# Research Notes in Mathematics

**Main Editors**
H. Brezis, Université de Paris
R. G. Douglas, State University of New York at Stony Brook
A. Jeffrey, University of Newcastle-upon-Tyne *(Founding Editor)*

**Editorial Board**

R. Aris, University of Minnesota
F. F. Bonsall, University of Edinburgh
W. Bürger, Universität Karlsruhe
J. Douglas Jr, University of Chicago
R. J. Elliott, University of Hull
G. Fichera, Università di Roma
R. P. Gilbert, University of Delaware
R. Glowinski, Université de Paris
K. P. Hadeler, Universität Tübingen
K. Kirchgässner, Universität Stuttgart

B. Lawson, State University of New York at Stony Brook
W. F. Lucas, Cornell University
R. E. Meyer, University of Wisconsin-Madison
J. Nitsche, Universität Freiburg
L. E. Payne, Cornell University
G. F. Roach, University of Strathclyde
J. H. Seinfeld, California Institute of Technology
I. N. Stewart, University of Warwick
S. J. Taylor, University of Virginia

## Submission of proposals for consideration

Suggestions for publication, in the form of outlines and representative samples, are invited by the Editorial Board for assessment. Intending authors should approach one of the main editors or another member of the Editorial Board, citing the relevant AMS subject classifications. Alternatively, outlines may be sent directly to one of the publisher's offices. Refereeing is by members of the board and other mathematical authorities in the topic concerned, throughout the world.

## Preparation of accepted manuscripts

On acceptance of a proposal, the publisher will supply full instructions for the preparation of manuscripts in a form suitable for direct photo-lithographic reproduction. Specially printed grid sheets are provided and a contribution is offered by the publisher towards the cost of typing. Word processor output, subject to the publisher's approval, is also acceptable.

Illustrations should be prepared by the authors, ready for direct reproduction without further improvement. The use of hand-drawn symbols should be avoided wherever possible, in order to maintain maximum clarity of the text.

The publisher will be pleased to give any guidance necessary during the preparation of a typescript, and will be happy to answer any queries.

## Important note

In order to avoid later retyping, intending authors are strongly urged not to begin final preparation of a typescript before receiving the publisher's guidelines and special paper. In this way it is hoped to preserve the uniform appearance of the series.

**Advanced Publishing Program**
**Pitman Publishing Inc**
**1020 Plain Street**
**Marshfield, MA 02050, USA**
**(tel (617) 837 1331)**

**Advanced Publishing Program**
**Pitman Publishing Limited**
**128 Long Acre**
**London WC2E 9AN, UK**
**(tel 01-379 7383)**

# Infinite dimensional analysis and stochastic processes

# S Albeverio (Editor)

Ruhr-Universität Bochum

# Infinite dimensional analysis and stochastic processes

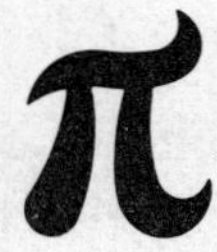

# Pitman Advanced Publishing Program
BOSTON · LONDON · MELBOURNE

PITMAN PUBLISHING INC
1020 Plain Street, Marshfield, Massachusetts 02050

PITMAN PUBLISHING LIMITED
128 Long Acre, London WC2E 9AN

*Associated Companies*
Pitman Publishing Pty Ltd, Melbourne
Pitman Publishing New Zealand Ltd, Wellington
Copp Clark Pitman, Toronto

© S Albeverio 1985

First published 1985

AMS Subject Classifications: (main) 60H, 60G, 80
                            (subsidiary) 81C20, 28C, 31C25

ISSN 0743-0337

Library of Congress Cataloging in Publication Data
Main entry under title:

Infinite dimensional analysis and stochastic processes.

   (Research notes in mathematics; 124)
   Seminars and lectures presented at a meeting held at
the University of Bielefeld in 1983, sponsored by the
Universitätsschwerpunkt Mathematisierung.
   Bibliography: p.
   1. Stochastic processes—Congresses.   2. Stochastic
analysis—Congresses.   3. Function spaces—Congresses.
I. Albeverio, Sergio.   II. Universität Bielefeld.
USP Mathematisierung der Einzelwissenschaften.
III. Series.
QA274.A1154   1985        519.2        84-26516
ISBN 0-273-08684-7

British Library Cataloguing in Publication Data

Infinite dimensional analysis and stochastic
   processes.—(Research notes in mathematics,
   ISSN 0743-0337; 124)
   1. Stochastic processes
   I. Albeverio, S.   II. Series
   519.2        QA274

   ISBN 0-273-08684-7

All rights reserved. No part of this publication may be reproduced,
stored in a retrieval system, or transmitted, in any form or by any
means, electronic, mechanical, photocopying, recording and/or
otherwise, without the prior written permission of the publishers.
This book may not be lent, resold, hired out or otherwise disposed
of by way of trade in any form of binding or cover other than that
in which it is published, without the prior consent of the publishers.

Reproduced and printed by photolithography
in Great Britain by Biddles Ltd, Guildford

# Contents

# Preface

The central topic of this volume is the theory of stochastic processes as developed by interactions with some of its applications to theoretical physics. In recent years these physical models have raised many deep mathematical problems stimulating fundamental research in stochastic processes, including those in infinite dimensional spaces. This area of interaction has been a continuing concern at seminars in Probability Theory and Mathematical Physics at the Universities of Bielefeld (Ph. Blanchard, L. Streit) and Bochum (S. Albeverio). We present some of the contributions to these seminars and other lectures at the Meeting "Infinite Dimensional Analysis and Stochastic Processes" held at ZiF, University of Bielefeld in 1983, sponsored by the Universitätsschwerpunkt Mathematisierung.

The contribution by A. Boukricha on "The Schrödinger equation with an isolated singularity" discusses, by methods of potential theory, continuous extensions of bounded solutions of the time independent Schrödinger equation at an isolated singularity of the potential assumed to be a positive Radon measure outside a given point. Singular Schrödinger operators and more generally Dirichlet forms are also discussed in the contribution by M. Röckner and N. Wielens, "Dirichlet forms – closability and change of speed measure". This paper contains in particular a review of results obtained in recent years on the problem of closability of minimally defined Dirichlet forms, as well as new results on this problem and the related one of the change of speed measure. These are fundamental to the approach to (symmetric) diffusions in locally compact spaces having infinitesimal generator with coefficients which are more general than measurable functions.

This theory of extended diffusions as processes associated with Dirichlet forms is the multidimensional analogue of Feller's one-dimensional theory and has been developed particularly by Beurling, Deny, Fukushima and Silverstein. It has been applied to quantum mechanics by S. Albeverio, R. Carmona, M. Fukushima, R. Høegh-Krohn, W. Karwowski, B. Simon, L. Streit and others. Some results of the theory and its relation to potential theory, Schrödinger operators and Green's functions are discussed in the

paper on "Capacity, Green's functions and Schrödinger operators" by
S. Albeverio, W. Karwowski, M. Röckner and L. Streit.

The paper by M. Fukushima and H. Kaneko on "(r,p)-Capacities for general
Markovian semigroups" is concerned with a problem related to the theory of
Dirichlet forms in infinite dimensional spaces. The authors associate to a
quite general Markovian semigroup on a metric space suitable Sobolev spaces
and associated capacities. This theory of capacity extends the theory for
capacity in Dirichlet spaces. In related work, M. Fukushima and M. Takeda
have given applications to the study of capacity on the infinite dimensional
Wiener space and fine properties of Brownian motion.

Schrödinger operators with stochastic or quasi-periodic potential have
received a lot of attention in recent years, by analysts, mathematical
physicists and probabilists, because of the interesting spectral phenomena
they exhibit. In physics they arise naturally in the study of disordered
systems such as amorphous solids. The paper by F. Martinelli and E. Scoppola,
"Absence of absolutely continuous spectrum in the Anderson model for large
disorder or low energy" is concerned with such problems. The authors give
in it the first proof of absence of the absolutely continuous component at
the lower end of the spectrum of a discrete random three-dimensional
Schrödinger operator (the Anderson tight-binding model of the physical
literature).

The paper by W. Craig gives a survey of results on "The Lyapounov index
and the integrated density of states for Schrodinger operators" with random
or almost periodic potentials. Schrödinger operators with random
potentials are also connected, via a Feynman-Kac formula, to certain
densities in Wiener space giving "polymer measures" and this provides a
mathematical realization of ideas developed by chemists and physicists in
the study of polymer models. Two papers by S. Kusuoka in this volume are
devoted to the mathematical study of such measures. The first, entitled
"Asymptotics of polymer measures in one dimension" contains a refinement of
Donsker-Varadhan and Schilder's asymptotic estimates for expectations of
functionals of Brownian motion, used to establish results on the asymptotics
in time of the mean absolute displacement of a polymer path, starting from
the origin. The second paper by S. Kusuoka in this volume, on "The path
property of Edward's model for long polymer chains in three dimensions",
contains the proof of the result that the sample paths of Edward's polymer

measure, as constructed in the three-dimensional case by Westwater, are those of a Dirichlet process in the sense of Föllmer, with its principal part a Brownian motion.

Another topic in stochastic analysis that has received a lot of attention in recent years is the study of the flows generated by diffusions. To this topic is devoted the contribution by Y. Le Jan, "Equilibrium state for a turbulent flow of diffusion". The author studies in it the asymptotics in time of the distribution of the stochastic flow given by a diffusion. In particular, a characterization of equilibrium states is given.

Limit theorems of various kinds are also central to statistical mechanics. This domain is represented in this volume by several contributions. The paper by J. Willms, on "Convergence of infinite particle systems to the Poisson process under the action of the free dynamic" studies the asymptotic time evolution of a system consisting of infinitely many classical particles. The paper relates to the work of Boldrighini, Dobrushin and Sukkov and the methods used are those of the theory of point processes. The subject of this paper is also related to that of the contribution by H. Zessin on "BBGKY hierarchy equations for Newtonian and gradient systems". In this, the time evolution of states describing systems of infinitely many interacting classical particles is studied and the asymptotic equilibrium states are characterized, by methods which are a synthesis of those used in the theory of point processes and in statistical mechanics.

Equilibrium states for systems with infinitely many degrees of freedom are also studied in two other contributions to these proceedings. In the paper "Gibbs random fields and symmetry breakdown" Ch.-E. Pfister analyzes the relation between the existence of several Gibbs states for lattice systems of classical statistical mechanics and the physical phenomenon of "symmetry breakdown". The subject and the methods belong to a theory originally developed by Dobrushin and Sinai. Related questions are also studied for a particular continuous system of classical statistical mechanics, the so-called "jellium model", by S. Albeverio, R. Høegh-Krohn and D. Merlini, in their contribution "Euler flows, associated generalized random fields and Coulomb systems". This article surveys theoretical and numerical results obtained in the study of the two-dimensional jellium model, in particular as a continuous model requiring a particular adaptation of Dobrushin's method for the construction of equilibrium states.

The contribution by A.S. Sznitman, "A fluctuation result for nonlinear diffusions", studies the asymptotic behaviour of infinitely many diffusion processes in "weak" interaction ("propagation of chaos" types of problem). In particular, the weak convergence of the fluctuation field to a Gaussian measure is proven. This paper is related to other recent work on the subject by Kusuoka-Tamura, Tanaka, Calderoni and Martinelli. The fundamental problem of constructing mechanical models of Brownian motion is discussed by D. Dürr in his paper "A limit theorem for stochastic acceleration with self-intersecting particle trajectories". The author gives a survey of results he has obtained in recent years, with S. Goldstein and J.L. Lebowitz, extending, in particular, work by Holley, Kesten, Papanicolaou and Spitzer. These results provide an explanation of basic stochastic phenomena starting from the classical Newtonian mechanics of infinitely many particles.

S. Albeverio

Bielefeld and Bochum
January 1985

# Acknowledgements

It is a pleasure to thank Prof. Dr Philippe Blanchard and Prof. Dr Ludwig Streit, who shared with me the organization of the Bielefeld-Bochum seminars on which this book is based.  I am also very happy to record here my gratitude to Professor Dr James S. Taylor for his kind interest and help with the publication.  He and the publishers should also be thanked for their understanding and the pleasant collaboration.

The support of the Universitätsschwerpunkt Mathematisierung of the University of Bielefeld (under the Research program "Infinite dimensional analysis"), of Research Projects of the University of Bochum and of the Deutsche Forschungsgemeinschaft ("Stochastic fields" and "Polymers and quantum fields") as well as of the Zentrum für Interdisziplinäre Forschung (Project No 2 Physics and Mathematics, directed by L. Streit) is also gratefully acknowledged.

S.A.

A BOUKRICHA

# The Schrödinger equation with an isolated singularity

In this paper we examine the continuous extension of the bounded solutions of the time-independent Schrödinger equation

$$\Delta u = cu$$

at an isolated singularity of the coefficient c in an open subset U of the space $\mathbb{R}^n$. $(n \geq 2)$. More generally, we consider the following equation

$$\Delta u = \mu u \qquad (*)$$

in the distributional sense where $\mu$ is a positive Radon measure on $\overline{U}\setminus\{x_o\}$ and $x_o$ is a given point in U.

In electrostatics $\mu$ is the distribution of charges on the open set U.

In many cases $\mu$ may not be absolutely continuous with respect to the Lebesgue measure. Take the case where the charges are distributed on a surface or consider the following case. Let V be a relatively compact regular set in U. By a theorem of Mrs. Hervé [8] there exists a positive Radon measure $\mu$ on $\mathbb{R}^n$ such that

$$\hat{R}_1^V(x) = \int \frac{1}{\| x-y \|^{n-2}} \, \mu(dy) \qquad (n \geq 3)$$

where $\hat{R}_1^V$ is a potential on $\mathbb{R}^n$. Since $\hat{R}_1^V$ is harmonic outside $V^*$, we get supp $\mu \subset V^*$ with $\mu \neq 0$. Hence $\mu$ cannot be absolutely continuous with respect to the Lebesgue measure.

For an open set U of $\mathbb{R}^n$ $(n \geq 2)$ and a given point $x_o \in U$, we consider a set of measures denoted by $M(U\setminus\{x_o\}$ which are defined on $\overline{U}\setminus\{x_o\}$ with the following property: for every relatively compact open set V in $\overline{U}\setminus\{x_o\}$ the function $p_V = \int G_t^V \, (dt)$ is continuous and real on V, where $G^V$ denotes the Green function of V with respect to the Laplace equation.

In Section 1 we study, for a given $\mu \in M(U\setminus\{x_o\})$, the existence of the limit of the bounded solution of $\Delta u = \mu u$ at $x_o$. Using the minimum principle, we prove the following result
(Prop. 1.5): if for a given $\nu \in M(U\setminus\{x_o\})$ and a relatively compact open set

V containing $x_o$ we have

$$\lim_{\substack{x \to x_o \\ x \neq x_o}} \int G_t^V(x)\nu(dt) \quad \text{exists and is real}$$

then

$$\lim_{\substack{x \to x_o \\ x \neq x_o}} \int G_t^W(x)\nu(dt) = \int G_t^W(x_o)\nu(dt)$$

for all open subsets $W$ of $U$ with $x_o \in W$. Then it follows that $\lim_{\substack{x \to x_o \\ x \neq x_o}} u(x)$

exists for every bounded solution of (*) if and only if there exists a
non-zero bounded solution of (*) which is continuously extendable at $x_o$.
Hence, if we denote by $H$ the set of functions u such that there exists $V$,
an open subset of $U$ containing $x_o$ with $u \in C_b(V \backslash \{x_o\})$ and $\Delta u = \mu u$ on
$V \backslash \{x_o\}$, we have the following result: either all functions of $H$ are
continuously extendable at $x_o$ or every non-zero element of $H$ has no limit
at $x_o$.

In Section 2 we take $U$ to be the unit ball in $\mathbb{R}^n$ and $x_o = 0$. For
every almost rotation-free measure (see §2) $\mu \in M(U \backslash \{x_o\})$ we show that all
functions of $H$ are continuously extendable at $0$. This is a generalization
of a result obtained by M. Nakai [8] in the case $n = 2$.

In the last section we define subNewtonian measures in a point $x_o$ as in
[2] and we give the following characterization for any measure $\mu \in M(U \backslash \{x_o\})$:
all functions of $H$ are continuously extendable at $x_o$ if and only if $\mu$ is
either not subNewtonian at $x_o$ or there exists an open set $V$ containing $x_o$
such that the function $x \to \int G_t^V(x)\mu(dt)$ is continuous and real at $x_o$.

From this, if the function $V$ in [1] is positive and is in $K_\nu^{loc}$ (see
[1, page 210]), it follows that $V$ defines a positive section of continuous
and real potentials, i.e., for every relatively compact open set $U$ in $\mathbb{R}^n$
the function $x \to \int G_t^U(x)V(t)dt$ is continuous on $U$ (see [6]).

From the results of Sections 1 and 3, it is easy to see that we can
replace the operator $\Delta$ by any elliptic differential operator of second order

2

which has a Green function:  for example, an operator of the kind
considered in $[9]$.  In Section 2 we can replace $\Delta$ by an elliptic
differential operator of second order which is rotation-free.  The
assumption of ellipticity may be to some extent avoided.

1  <u>THE BEHAVIOUR OF THE BOUNDED SOLUTIONS OF THE SCHRODINGER EQUATION</u>
   <u>AT AN ISOLATED SINGULARITY</u>

In the following we consider a relatively compact open set  U  of $\mathbb{R}^n$  and we
fix a point $x_o \in U$.

For $A \subset \mathbb{R}^n$,  $C(A)$ (resp. $B(A)$) denotes the set of continuous (resp. Borel
bounded) functions on A.  Let $M(U \setminus \{x_o\})$ be the set of measures $\mu$ on
$U \setminus \{x_o\}$ such that, for every open set  V  which is relatively compact in
$\overline{U} \setminus \{x_o\}$, the function $\int G_t^V \mu(dt)$ is continuous and real on  V.  $G^V$ is the
Green function on  V  for the Laplace equation.  The aim of this section is
the following result:  every bounded continuous solution of $\Delta u = \mu u$ on
$U \setminus \{x_o\}$ in the distributional sense is continuously extendable at the point
$x_o$ if and only if there exists one non-identically zero-bounded solution of
this equation which is continuously extendable at the point $x_o$.

For $0 \subset \overline{0} \subset \overline{U} \setminus \{x_o\}$ regular and $g \in C(\partial 0)$, $^{\mu}H_0 g$ denotes the solution of
the Dirichlet problem for 0 and f according to the equation $\Delta u = \mu u$.  As
proved in $[2]$, we have

$$H_0 g = {}^{\mu}H_0 g + \int G_t^0 \, {}^{\mu}H_0 g(t) \mu(dt).$$

In order to eliminate notation such as $\int_{V \setminus \{x_o\}} \phi(t) \mu(dt)$ for a function $\phi$

on $V \setminus \{x_o\}$, we shall suppose $\mu(\{x_o\}) = 0$.


1.1.  <u>Proposition.</u>  Let  V  be a regular open set with $x_o \in V$.  Then,
for all $f \in C(\partial V)$, there exists on $V \setminus \{x_o\}$ a unique continuous bounded
solution u of $\Delta u = \mu u$ such that $u_{|\partial V} = f$.


<u>Proof.</u>  Let $n_o \in \mathbb{N}$ be such that the ball $B(x_o, \frac{1}{n_o})$ with centrum $x_o$ and
radius $\frac{1}{n_o}$ is in the set  V.  For all $n \geq n_o$ let

$$W_n := V \backslash B(x_o, \frac{1}{n}).$$

3

An easy calculation shows that the sequence $({}^{\mu}H_{W_n}(H_V f))_{n \geq n_o}$ is decreasing and its limit u verifies $\Delta u = \mu u$ and $u_{|\partial V} = f$. The uniqueness derives from the minimum principle [see [2], Lemma II.1.2).

1.2. **Notation.**  Let $r > 0$ such that $B(x_o,r) \subset U$. $e_r$ denotes the bounded solution of (*) on $B(x_o,r) \setminus \{x_o\}$ with $e_r(x) = 1$ for all x with $\| x \| = r$.

1.3. **Lemma.**  We have

$$1 = e_r + \int G_t^{B(x_o,r)} e_r(t)\mu(dt) \quad \text{on} \quad B(x_o,r) \setminus \{x_o\}.$$

**Proof.**  See [2,II.1.1.].

1.4. **Proposition.**  Let W be open with $x_o \in W \subset \overline{W} \subset U$; then the function

$$\phi(x) = \int_{C_W} G_t^U(x)\mu(dt) \quad \text{is continuous on} \quad U.$$

**Proof.**  Since $\mu \in M(U \setminus \{x_o\})$, we have $\mu(U \setminus W) < \infty$.

Let $y \in U \setminus \{x_o\}$, then there exists $V \subset \overline{V} \subset U \setminus \{x_o\}$ such that $y \in V$ and V is regular.  Whence

$$G_t^U = G_t^V + H_V G_t^U(x) \quad \text{for all} \quad x \in V,$$

thus

$$\phi(x) = \int_{C_W} G_t^V(x)\mu(dt) + \int_{C_W} H_V G_t^U(x)\mu(dt) \quad \text{for } x \in V.$$

Since $\mu \in M(U \setminus \{x_o\})$, it follows that

$$x \to \int_{C_W} G_t^V(x)\mu(dt)$$

is continuous on  V  and hence in y.  The continuity of $x \to \int_{C_V} H_V G_t^U(x)\mu(dt)$ in y follows from [2, II, 3.8] and the lemma of Fatou.

The continuity at $y = x_o$ results from the dominated convergence theorem, since the function $x \to G_t^U(x)\big|_{C_V}(t)$ is bounded uniformly in t in a neighbourhood of $x_o$.

4

__1.5. Proposition.__  Let $r > 0$ such that $B(x_o, r) \subseteq U$.

If $\lim\limits_{\substack{x \to x_o \\ x \neq x_o}} e_r(x)$ exists then $\lim\limits_{\substack{x \to x_o \\ x \neq x_o}} \int G_t^{B(x_o, r)}(x) e_r(t) \mu(dt)$ exists and we have

$$\lim_{\substack{x \to x_o \\ x \neq x_o}} \int G_t^{B(x_o, r)}(x) e_r(t) \mu(dt) = \int G_t^{B(x_o, r)}(x_o) e_r(t) \mu(dt).$$

__Proof.__  By Lemma 1.3 we have the existence of

$$\lim_{\substack{x \to x_o \\ x \neq x_o}} \int G_t^{B(x_o, r)}(x) e_r(t) \mu(dt).$$

Let $p(x) = \int G_t^{B(x_o, r)}(x) e_r(t) \mu(dt)$ if $x \neq x_o$ and with

$$p(x_o) = \lim_{\substack{x \to x_o \\ x \neq x_o}} \int G_t^{B(x_o, r)}(x) e_r(t) \mu(dt).$$

Let $n_o \in \mathbb{N}$ be such that $\dfrac{1}{n_o} < r$ and let

$$P_n(x) := \int G_t^{B(x_o, r)}(x) e_r(t) \mu(dt) \qquad n \geq n_o$$

$$B(x_o, r) \setminus B(x_o, \tfrac{1}{n})$$

Applying Proposition 1.4, we get that $p_n \in C(U)$ for all $n \geq n_o$.
Furthermore, $p_n$ is an increasing sequence with $\lim\limits_{n \to \infty} p_n = p$ on $U \setminus \{x_o\}$.

By Fatou's lemma we obtain

$$\int G_t^{B(x_o, r)}(x_o) e_r(t) \mu(dt) = \lim_{n \to \infty} p_n(x_o) \leq p(x_o).$$

From the continuity of $p$ we have that, for all $\varepsilon > 0$, there exists $\delta > 0$ such that $\| x - x_o \| \leq \delta$ implies $p(x_o) - \varepsilon \leq p(x)$. Without loss of generality we can suppose $\delta < r$.

By Dini's theorem we get the uniform convergence of $p_n$ on $\| x - x_o \| = \delta$. Then there exists $n_1 \in \mathbb{N}$ such that

$$p(x) \leq p_n(x) + \varepsilon \quad \text{on } \{x \mid \| x - x_o \| = \delta\} \quad \text{for } n \geq n_1$$

and

$$p(x_o) - \varepsilon \leq p_n(x) + \varepsilon \quad \text{on} \quad \{x \mid \|x - x_o\| = \delta\}.$$

Since $p_n + 2\varepsilon - p(x_o)$ is a superharmonic function on $B(x_o, \delta)$ which is positive on the boundary of this ball, we get, by the minimum principle,

$$p_n(x) + 2\varepsilon - p(x_o) \geq 0 \quad \text{on} \quad B(x_o, \delta) \quad \text{for} \quad n \geq n_1.$$

This gives us

$$\lim_{n \to \infty} p_n(x_o) + 2\varepsilon - p(x_o) \geq 0 \quad \text{for all} \quad \varepsilon > 0,$$

thus we obtain

$$\lim p_n(x_o) - p(x_o) \geq 0$$

and

$$p(x_o) = \int G_t^{B(x_o, r)}(x_o) e_r(t) \mu(dt).$$

<u>1.6. Corollary.</u>   If for some $\rho > 0$ there exists

$$\lim_{\substack{x \to x_o \\ x \neq x_o}} e_\rho(x)$$

then

$$\lim_{\substack{x \to x_o \\ x \neq x_o}} e_r(x) \quad \text{exists for all } r > 0 \text{ with } B(x_o, r) \subset U.$$

<u>Proof.</u>   Let $r > 0$ with $B(x_o, r) \subset U$.  By the minimum principle $[2, \text{II}.1.2]$, there exists $\alpha > 0$ such that $e_r \leq \alpha e_\rho$ on $B(x_o, \min(\rho, r)) =: B$.

If $r < \rho$ we have

$$G_t^{B(x_o, r)} e_r(t) \leq \alpha G_t^{B(x_o, \rho)} e_\rho(t).$$

Then it follows from the proposition 1.5 and Fatou's lemma that

$$\lim_{\substack{x \to x_o \\ x \neq x_o}} \int G_t^{B(x_o, r)}(x) e_r(t) \mu(dt) = \int G_t^{B(x_o, r)}(x_o) e_r(t) \mu(dt)$$

If $r < \rho$ we have

$$\int G_t^{B(x_o,r)} e_r(t)\mu(dt) = \int_{B(x_o,\rho)} G_t^{B(x_o,r)} e_r(t)\mu(dt) + \int_{{}^cB(x_o,\rho)} G_t^{B(x_o,r)} e_r(t)\mu(dt).$$

Then it is enough to show that

$$\lim_{\substack{x \to x_o \\ x \neq x_o}} \int_{B(x_o,\rho)} G_t^{B(x_o,r)}(x)e_r(t)\mu(dt) = \int_{B(x_o,\rho)} G_t^{B(x_o,r)}(x_o)e_r(t)\mu(dt)$$

This follows from Fatou's lemma and $[2,\text{I, Lemma } 3.8]$. Hence we have for all $r > 0$ with $B(x_o,r) \subset U$

$$\lim_{\substack{x \to x_o \\ x \neq x_o}} \int G_t^{B(x_o,r)}(x)e_r(t)\mu(dt) = \int G_t^{B(x_o,r)}(x)e_r(t)\mu(dt).$$

The result follows then from Lemma 1.2.

<u>1.7. Theorem.</u>    Let $\mu \in M(U \smallsetminus \{x_o\})$; then the following properties are equivalent:

(1) all bounded continuous solutions of (*) in an open subset $V$ containing $x_o$ are continuously extendable at $x_o$;

(2) there exists a subset $V$ of $U$ containing $x_o$ and a non-zero bounded continuous solution of (*) on $V \smallsetminus \{x_o\}$ which has a limit at $x_o$;

(3) there exists $r > 0$ such that $e_r$ is continuously extendable at $x_o$.


<u>Proof.</u>    (1) $\Rightarrow$ (2) is clear.

(2) $\Rightarrow$ (3)  has the same proof as Corollary 1.6.

(3) $\Rightarrow$ (1).    Let $V \subset U$ with $x_o \in V$ and $u \in C(V \smallsetminus \{x_o\})$ such that $u$ is bounded on $V$ and

$$\Delta u = \mu u \quad \text{on} \quad V \smallsetminus \{x_o\}.$$

Let $\rho > 0$ such that $\overline{B}(x_o,\rho) \subset V$; without loss of generality we can suppose that $u \geq 0$ on $B(x_o,\rho)$. Then there exists $M > 0$ such that $u \leq M$ on $B(x_o,\rho)$ and hence we have

$$u \leq Me_\rho \quad \text{on} \quad B(x_o,\rho).$$

It is easy to see that the function $v = u + \int G_t^{B(x_o,\rho)} u(t)\mu(dt)$ is a bounded

harmonic function on $B(x_o,\rho)\setminus\{x_o\}$. By $\left[5,\text{ Th. }7.7\right]$, $v$ is extendable to a

continuous function on $B(x_o,\rho)$. Since $G_t^{B(x_o,\rho)} u(t) \leq MG_t^{B(x_o,\rho)} e_\rho(t)$, we

obtain by Corollary 1.6 and Fatou's lemma that

$$\lim_{\substack{x \to x_o \\ x \neq x_o}} \int G_t^{B(x_o,\rho)}(x)u(t)\mu(dt) = \int G_t^{B(x_o,\rho)}(x_o)u(t)\mu(dt).$$

Thus we have proved the existence of $\lim_{\substack{x \to x_o \\ x \neq x_o}} u(x)$.

1.8. Corollary. We have the following alternatives: either all bounded continuous solutions of $\Delta u = \mu u$ are continuously extendable at $x_o$ or every non-zero bounded solution has no limit in $x_o$.

1.9. Corollary. Let c be a locally bounded measurable function on $U\setminus\{x_o\}$. Then either all bounded continuous solutions of $\Delta u = cu$ in the distribution sense are continuously extendable at $x_o$ or every non-zero bounded solution has no limit in $x_o$.

## 2  ROTATION-FREE MEASURES

In this section we consider $U = \{x \in \mathbb{R}^n \mid \|x\| < 1\}$ and a measure $\mu \in M(U\setminus\{0\})$ such that, for all f measurable with compact support in $\overline{U}\setminus\{0\}$, we have $\mu(f \circ g) = \mu(f)$ for all $g \in SO(n)$.

Example. Let $\mu = c\lambda$ where $\lambda$ is the Lebesgue measure and c is a measurable function on $\overline{U}\setminus\{0\}$ with $(c \circ g)(x) = c(x)$ for all $x \in \overline{U}\setminus\{0\}$ and $g \in SO(n)$; i.e., there exists a measurable function F defined on the interval $]0,1]$ such that

$$e(x) = F(\|x\|).$$

2.1. Lemma. e is rotation-free where $e = e_1$.

_Proof._    By Lemma 1.3 we have

$$1 = e + \int G^U_t e(t) \mu(dt)$$

Let $g \in SO(n)$ and $x \in U \setminus \{0\}$, then

$$1 = e(g(x)) + \int G^U_t (g(x)) e(t) \mu(dt).$$

By the hypothesis it follows that

$$\int G^U_t (g(x)) e(t) \mu(dt) = \int G^U_{g(t)} (g(x)) e(g(t)) \mu(dt).$$

By $[7]$ we can easily see that

$$G^U_{g(t)} g(x)) = G^U_t (x),$$

hence we get

$$1 = e(g(x)) + \int G^U_t (x) e(g(t)) \mu(dt)$$

$$= e(x) + \int G^U_t (x) e(t) \mu(dt).$$

By $[2, \text{I. Lemma } 1.1]$ we get

$$(e \circ g)(x) = e(x).$$

_2.2. Corollary._    $\lim_{x \to 0} e(x)$ exists.

_Proof._    Let $\alpha = \liminf_{x \to x_o} e(x)$ and let $F$ be such that $F(\| x \|) = e(x)$. By the definition of the lim inf there exists a decreasing sequence $(r_n)$ with $\inf r_n = 0$ and

$$\alpha = \lim F(r_n).$$

By the minimum principle $[2, \text{II}.1.2]$, we get

$$e(x) \leq f(r_n) \quad \text{for all} \quad x \in B(0, r_n).$$

Hence we have

$$\limsup_{x \to 0} e(x) \leq f(r_n) \quad \text{for all} \quad n \in \mathbb{N}$$

and then

$$\limsup_{x \to 0} e(x) = \liminf_{x \to 0} e(x).$$

**2.3. Theorem.** Let $\mu \in M(U \setminus \{0\})$ be rotation-free on $U \setminus \{0\}$. Then, for every open set $V$ containing $0$, all bounded continuous solutions (in the distributional sense) on $V \setminus \{0\}$ of the equation $\Delta u = \mu u$ are continuously extendable at $0$.

**Proof.** Corollary 3.2 and Theorem 1.7.

**2.4. Corollary.** Let $c$ be a locally bounded function on $U \setminus \{0\}$ such that there exists $r > 0$ and a measurable function $F$ defined on $]0,r]$ with $c(x) = F(\|x\|)$ for all $\|x\| \leq r$; then every bounded solution of the equation $\Delta u = cu$ is continuously extendable at the point $x_o$.

**2.5. Corollary.** Let $c$ be a locally bounded function on $\overline{U} \setminus \{0\}$ which is rotation-free on $U \setminus \{0\}$ and satisfies the Picard principle on $U$. Then all positive solutions of the equation $\Delta u = cu$ in the distributional sense are continuously extendable at $0$ (eventually we have $\lim_{x \to 0} u(x) = +\infty$).

**Proof.** Let $V \subset U$ with $0 \in V$ and $u \in (V \setminus \{0\})$ and $\Delta u = cu$. We can suppose that $u$ is unbounded. Let $r > 0$ such that $\overline{B(0,r)} \subset V$.

Let $v$ be the bounded solution of

$$\Delta u = cu \quad \text{with} \quad v_{|\partial B(0,r)} = u.$$

Then by the construction of $v$ we get $v \leq u$ and hence $u - v \geq 0$; $u - v$ is a positive solution of $\Delta u - cu = 0$ which is zero at the boundary of $B(0,r)$.

Since $c$ is rotation-free and verifies the Picard principle, we obtain

$$\lim_{x \to x_o} (u-v)(x) = +\infty$$

and hence

$$\lim_{x \to x_o} u(x) = +\infty.$$

2.6. <u>Definition</u>   (cf. $[8]$).  We say that a function c on $U \setminus \{0\}$ is almost rotation-free if there exists a constant $k \in [1, \infty[$ and a real function F on $]0,1[$ such that

$$k^{-1} F(\|x\|) \leq c(x) \leq k F(\|x\|) \quad \text{for all} \quad x \in U \setminus \{0\}.$$

2.6. <u>Theorem.</u>   Let c be locally bounded on $\overline{U} \setminus \{0\}$ which is almost rotation-free; then all bounded continuous solutions cf $\Delta u = cu$ in the distributional sense are continuously extendable in $0$.

<u>Proof.</u>   It is enough to show that

$$\lim_{\substack{x \to x_o \\ x \neq x_o}} \int G_t^{B(0,1)}(x) e(t) c(t) \, dt = \int G_t^{B(0,1)}(x_o) e(t) c(t) \, dt.$$

Let e' and e" be respectively the bounded solutions of

$$\Delta u = k^{-1} F(\| \cdot \|) u \quad \text{and} \quad \Delta u = k F(\| \cdot \|).$$

By Corollary 2 we have

$$\lim_{\substack{x \to x_o \\ x \neq x_o}} \int G_t^{B(0,1)}(x) e'(t) F(\|t\|) \, dt = \int G_t^{B(0,1)}(x_o) e'(t) F(\|t\|) \, dt.$$

Since $e \leq e'$, we have

$$G_t^{B(0,1)}(x) e(t) c(t) \leq k G_t e'(t) F(\|t\|).$$

Fatou's lemma gives us the equality and the result follows by Lemma 1.3 and Theorem 1.7.

## 3  SUBNEWTONIAN MEASURES

Let U be a relatively compact open set in $\mathbb{R}^n$.  We fix a point $x_o \in U$.  By means of the following definition we give a characterization of the measures $\mu \in M(U \setminus \{x_o\})$ such that, for all open sets V containing $x_o$ and every bounded continuous solution on $V \setminus \{x_o\}$ of $\Delta u = \mu u$  (in the distributional sense), $\lim_{\substack{x \to x_o \\ x \neq x_o}} u(x)$  exists.

3.1. <u>Definition</u> (cf. [2]). A measure $\mu$ in $M(U \setminus \{x_o\}$ is said to be subNewtonian at $x_o$ if there exists on $U \setminus \{x_o\}$ a strictly positive continuous function h such that $\Delta h = \mu h$ and $h \leq G^U_{x_o}$ .

3.2. <u>Example.</u> Every positive measure $\mu$ on U which has a locally bounded density relative to the Lebesgue measure is subNewtonian at every point of U.

3.3. <u>Definition.</u> Let $y_o \in U$ and $E \subset U$. We say that E is thin at $y_o$ if and only if there exists a positive bounded Radon measure $\mu$ on U such that

$$\lim_{\substack{x \to y_o \\ x \in E \setminus \{y_o\}}} \int G^U_t(x)\mu(dt) > \int G^U_t(x_o)\mu(dt).$$

3.4. <u>Examples.</u>
  (a) (Theorem of Zaremba). If $y_o$ is a boundary point of the bounded open
      set $E \subset \mathbb{R}^n$, $n \geq 2$, and there is a truncated closed solit cone in $c_E$
      with vertex at $y_o$, then E is not thin at $y_o$.

  (b) Let $n = 2$. If E is thin at $y_o$ then one can find arbitrary small $\rho$
      such that $\{x \in \mathbb{R}^2 \mid \| x - x_o \| = \rho\} \subset E$. For more details see [7].

  (c) Probabilistic interpretation: E is thin at a point $y_o$ if a particle
      undergoing Brownian motion and starting from $y_o$ at $t = 0$ almost
      certainly hits the set E after a strictly positive time.

3.5. <u>Theorem.</u> Let $\mu \in M(U \setminus \{x_o\})$, then the following conditions are equivalent.
  (a) For every bounded continuous solution u of $\Delta u = \mu u$ on $V \setminus \{x_o\}$ with
      $x_o \in V \subset U$ we have

      $$\lim_{\substack{x \to x_o \\ x \neq x_o}} u(x) = 0.$$

  (b) $\mu$ is not subNewtonian at $x_o$.

  (c) For every set E thin at $x_o$ we have

$$\int_{U \setminus E}^* G^U_t(x_o)\mu(dt) = +\infty.$$

<u>Proof.</u>   See $[2]$, Theorem 2.4 and 4.2.

<u>3.6. Example.</u>   If there exists a neighbourhood $V$ of $x_o$ and $p \geq 2$ such that

$$c(x) \geq \alpha \, \frac{1}{\| x - x_o \|^p} \, , \qquad \alpha > 0,$$

then the measure $\mu = c\lambda$ is not subNewtonian at $x_o$ and hence, for all bounded
continuous solutions $u$ of $\Delta u = cu$ on $U \setminus \{x_o\}$, we have

$$\lim_{\substack{x \to x_o \\ x \neq x_o}} u(x) = 0.$$

<u>3.7. Theorem.</u>   Let $\mu \in M(U \setminus \{x_o\})$. Then all bounded continuous solutions of
$\Delta u = \mu u$ are continuously extendable at $x_o$ if and only if either $\mu$ is not
subNewtonian at $x_o$ or there exists an open set $V$, with $x_o \in V \subset \overline{V} \subset U$, such
that

$$\lim_{\substack{x \to x_o \\ x \neq x_o}} \int G_t^V(x)\mu(dt) = \int G_t^V(x_o)\mu(dt)$$

and the limit is real.

<u>Proof.</u>   "$\Rightarrow$" Let $r > 0$ such that $B(x_o, r) \subset U$.   If

$$\lim_{\substack{x \to x_o \\ x \neq x_o}} e_r(x) = 0$$

it follows then that $\mu$ is not subNewtonian at $x_o$.   If

$$\lim_{\substack{x \to x_o \\ x \neq x_o}} e_r(x) > 0$$

then there exists $\alpha > 0$ such that $\alpha < e_r$ on $B(x_o, r)$.

Since
$$\lim_{x \to x_o} \int G_t^{B(x_o,r)}(x) e_r(t)\mu(dt) = \int G_t^{B(x_o,r)}(x_o) e_r(t)\mu(dt) < 1$$

and

$$G_t^{B(x_o,r)}(x) \le \frac{1}{\alpha} G_t^{B(x_o,r)}(x) e_r(t), \qquad x,t \in B(x_o,r),$$

we obtain by Fatou's lemma that

$$\lim_{\substack{x \to x_o \\ x \ne x_o}} \int G_t^{B(x_o,r)}(x)\mu(dt) = \int G_t^{B(x_o,r)}(x_o)\mu(dt)$$

and thus we get the result.

"$\Leftarrow$" follows from Lemma 1.3 and Theorem 3.5.

3.8. Corollary (cf. [1] Theorem 4.15). Let c be a real function on $\mathbb{R}^n$. If $c \in K_\nu^{loc}$ (see [1, page 210]), then c defines a positive section of continuous and real potentials on $\mathbb{R}^n$.

Proof. It is enough to show that for every relatively compact open set U of $\mathbb{R}^n$ the function $x \to \int G_t^U(x)c(t)dt$ is continuous and real on U. This results from [1, Theorem 1.5] and Theorem 3.7.


REFERENCES

[1]  Aizenman, M. and Simon, B. *Brownian Motion and Harnack Inequality for Schrödinger Operators*. Communications on Pure and Applied Mathematics, Vol XXXV 209-273, Wiley (1982).

[2]  Boukricha, A. Das Picard-Prinzip und verwandte Fragen bei Störung von harmonischen Räumen. *Math. Ann. 239*, 247-270 (1979).

[3]  Boukricha, A. and Hueber, H. The Poisson space $^cP_x$ for $\Delta u = cu$ with rotation free c. Academie Royale de Belgique. Classe des Sciences. $5^o$ série-Tome LXIV 1978-10.

[4]  Brelot, M. Etude de L'equation de la chaleur $\Delta u = c(M)u(M)$, $c(M) \ge 0$ au voisinage d'un point singulier du coefficient. *Ann. Sci. Ecole Norm. Sup. 48*, 153-246 (1931).

[5]  Constantinecsu, C. and Cornea, A. *Potential theory on harmonic spaces*. Berlin-Heidelberg-New York: Springer (1972).

[6]  Hansen, W.  Cohomology in harmonic spaces.  In: *Seminar on potential theory II*, 63-101.  Lecture Notes in Mathematics 226.  Berlin-Heidelberg-New York: Springer (1971).

[7]  Helms, L.L.  *Introduction to Potential Theory*, Berlin-New York: Wiley-Interscience (1969).

[8]  Hervé, R.M.  Recherches axiomatiques sur la théorie des fonctions surharmoniques et du potentiel.  *Ann. Inst. Fourier 12*, 415-571 (1962).

[9]  Hervé, R.M. and Hervé, M.  Les fonctions surharmoniques associés à un operateur elliptique du second ordre à coefficients discontinus.  *Ann. Inst. Fourier 19*, 305-359 (1969).

[10]  Nakai, M.  Martin boundary over an isolated singularity of rotation free density.  *J. Math. Soc. Japan 56*, 483-507 (1974).


Abderrahman Boukricha,
Département de Mathématiques,
Faculté des Sciences,
Campus Universitaire,
1060 - Tunis,
Tunisie.

W CRAIG
# The Lyapunov index and the integrated density of states for stochastic Schrödinger operators

There has been a lot of recent research interest in the spectral properties of Schrödinger operators endowed with potentials with 'stochastic' or 'ergodic' properties. Two principal classes of potentials which share this property are random potentials and almost periodic potentials. These are usually treated separately, but the results of this article mostly apply to both. Instead of discussing the wealth of recent work, I shall consider a few details of spectral theory, concerning the Lyapounov index $\gamma(\lambda)$ and the integrated density of stated $k(\lambda)$. Central to the discussion is a useful formula derived (formally) by Thouless [10], giving a relationship between these quantities, and having implications on their regularity. At least in the discrete case, an analog of the Thouless formula holds for strip problems, giving higher dimensional results of the same character. I am reporting on many people's work; all of my contributions to the results summarized here have been in collaboration with Barry Simon [1][2].

INTRODUCTION

The spectral problems that we consider here are the following Schrödinger operators:

$$H\psi(x) = -\frac{d^2}{dx^2}\psi + q(x)\psi = \lambda\psi, \qquad \psi \in L^2(\mathbb{R}), \tag{1}$$

and the discretized version

$$(h\psi)_j = \psi_{j+1} + \psi_{j-1} + q_j\psi_j = \lambda\psi_j, \qquad \psi \in \ell^2(\mathbb{Z}).$$

Of course in several dimensions the discrete operator takes the form

$$(h\psi)_j = \sum_{|k|=1} \psi_{j+k} + q_j\psi_j = \lambda\psi_j, \qquad \psi \in \ell^2(\mathbb{Z}^n),$$

where now $j = (j_1,\ldots,j_n)$ an integer vector, with $|j| = \sum_{r=1}^{n} |j_r|$.

The stochastic potentials considered here are given by bounded measurable functions on an underlying probability space $\Omega$ which admits an ergodic transformation T. That is, in the discrete problem

16

$$T; \Omega \to \Omega \quad \text{ergodic}$$

$$f; \Omega \to \mathbb{R} \quad \text{measurable, bounded}$$

then

$$q_j(\omega) = f(T^j \omega).$$

In the continuous case (1) we ask instead that $T_t$, $t \in \mathbb{R}$ be an ergodic one-parameter family of transformations. In the several-dimensional discrete problem we write $T^j = T_1^{j_1} \circ T_2^{j_2} \circ \ldots \circ T_n^{j_n}$, where all $T_r$ commute.

<u>Examples.</u>  (1) Let

$$\Omega = \prod_{j \in \mathbb{Z}^n} (-\infty, \infty)$$

with product measure

$$dP = \prod g(q_j) dq_j$$

This of course gives the independent, identically distributed random potention model on $\mathbb{Z}^n$.

(2)  Set

$$\Omega = T^d$$

the d-dimensional standard torus, endowed with Haar measure. The ergodic transformation is given by $T\omega = \omega + \theta$, where $\theta$ is a d-vector, and $1$, $\theta_1, \ldots, \theta_d$ are required to be rationally independent. A function $f; T^d \to \mathbb{R}$ gives rise to a quasiperiodic potential.

<u>Remark.</u>  The hypothesis on the boundedness of $f(\omega)$ can be weakened to $\log(1 + |f(\omega)|) \in L^1(\Omega)$ in much of the following.

Some one-dimensional results are considered first. We introduce several functions relevant to the spectral problems; let $\Phi_\omega(x; \lambda)$ be the fundamental solution matrix of (1) for $\lambda \in \mathbb{C}$.

Define

$$\gamma(\lambda, \omega) = \lim_{T \to \infty} \frac{1}{T} \log \| \Phi_\omega(T; \lambda) \|.$$

To dispense with anomolous behaviour of $\gamma(\lambda,\omega)$ for atypical $(\omega,\lambda)$, we further define

$$\gamma(\lambda) = \int_{\Omega} \gamma(\lambda,\omega)\,dP$$

to be the Lyapounov index. Note that the quantity $\gamma(\lambda)$ is inherently *nonnegatve*, since the matrix $\Phi_{\omega}(x;\lambda)$ is symplectic. The Lyapounov index for the discrete problem is defined by analogy.

Secondly, define the integrated density of states for $\lambda \in \mathbb{R}$ as the following limit:

$$k(\lambda) = \lim_{T\to\infty} \frac{1}{T}\, N(T,\lambda,\omega) \tag{3}$$

where $N(T,\lambda,\omega) = \#\{\text{Dirichlet eigenvalues of H on } [0,T] \leq \lambda\}$. It turns out that $k(\lambda)$ exists and is a.e. independent of $\omega$ for *all* $\lambda$, and this is reflected in the notation. Here as well the quantity $k(\lambda)$ is inherently *nonnegative*, and is increasing in $\lambda$. Of course $k(\lambda)$ is related to the rotation numbers via the Sturm comparison theorem,

$$\pi k(\lambda) = \alpha(\lambda).$$

We define $k(\lambda)$ for $\lambda$ complex as well. Referring to Johnson, Moser [7], if $\theta(x;\lambda)$ is the Prüfer transform of (1), and $\lambda \in \mathbb{R}$,

$$\alpha(\lambda) = \lim_{T\to\infty} \frac{1}{T}\, \theta(T;\lambda),$$

while for $\lambda \in \mathbb{C} - \mathbb{R}$ we may define

$$\alpha(\lambda) = \lim_{T\to\infty} \frac{1}{T}\, \arg \phi(T;\lambda)$$

where $\phi(T;\lambda)$ is a complex solution of (1).

Finally, to motivate the study of these quantities, I quote several relations between $k(\lambda)$ and $\gamma(\lambda)$, and the spectrum of the operator (1).

<u>Theorem 1.</u>    $\sigma(H) = $ points of increase of $k(\lambda)$.

This theorem is stated in Johnson, Moser [7], as well as being explicit or implicit in much other work. If we define the set $S = \{\lambda;\gamma(\lambda) = 0\}$, then:

Theorem 2 [Ishii] [Pastur].   For a typical $\omega$, $\displaystyle\int_{\mathbb{R}-S} d\rho_{ac}(\lambda) = 0$.

Theorem 3 [Kotani].   For a typical $\omega$, for a.e. $\lambda \in S$, then $\lambda \in \sigma_{ac}(H)$.  In fact, if S contains an interval I, then $I \subseteq \sigma_{ac}(H)$ and $-\gamma(\lambda) + ik(\lambda)$ can be continued analytically through I.

## 2.  THE THOULESS FORMULA

We go now to the relationship between $k(\lambda)$ and $\gamma(\lambda)$ known as the Thouless formula.  From this we demonstrate the continuity of $k(\lambda)$ (a different proof than that of Johnson, Moser [7]) and then obtain some consequences and extensions of the formula.  The exposition will be carried out for the discrete problem (1) for illustration; for the continuous version, some modifications are necessary.  We have defined the quantity

$$\gamma(\lambda,\omega) = \lim_{T\to\infty} \frac{1}{T} \log\| \Phi_\omega(T;\lambda) \| .$$

In the discrete case the fundamental solution matrix is a product

$$\Phi_\omega(T;\lambda) = M_T(\lambda)\, M_{T-1}(\lambda)\, \ldots\, M_1(\lambda)$$

where each transfer matrix has the form

$$M_j(\lambda) = \begin{pmatrix} q_j(\omega)-\lambda & -1 \\ 1 & 0 \end{pmatrix}.$$

Thus it is clear that each matrix element $m_{ij}(T;\lambda)$ of $\Phi_\omega(T;\lambda)$ is a monic polynomial in $\lambda$ of degree $T$, $T-1$ or $T-2$.  Focus, for example, on the matrix element $m_{11}(T;\lambda)$ and denote its roots $\lambda_k(T;\omega)$,

$$m_{11}(T;\lambda) = \prod_{k=1}^{T} (\lambda-\lambda_k(T;\omega)).$$

This suggests definition of an approximation

$$\begin{aligned}
\gamma_T(\lambda,\omega) &= \frac{1}{T} \log \prod_{k=1}^{T} |\lambda-\lambda_k(T,\omega)| \\
&= \frac{1}{T} \sum_{k=1}^{T} \log|\lambda-\lambda_k(T,\omega)| \\
&= \mathrm{re} \int \log(\lambda-\lambda')\, dk_T(\lambda')
\end{aligned}$$

for an appropriate step function $k_T(\lambda)$. Observe that $dk_T(\lambda)$ is an approximating density of states measure defined by restricting the operator $h$ to the interval $[-1, T+1]$ with Dirichlet boundary conditions, for the vanishing of the matrix element $m_{11}$ for some $\lambda$ corresponds to the existence of an eigenfunction. Other matrix elements $m_{ij}$ correspond to similar Dirichlet problems, and it is not hard to show that any matrix element will give in the limit the same growth rate $\gamma(\lambda, \omega)$. For $\lambda \in \mathbb{C} - \sigma(h)$, $\log(\lambda-\lambda')$ is continuous on $\mathrm{supp}(dk)$, and the convergence of $dk_T(\lambda) \to dk(\lambda)$ weakly gives us

$$\gamma(\lambda, \omega) = \lim_{T \to \infty} \gamma_T(\lambda, \omega) = \mathrm{re} \int \log(\lambda-\lambda') dk(\lambda')$$

$$= \int \log|\lambda-\lambda'| dk(\lambda').$$

The hard part is to verify the formula for $\lambda \in \mathbb{R}$ in the spectrum. This is exactly where we want to use it.

<u>Definition.</u>  A function $\mu(\lambda)$ is *subharmonic* if both

    (i) $\mu(\lambda)$ is upper semicontinuous and

    (ii) for any disc $B_r(\lambda_o)$

$$\mu(\lambda_o) \leq \frac{1}{\pi r^2} \int_{B_r(\lambda_o)} \mu(\lambda) d\lambda.$$

This second property is called *submean*. An easy consequence of subharmonicity is the following:

<u>Lemma 1.</u>  If $\mu(\lambda)$ is subharmonic, then

$$\mu(\lambda_o) = \lim_{r \to 0} \frac{1}{\pi r^2} \int_{B_r(\lambda_o)} \mu(\lambda) d\lambda.$$

If we define both $\gamma(\lambda)$ and $\int \log|\lambda-\lambda'| dk(\lambda')$ to be $-\infty$ at values of $\lambda$ such that the expressions diverge to $-\infty$, then we have:

<u>Theorem 4.</u>  Both $\gamma(\lambda)$ and $\int \log|\lambda-\lambda'| dk(\lambda')$ are subharmonic.

This is the key observation. Use the fact that $\gamma(\lambda) = \int \log|\lambda-\lambda'| dk(\lambda')$ for a.e. $\lambda \in \mathbb{C}$ and the conclusion of Lemma 1 to conclude that:

Corollary.   The Thouless formula

$$\gamma(\lambda) = \int \log|\lambda-\lambda'| \, dk(\lambda')$$

holds for *all* $\lambda \in \mathbb{C}$.

Again remark that it is $\gamma(\lambda) = \int_\Omega \gamma(\lambda,\omega) \, dP$, the averaged quantity, that is being considered here.  I would like to indicate the ideas behind the proof of Theorem 4.

(1)  $\log|\lambda-\lambda'|$ is subharmonic, so by Fatou's lemma $\int \log|\lambda-\lambda'| \, dk(\lambda')$ is submean.

(2)  Furthermore, we can write $\int \log|\lambda-\lambda'| \, dk(\lambda') = \inf_N \int \max(-N, \log|\lambda-\lambda'|) \, dk(\lambda')$, an infimum of continuous functions.  This implies upper semicontinuity.

(3)  Consider the function $\gamma(\lambda)$ next.   $\Phi_\omega(T;\lambda)$ has elements which are polynomials in $\lambda$, hence it is classical that $\gamma(T,\lambda,\omega) = T^{-1} \log\|\Phi_\omega(T;\lambda)\|$ is subharmonic.   Since $\Phi_\omega(s_1 + s_2;\lambda) = \Phi_{T\omega^{s_1}}(s_2;\lambda) \Phi_\omega(s_1;\lambda)$, we have

$$\| \Phi_\omega(s_1 + s_2;\lambda) \| \;\leq\; \| \Phi_{T\omega^{s_1}}(s_2;\lambda) \| \; \| \Phi_\omega(s_1;\lambda) \|$$

and thus

$$(s_1 + s_2)\gamma(s_1 + s_2;\lambda,\omega) \;\leq\; s_2\gamma(s_2,\lambda, T_\omega^{s_1}) + s_1\gamma(s_1,\lambda,\omega).$$

Average over $\Omega$ to find that

$$s \int_\Omega \gamma(s,\lambda,\omega) \, dP = s\gamma(s,\lambda)$$

is subadditive, and hence

$$\gamma(\lambda) = \lim_{T\to\infty} \gamma(T,\lambda) = \varlimsup_{T\to\infty} \gamma(T,\lambda).$$

Thus the limit can be taken through a monotone decreasing sequence of subharmonic functions, and is consequently subharmonic.  The proof is concluded.   $\square$

Using the Thouless formula I would like to address the continuity of $k(\lambda)$. It follows from the following lemma.

<u>Lemma 2.</u>    Suppose that $\mu(\lambda)$ is monotone increasing, and that there exist constants $c_o > 0$, $c_1 > -\infty$ such that

$$\text{(i)} \quad \int d\mu(\lambda) < c_o$$

$$\text{supp } d\mu = \left[-A, A\right]$$

$$\text{(ii)} \quad c_1 \leq \int \log|\lambda - \lambda'| d\mu(\lambda').$$

Then $\mu(\lambda)$ is continuous, with a uniform logarithmic modulus of continuity.

<u>Proof.</u>    Consider $|\lambda_o - \lambda_1| < \frac{1}{2}$, with $\lambda_o < \lambda_1$,

$$c_1 \leq \int \log|\lambda_o - \lambda'| d\mu(\lambda')$$

$$= \int_{\lambda_o}^{\lambda_1} \log|\lambda_o - \lambda'| d\mu(\lambda') + \int_{\substack{\lambda' \notin (\lambda_o, \lambda_1) \\ |\lambda' - \lambda_o| \geq 1}} \log|\lambda_o - \lambda'| d\mu(\lambda')$$

$$+ \int_{\substack{\lambda' \notin (\lambda_o, \lambda_1) \\ |\lambda' - \lambda_o| < 1}} \log|\lambda_o - \lambda'| d\mu(\lambda').$$

The last term is negative, by the positivity of $d\mu(\lambda)$, and the first term is easily bounded from above:

$$c_1 \leq \log|\lambda_o - \lambda_1| \int_{\lambda_o}^{\lambda_1} d\mu(\lambda) + \int_{|\lambda' - \lambda_o| > 1} \log|\lambda_o - \lambda'| d\mu(\lambda').$$

Hence

$$-\log|\lambda_o - \lambda_1|(\mu(\lambda_1) - \mu(\lambda_o)) \leq -c_1 + c_o \log^+ 2A$$

That is,

$$|\mu(\lambda_1) - \mu(\lambda_o)| \leq \frac{c_2}{-\log|\lambda_1 - \lambda_o|}. \qquad \square$$

In this case we say that $k(\lambda)$ is *log-Hölder continuous*.

<u>Remark.</u>    In the continuous version of equation (1), the Thouless formula requires a convergence factor.  That is, one controls the integral

$$\int \log|\lambda-\lambda'| \, (dk(\lambda') - dk_o(\lambda'))$$

where $k_o(\lambda)$ is the 'free' integrated density of states. The Sturm comparison theorem implies that $|k(\lambda) - k_o(\lambda)| \le \lambda^{-\frac{1}{2}} \|q\|_{L_\infty}$ for $\lambda$ large, so that the integral is convergent

I conclude this section with some remarks on control of the Lyapounov exponent. In contrast to the function $k(\lambda)$, not a great deal is known. The following are several facts; I think it is an interesting open question to know more.

(1) Johnson $\begin{bmatrix}5\end{bmatrix}$ has examples (almost periodic-limit periodic) in which $\gamma(\lambda)$ is not continuous, so that one must expect less regular behaviour than $k(\lambda)$.

(2) However, $\gamma(\lambda) \ge 0$ and is upper semicontinuous; hence, on $S = \{\lambda; \gamma(\lambda) = 0\}$, $\gamma(\lambda)$ is continuous.

(3) The quantity $\log|\lambda'-(\lambda+i\epsilon)| \to \log|\lambda'-\lambda|$ monotonically as $\epsilon \to 0$, hence

$$\gamma(\lambda+i\epsilon) = \int \log|\lambda'-(\lambda+i\epsilon)| \, dk(\lambda') \to \gamma(\lambda)$$

for *all* $\lambda \in \mathbb{R}$. This can be strengthened to show that nontangential limits of $\gamma(\lambda)$ exist for *all* $\lambda$. Again remember that $\gamma(\lambda)$ is the averaged quantity.

(4) For $\lambda \in \mathbb{R}$, $\lambda < \lambda'$, then $\log|\lambda'-(\lambda+\epsilon)| \to \log|\lambda'-\lambda|$ monotonically as $\epsilon \to 0$. Hence for any interval such that $(a,b) \cap \sigma(H) = \phi$, but $b \in \sigma(H)$, we have

$$\gamma(b) = \lim_{\epsilon \to 0} \gamma(b-\epsilon).$$

See Johnson $\begin{bmatrix}6\end{bmatrix}$ for this result for almost periodic potentials.


## 3.  MORE THAN ONE-DIMENSION

In higher dimensional cases relatively little is known in general. Here I will discuss the discrete operator (1) alone. We first define the integrated density of states.

<u>Definition.</u>  Let $\Lambda_L \subseteq \mathbb{Z}^n$ be hypercubes of width L.  Then

$$k(\lambda) = \lim_{\Lambda_L \to \mathbb{Z}^n} \frac{1}{|\Lambda_L|} \#\{\text{Dirichlet eigenvalues of } h \text{ on } \Lambda_L \le \lambda\}. \tag{4}$$

<u>Remark.</u>  The quantity $k(\lambda)$, as before, is inherently nonnegative, and increasing in $\lambda$.

A very nice result of Wegner [11] is repeated here.  It applies to the
independent random potential case.

<u>Theorem 5.</u>    Suppose that the potential $q_j$ is given by independent random
variables.  If there is a uniform bound on the distribution densities $g(q)$,
then $k(\lambda)$ is Lipschitz.

<u>Proof.</u>    Consider a large cube $\Lambda$, and set $N_\Lambda(\lambda,q) = \#\{$Dirichlet eigenvalues
of $h(q)$ on $\Lambda \leq \lambda\}$.  We will bound $|\Lambda|^{-1}\partial N_\Lambda/\partial\lambda$ uniformly in $|\Lambda|$.  The first
remark is that $\lambda$-differentiation may be exchanged for differentiation with
respect to $q$,

$$\frac{\partial N_\Lambda}{\partial\lambda} = \sum_{j\in\Lambda} - \frac{\partial}{\partial q_j} N_\Lambda(\lambda,q).$$

Average over all potentials within $\Lambda$,

$$\int \cdots \int \frac{\partial N_\Lambda}{\partial\lambda} \cdot \prod_{j\in\Lambda} g_j(q_j)dq_j = - \sum_{j\in\Lambda} \int \cdots \int \frac{\partial N_\Lambda}{\partial q_j}(\lambda,q) \prod_{j\in\Lambda} g_i(q_j)dq_j.$$

We note that $-\partial N_\Lambda/\partial q_j \geq 0$ since $N_\lambda$ is decreasing in $q_j$; hence,

$$- \frac{\partial N_\Lambda}{\partial q_j} \leq \sup_{\substack{s\in\mathbb{R}\\k\in\Lambda}} g_k(s) \cdot \sum_{j\in\Lambda} \int \cdots \int \left(-\int \frac{\partial N_\Lambda}{\partial q_j} dq_j\right) \prod_{i\neq j} g_i(q_i)dq_i$$

$$\leq |\Lambda| \sup g_k(s) \int - \frac{\partial N_\Lambda}{\partial q_j} dq_j.$$

Since a change in $q_j$ is a rank one perturbation of $h$, this last integral is
at most one, and the estimate is complete.    $\square$

For the general stochastic potential this result is of course not true,
the regularity cannot be so good.  For example, if the potential is
Bernouilli,

$$q_j = \begin{cases} -c & \text{with probability } p \\ c & \text{with probability } (1-p) \end{cases}$$

then $k(\lambda)$ is not expected to be Lipschitz.  And in the almost periodic case
there exist examples [3] where $k(\lambda)$ is not Hölder $\alpha$ continuous for any
$0 < \alpha \leq 1$.  However an approach through a Thouless-like formula still works

to obtain the continuity (with a logarithmic modulus) of $k(\lambda)$.  I give  here
a brief sketch of the proof.

The goal is to show that there is a constant such that

$$c \leq \int \log|\lambda-\lambda'| \, dk(\lambda') \tag{5}$$

and then to invoke Lemma 2.  In fact $c$ can be chosen to be zero.  One
approximates $k(\lambda)$ by the integrated density of states for strips of width L.
Strip problems are in some sense one-dimensional, and there is an analog of
the Thouless formula which holds.  That is, we can define $2L \times 2L$ transfer
matrices for the initial value problem in the strip

$$\Phi_\omega(T;\lambda) = M_T(\lambda) \cdot M_{T-1}(\lambda) \ldots M_1(\lambda)$$

where each

$$M_j(\lambda) = \begin{pmatrix} A_j(\lambda) & -I \\ I & 0 \end{pmatrix} .$$

$A_j(\lambda)$ is the transverse operator, a Schrödinger operator on an $(n-1)-$
dimensional cube of width L, with Dirichlet boundary conditions.  Now define

$$\gamma(\lambda,L,\omega) = \lim_{T\to\infty} \frac{1}{L^{n-1}} \frac{1}{T} \log \| \Lambda^L \Phi_\omega(T;\lambda) \| \tag{6}$$

where $\Lambda^L$ denotes the usual 'wedge' or antisymmetric tensor product.  If
$\gamma(\lambda,L)$ is defined to be the averaged quantity, again we can associate it and
$k(\lambda,L)$ (the strip integrated density of states) via a Thouless-type formula,
to conclude:

Theorem 6.    (i)  $\gamma(\lambda,L)$  is subharmonic;

           (ii)  $k(\lambda,L)$  is log-Hölder continuous.

Finally, using the ergodic theorem, $k(\lambda,L) \to k(\lambda)$ as $L \to \infty$, while (5) holds
uniformly in L for each $k(\lambda,L)$.

I finish this article with several questions.

(1)  For strip problems we may define

$$L_\gamma(\lambda,L) = \gamma_1(\lambda) + \ldots + \gamma_L(\lambda)$$

where recursively

$$\gamma_j(\lambda) + \gamma_{j-1}(\lambda) + \ldots + \gamma_1(\lambda) = \lim_{T \to \infty} \frac{1}{T} \log \| \Lambda^j \Phi_\omega(T;\lambda) \| \, .$$

Each $\gamma_j(\lambda)$ is harmonic on $\mathbb{C}^+$. Are they subharmonic on  ? If so, what meaning does the associated measure $d\rho_j(\lambda)$ have?

(2) Is there a connection between $\gamma_L(\lambda)$ (the smallest) and $\sigma_{ac}(h)$? For example, are there analogs of Theorems 2 and 3?

(3) What is the distribution of $\gamma_j(\lambda)$, $1 \leq j \leq L$ as $L \to \infty$? Does this give an indication of the spectral multiplicity of $\sigma_{ac}(h)$ in several dimensions? It has been conjectured that the spectral multiplicity of $\sigma_{ac}(h)$ in the random case should be only 1. As well, differences may appear between the dimensions 2 and $n \geq 3$.

(4) As a final remark, I want to mention that the continuous case of Theorem 6 in $\mathbb{R}^n$ $n > 1$ seems more difficult.

REFERENCES


[1] Craig, W. and Simon B., Subharmonicity of the Lyapounov index, *Duke Math. Journal, 50 (2)* (1983) 551–560.

[2] Craig, W. and Simon B., Log-Hölder continuity of the integrated density of states for stochastic Jacobi matrices, *Commun. Math. Phys., 90* (1983) 207–218.

[3] Craig, W., Pure point spectrum for discrete almost periodic Schrödinger operators, *Commun. Math. Phys. 88,* 113–131 (1983).

[4] Ishii, K., Localization of eigenstates and transport phenomena in the one dimensional disordered system, *Supp. Theor. Phys. 53,* 77–138 (1973).

[5] Johnson, R., The recurrent Hill's equation, *Journal Diff. Eq. 46(2),* 165–193 (1982).

[6] Johnson, R., Lyapunov exponents for the almost periodic Schrödinger equation, *Ill. J. Math., 28 (3)* (1984) 397–419.

[7] Johnson, R. and Moser, J., The rotation number for almost periodic potentials, *Commun. Math. Phys. 84,* 403–438 (1982).

[8] Kotani, S., Lyapounov indices determine absolutely continuous spectra of stationary random one-dimensional Schrodinger operators, *Proc. Kyoto Stoch. Conf.* (1982).

[9]  Pastur, L.,  Spectral properties of disordered systems in the one
     body approximation, *Commun. Math. Phys.* 75, 179-196 (1980).

[10] Thouless, D.,  A relation between the density of states and the range
     of localization for one dimensional random systems, *J. Phys. C.*
     5, 77-81 (1972).

[11] Wegner, F.,  *Z. Physik B* (1983) (to appear).

Walter Craig,
Department of Mathematics,
California Institute of Technology,
Pasadena,
California  91125,
U.S.A.

present address:

Department of Mathematics,
Stanford University,
Stanford, California 94305,
U.S.A.

D DÜRR

# A limit theorem for stochastic acceleration with self-intersecting particle trajectories

## INTRODUCTION

Rigorous work in non-equilibrium statistical mechanics is mainly concerned with the derivation of the macroscopic (phenomenological) behaviour of a complex system from the underlying microscopic equations of motion of its constituents.

Often this is done via the so-called reduction procedure where only a few relevant variables are singled out and their time evolution is considered under the "noisy" influence of the many remaining variables. A typical example is that of an infinite ideal gas system of identical point particles in one dimension where one considers the motion of a tagged particle under the influence of elastic collisions of the other ideal gas particles. Suppose that the initial condition of the system is the following. The tagged particle starts at the origin with velocity ±1 each with probability 1/2. The other particles start at the even integers also with velocities ±1 with probability 1/2. The time evolution of the position of the tagged particle, which we assume to have the same mass as the other particles, is determined by elastic collisions between the tagged particle and the other particles. One can convince oneself rather easily that the tagged particle will perform a simple symmetric random walk where at each time $t_i = \frac{i}{2}$ , $i \in \mathbb{N}$ , the velocity of the tagged particle will either change its sign or keep its sign, each with probability 1/2. Call $v_i$ the velocity of the tagged particle at time $t_i$, then the $v_i$'s are all i.i.d. random variables and

$$X(t) = \sum_{i=0}^{[t]} v_i + v_{[t]} (t - [t])$$

is the position of the tagged particle on the microscopic scale, i.e. the effect of each single collision is visible in the graph of $X(t)$.

The important step in the reduction procedure is now the scale transformation. One describes the motion of the tagged particle on a macroscopic scale. This is most easily understood in the example above by introducing

$$X_A(t) = \frac{1}{\sqrt{A}} X(At)$$

where A is a parameter which should be considered to have a high value. Then, using the formula for X(t), we obtain

$$X_A(t) = \frac{1}{\sqrt{A}} \sum_{i=0}^{[At]} 1/2 \; v_i + v_{[At]} (At - [At])$$

which is the standard sum one looks at in the central limit theorem for i.i.d. random variables; moreover, we know $\boxed{4}$ that, as A goes to infinity, $X_A(t)$ converges in distribution to a Wiener process. The Wiener process is, so to speak, the macroscopic description of the simple random walk model. Note that for very large A the single collisions are no longer visible in the graph of $X_A(t)$.

The scaling one uses depends on the model. In the central limit theorem type of scaling, which is often called hydrodynamical scaling, space and time are scaled differently (as in diffusion, since the mean square displacement of the Wiener process $< W(t)^2 > \sim t$). In the model we describe next, space and time are scaled in the same way.


1. <u>DESCRIPTION OF PARTICLE MOTION</u>

The problem of describing the motion of a particle in a random medium arises naturally in many areas of physics: there is the motion of a charged particle in a plasma, the motion of a pollen in a liquid (Brownian motion) or, in general, a particle moving in a turbulent fluid. The medium, which we shall call the fluid, will be modelled by a random force field $\underline{F}(x,t) \in \mathbb{R}^\nu$, which for fixed $\underline{x} \in \mathbb{R}^\nu$ and $t \in \mathbb{R}^+$ is a random variable on some probability space $(\Omega, F, P)$. For each $\omega \in \Omega$, $\underline{F}(x,t,\omega)$ is then a realization of the force field.

The equations of motion for a point particle moving in the force field are the Newtonian equations

$$\frac{d\underline{x}(t)}{dt} = \underline{v}(t), \qquad \underline{x}(0) = \underline{x}_o \tag{1}$$

$$\frac{d\underline{v}(t)}{dt} = \underline{F}(x(t),t,\omega), \qquad \underline{v}(0) = \underline{v}_o \tag{2}$$

where $\underline{x}(t)$ is the position of the particle at time t and $\underline{v}(t)$ is its velocity. It is immediate from (1) and (2) that $\underline{x}(t)$ are also random variables on $(\Omega, F, P)$ and hence $(\underline{x}_t, \underline{v}_t)_{t \in [0, \infty)}$ is a stochastic process, if $F$ is well behaved. It is of course clear that in general not much more can be said about this; physicists (and thus mathematicians) therefore tend to look at the problem from a perturbation point of view, i.e., when the influence of the force field on the particle is very small so that the effect is only visible when the motion of th particle is considered on appropriate time (long) and space (large) scales. To get to some level of rigour one eventually ends up considering limit situations, such as the so-called van Hove limit [1]. This limit is described in detail later in connection with the special fluid model we are concerned with in this paper.

First, however, I wish to remark that the limit contemplated here is likewise a transition from a microscopic (detailed) description to a macroscopic one, in which the detailed microscopic events are rendered out of focus. It is quite natural to expect on the macroscopic level the motion of the particle to be of much simpler form, for example that the position process or velocity process is a diffusion process. In our introductory example we obtained a Wiener process for the position. The scaling considered here is a special one since it goes hand in hand with the rescaling of the force field. The limit procedure is therefore often called the *weak coupling limit*, which is to be distinguished from the *hydrodynamical limit* in which the force field (i.e. the physics) remains unchanged, while the motion of the particle is looked upon on a macroscopic scale.


2. <u>CONSTRUCTION OF THE FORCE FIELD</u>

The force field will be generated by identical scatterers, which are spherically symmetric potentials each of which creates a force field of finite range, and which are uniformly distributed (randomly) in space $\mathbb{R}^\nu$. Technically this is done as follows.

Let $P_\rho = (\Omega, F, P)$ be a Poisson point process on $\mathbb{R}^\nu$ ($\nu \geq 2$) with density $\rho$, i.e., for a Borel set B in $\mathbb{R}^\nu$ the number $N_B$ of points in B is Poisson distributed with mean $P(N_B) = |B|\rho$, where $|B|$ denotes the Lebesgue measure of B. The uniform (Poisson) distribution has the important property that the numbers of points in distinct regions of space are independent (random variables). Let $V(\underline{r}) = V(|\underline{r}|) \in C_o^3(\mathbb{R}^\nu)$ represent a potential of compact support and let $U(\underline{x})$ denote the random potential field:

$$U(\underline{x}) = \sum_{\underline{r}_i \in \omega} V(|\underline{x} - \underline{r}_i|) \tag{3}$$

where $(\underline{r}_i)_{i \in \mathbb{N}} =: \omega \in \Omega$ is a realization of random points of $P_\rho$. The negative gradient of $U(\underline{x})$ gives the desired force field. (The centres $\underline{r}_i$ of the scatterers are uniformly distributed.)

For $\varepsilon > 0$ let us introduce the scaled Poisson field

$$P_{\rho\varepsilon} = (\Omega, F, P^\varepsilon) \text{ with density } \rho^\varepsilon = \rho\varepsilon^{-2\nu} \tag{4}$$

and let us consider the scaled potential field $(U^\varepsilon(\underline{x}), P_{\rho\varepsilon})$ where

$$U^\varepsilon(\underline{x}) = \sum_{\underline{r}_i \in \omega} \varepsilon V\left(\frac{|\underline{x} - \underline{r}_i|}{\varepsilon^2}\right) =: \sum_{\underline{r}_i \in \omega} V^\varepsilon(|\underline{x} - \underline{r}_i|). \tag{5}$$

Note that the support $\kappa^\varepsilon$ of the scaled scatterer $V^\varepsilon(|\underline{x}|)$ shrinks as $\varepsilon \to 0$, namely $|\kappa^\varepsilon| = \varepsilon^{2\nu}|\kappa|$, but by virtue of (4) the fraction of volume occupied by the scatterers is independent of $\varepsilon$.

Note also that $U^\varepsilon(\underline{x}) \sim \varepsilon$ (weak coupling) and that the shrinking of the support of the scatterers and the rescaling of the density reflects a rescaling of space. The reader should therefore not be surprised to find the problem already posed in its final form when we consider now the motion of a point particle in the potential field (5). (We are really looking then at the motion of the particle on a time-scale $\varepsilon^{-2}$ and space-scale $\varepsilon^2$ $(x_\varepsilon(t) = \varepsilon^2 x(t\varepsilon^{-2}))$ in a potential field $U \sim \varepsilon$, with otherwise unscaled scatterers and unscaled density $\rho$: von Hove limit.)

We have, with $\underline{F} = -\nabla V$ and (1), (2) and (5),

$$\frac{d\underline{x}^\varepsilon(t)}{dt} = \underline{v}^\varepsilon(t), \qquad \underline{x}^\varepsilon(0) = \underline{x}_o \tag{6}$$

and

$$\frac{d\underline{v}^\varepsilon(t)}{dt} = \sum_{\underline{r}_i \in \omega} \frac{1}{\varepsilon} \underline{F}\left(\frac{|\underline{x}^\varepsilon(t) - \underline{r}_i|}{\varepsilon^2}\right), \qquad \underline{v}^\varepsilon(0) = \underline{v}_o. \tag{7}$$

Of course, $(\underline{x}^\varepsilon(t), \underline{v}^\varepsilon(t))_{t \in [0, \infty)}$ is a stochastic process on $P_{\rho\varepsilon}$, provided that a unique solution of (7) almost certainly exists, i.e., if the sum in

(7) is almost certainly finite for all $\varepsilon > 0$. To show this requires merely a minor technical argument. We henceforth assume that a stochastic process as solution of (6) and (7) exists and denote it by $(\underline{x}_{-t}^{\varepsilon}, \underline{v}_{-t}^{\varepsilon})$.

## 3. LIMIT OF DISTRIBUTION OF $\underline{v}_{-t}^{\varepsilon}$ AS $\varepsilon \to 0$

Our theorem, which of course reflects the spirit of our introductory remarks, establishes and describes the limit of the distribution of the process $\underline{v}_{-t}^{\varepsilon}$ as $\varepsilon \to 0$. The rest of the paper is then devoted to outlining the proof and the difficulties involved. We shall see that a special role is played by the dimensionality  of the space. The theorem has been proved before for more general force fields by Kesten and Papanicolaou [2] for dimension $\nu \geq 3$. Our contribution is to show that the result, at least in the above model, holds true in dimension $\nu = 2$. The complete proof appears in a paper by S. Goldstein, J.L. Lebowitz and the author [3].

<u>Theorem.</u>    For $\nu \geq 2$ let $\underline{v}_t$ be a Wiener process* on the $\nu-1$-sphere of radius $v_o$ with diffusion coefficient

$$D = \frac{\pi}{\nu-1} \frac{\rho}{v_o} \int d^{\nu-1}\underline{k} \; k^2 |V(k)|^2$$

where $V(k)$ is the Fourier transform of $V(x)$. Then $\underline{v}_{-t}^{\varepsilon}$, the solution of (6) and (7), converges in distribution to $\underline{v}_t$ as $\varepsilon \to 0$.

Note that we may define the Wiener process on the sphere via its generator

$$A = D.\nabla_v.\underline{\underline{P}}.\nabla_v$$

where

$$\underline{\underline{P}} = \frac{\underline{v}\underline{v}}{v_o^2} \; , \qquad \underline{v}\,\underline{v} = \text{the tensor produce}$$

and

$$\nabla_v = ( \frac{\partial}{\partial v_1} , \frac{\partial}{\partial v_2} , \frac{\partial}{\partial v_3} ).$$

---

*Its transition probability p solves $\partial p/\partial t = D\Delta_B p$, where $\Delta_B$ is the Laplace Beltrami operator on the $\nu-1$-sphere of radius $v_o$.

It follows that also $\underline{x}_t$ converges in distribution to $\underline{x}_t = \underline{x}_o + \int_0^t \underline{v}(s)\,ds$.

The convergence in distribution is a synonym for the weak convergence of measures, i.e., the convergence of expectations of bounded continuous functions on the path space of the stochastic processes. Here $\underline{v}_t^\varepsilon$ induces a measure $\mu^\varepsilon$ on $C([0,\infty), \mathbb{R}^\nu)$, the space of continuous functions on $[0,\infty)$ with values in $\mathbb{R}^\nu$, and likewise $\underline{v}_t$ induces a measure $\mu$ on the same space. Hence the theorem states that $\mu^\varepsilon(f) \to \mu(f)$ as $\varepsilon \to 0$ for all bounded $f \in C(C([0,\infty), \mathbb{R}^\nu), \mathbb{R})$. In the following I shall try to explain on a heuristic level why the theorem holds, thereby indicating the sharp edges which have to be smoothed out in a rigorous approach.

We start with a crucial observation about the process $(\underline{x}_t^\varepsilon, \underline{v}_t^\varepsilon)$. The fact that the scatterers are extended means that this process is not Markovian. To be more specific, we give two reasons:

(1) A scatterer affects the particle over a time period of length $\varepsilon^2$. Due to the force acting on it, its path curves smoothly in the scatterer during this time. Thus the future path of the particle is not independent of the past trajectory, given the present data, because the curve gives information about the position of the particle relative to the scatterer.

(2) The particle can return to places it has previously visited. Since the curve of the past trajectory contains information about the location of scatterers, this indicates its future evolution as affected by real recollisions (when the particle enters a "known" scatterer again) and virtual recollisions (when the particle traverses again a region which is free from scatterers). It will be intuitively clear to the reader that the first cause of "non-Markovianness" is independent of the space dimension, whereas the second cause seems very dependent on this dimension since it has to do with the self-intersection of a path $\underline{x}(t,\omega)$: a path wiggling around in two dimensions is more likely to intersect itself (in the present problem, with probability one) than it is in higher dimensions (with probability zero). The statements in parenthesis hold for the limit process $\underline{x}_t = \int_0^t v(s)\,ds + \underline{x}_o$, which of course, for very small $\varepsilon$, should be near $\underline{x}_t^\varepsilon$ and therefore contains the relevant information about the self-intersection properties of $\underline{x}_t^\varepsilon$. Roughly speaking, in dimension $\nu \geq 3$ one need only get rid of the first

cause, since the second vanishes automatically for "topological" reasons as
$\varepsilon \to 0$. In two dimensions, however, the second cause is always present and
is rendered unimportant only by the fact that the effect of a finite number
of scatterers on the motion of the particle vanishes as $\varepsilon \to 0$, which again
is due to the shrinking of the support of the scatterer and the weakness of
the potential. For this very reason the first cause will not persist in the
limit; thus both obstacles to being Markovian are removed at one stroke.
Nevertheless, in the rigorous proof a separate argument from that which
overcomes the first obstacle is used to obliterate the second. We shall
come back to this point. Now that the reader is warned of the non-Markovian
character of the motion, a simple heuristic argument is presented for the
existence and structure of the limit process $\underline{v}_t$.

According to (5) the potential energy of the particle is of order $\varepsilon$ (if
the number of overlapping scatterers stays bounded), hence the speed of the
particle (by conservation of energy) will be roughly $v_o$. The cross-section
of a scatterer is of order $\varepsilon^{2(\nu-1)}$ (again by (5)) and hence the rate $R^\varepsilon$ for
"collisions" of the particle with a scatterer is $R^\varepsilon \sim v_o \rho^\varepsilon \varepsilon^{2(\nu-1)} \sim 1/\varepsilon^2$
by (4). The effect $\Delta\underline{v}$ of one scatterer on the velocity of the particle is

$$\Delta\underline{v} = \frac{1}{\varepsilon} \int_{-\infty}^{\infty} dt\ \underline{F}(\frac{\underline{x}(t)}{\varepsilon^2}) \sim \frac{1}{\varepsilon} \int_{-\infty}^{\infty} dt\ \underline{F}(\frac{\underline{r}_s}{\varepsilon^2} + \underline{v}\,\frac{t}{\varepsilon^2}) \tag{9}$$

$$\sim \varepsilon \int_{-\infty}^{\infty} dt\ \underline{F}(\frac{\underline{r}_s}{\varepsilon^2} + \underline{v}t) \sim \varepsilon$$

where $\underline{r}_s$, $\underline{v}$ are the position and velocity of the particle at the moment of
entrance ($t = 0$) into the scatterer. Hence the variance per scattering
event is $<\Delta\underline{v}^2> \sim \varepsilon^2$. If we multiply this by the rate $R^\varepsilon$ times t we get
the total variance in time t, which is of order t. In doing this we assume
of course that the scattering events become independent as $\varepsilon \to 0$; this means
in particular that the process becomes Markovian.

From the symmetry of the scatterer field it is also intuitively clear
that the expected change in angle $< \Delta\Psi > = 0$, where $\Delta\Psi = $ angle $(\underline{v}, \underline{v} + \Delta\underline{v})$
(note that by the law of conservation of energy a scatterer can only cause
a change of the direction of the particle velocity). So far we can conclude,
recalling, for example, the standard central limit theorems, that the
velocity under the above circumstances will describe a Wiener process on the
sphere.

34

The averaging $< \ >$ is an integration over the surface points $\underline{r}_s$ (cf. (9)) for which $\underline{r}_s \cdot \underline{v} \leq 0$, assuming that the scatterer is centred at the origin. Noting this and using (9), one computes with little effort the diffusion coefficient in the theorem from $< \Delta\underline{v}^2 >$.

So far we have dealt with the heuristics, which are useful in that they will guide us through the rigorous proof. We shall henceforth restrict ourselves to dimension $\nu = 2$.

Our proof is a lengthy exercise on the (elementary) theory of weak convergence of measures [4], but the fact central to our argument is the following.

Suppose that we have an abstract process $\underline{\tilde{v}}_t^\varepsilon$ (this means that the process is primarily defined not by its realization on a particular probability space but, for example, through a transition probability – if it is a Markov process) for which we know that $\underline{\tilde{v}}_t^\varepsilon \to \underline{v}_t$ (convergent in distribution). Then we may conclude that $\underline{v}_t^\varepsilon \to \underline{v}_t$, provided that we can realize $\underline{v}_t^\varepsilon$ simultaneously with $\underline{\tilde{v}}_t^\varepsilon$ on the same probability space say $(\overline{\Omega}, \overline{F}, \overline{P}^\varepsilon)$ such that, for any $T < \infty$ and $\delta > 0$,

$$\lim_{\varepsilon \to 0} \overline{P}^\varepsilon \{ (\overline{\omega}; \sup_{t \in [0,T]} |\underline{\tilde{v}}_t^\varepsilon(\overline{\omega}) - \underline{v}_t^\varepsilon(\overline{\omega})| > \delta \} ) = 0. \tag{10}$$

A simultaneous realization of two processes is called a coupling and if (10) holds the coupling is said to be good [5].

The next observation gives the clue to how to proceed. We normalize the scatterers to be discs of radius $\varepsilon^2$. Whenever the path $\underline{x}_t^\varepsilon(\omega)$ stays away from its past $2\varepsilon^2$-neighbourhood* the particle encounters "fresh", i.e., Poisson-distributed scatterers. There are three steps in the proof of the theorem.

<u>Step 1:</u>  Define a Markov process $\underline{\tilde{v}}_t^\varepsilon$. This will be a jump process with exponential waiting time, determined by the "collision rate" at which a particle with speed $v_o$ encounters uniformly distributed scatterers. If a jump occurs, its magnitude is computed by the r.h.s. of (9), where $\underline{v}$ is the value of the process $\underline{\tilde{v}}_t^\varepsilon$ just before the time of the jump. Since the size of

---

* Consider the motion of a sphere of radius $2\varepsilon^2$ centred at $\underline{x}_t^\varepsilon(\omega)$. In this sphere lie all the centres of scatterers which affect the motion of the particle at time t. The region traced by the sphere is a tube: the $2\varepsilon^2$-neighbourhood.

the jump depends on $\underline{r}_s$, the transition probability is computed from the probability that a particle enters a scatterer at $\underline{r}_s$ while the centre of the scatterer is Poisson-distributed. Note that by the rotational symmetry of the scatterer field the transition rate of the Markov process $\underline{v}_t$ will be rotationally invariant.

It is a matter of routine computation to establish that $\underline{\tilde{v}}_t^{\varepsilon} \to \underline{v}_t$. What one does is to check the strong convergence of the family of generators of the processes $\underline{v}_t^{\varepsilon}$ to the generator $A$ of $\underline{v}_t$, on a core of $A$. The computation involved is of the kind suggested in the heuristic argument.

<u>Step 2:</u>  Define a semimechanical process $\underline{\tilde{u}}_t^{\varepsilon}$ which behaves very much like $\underline{v}_t^{\varepsilon}$ except that, in regions where $\underline{\tilde{y}}_t^{\varepsilon} = x_o + \int_0^t \underline{\tilde{u}}_s^{\varepsilon} \, ds$ comes back to its past $2\varepsilon^2$-neighbourhood, a particle following the $\underline{\tilde{u}}_t^{\varepsilon}$ process encounters fresh scatterers. The fluid, so to speak, is always kept in equilibrium. This process may be constructed by replacing the scatterer field around the scatterers which overlap the particle, at each instant of time, by a "fresh" one so that nothing can be learned from the past particle trajectory. But still the path curves smoothly in a scatterer which prevents the motion from being Markovian. We show that $\underline{\tilde{u}}_t^{\varepsilon} \to \underline{v}_t$ by showing that there exists a good coupling of $(\underline{\tilde{u}}_t^{\varepsilon}, \underline{\tilde{v}}_t^{\varepsilon})$.

Ignoring some minor technical details, the coupling is done as follows. $\underline{\tilde{v}}_t^{\varepsilon}$ will be represented on the probability space $(\Omega_u, F_u, P_u^{\varepsilon})$ on which $\underline{\tilde{u}}_t^{\varepsilon}$ is defined. We prescribe the motion of a particle Ma, the velocity of which is $\underline{\tilde{v}}_t^{\varepsilon}$. We call Me the particle which follows $\underline{\tilde{u}}_t^{\varepsilon}$.

Ma and Me have the same initial conditions. The motions of Ma and Me are the same until, at time $\tau$, say, Me enters a scatterer, say at $\underline{r}_s$, with direction given by $\underline{\tilde{u}}^{\varepsilon} = \underline{v}$. At this moment Ma changes its velocity instantaneously by an amount determined by (9). Ma keeps its new velocity, say $\underline{v}'$, until Me enters the next scatterer, at which time the velocity of Ma changes again with the "collision" parameters $\underline{r}_s'$ and $\underline{v}'$, where $\underline{r}_s'$ is determined by the requirement that the angle between $\underline{r}_s'$ and $\underline{v}'$ is the same as for the corresponding parameters of Me. One then proceeds in the same way.

A word of warning: The speeds in the two processes are different and Ma has collisions at the rate given by the speed of Me. One really obtains $\overset{\sim}{\underset{-t}{v}}{}^{\varepsilon}$ on a time-scale that is random but is very close to the correct one (recall that $|\overset{\sim}{\underset{-t}{u}}{}^{\varepsilon}| \sim v_{o}$). It needs some moments of reflection to convince oneself that one obtains from this construction a true copy of $\overset{\sim}{\underset{-t}{v}}{}^{\varepsilon}$.

There is still a long way to go to show that this coupling is good (i.e., to prove (10)), but having done so one has nearly proved the theorem in dimensions $\nu \geq 3$, where self-intersection in the limit position process does not occur with probability one. Kester and Papanicolaou used a different technique which is not so down-to-earth as ours; theirs is analytic and handles also non-isotropic scatterer fields, but fails in two dimensions.

<u>Step 3:</u>   A new process $\underset{-t}{u}^{\varepsilon}$ is now used to handle the self-intersections. Recall that, in between self-intersections of the process $\underset{-t}{x}^{\varepsilon}$, $\underset{-t}{v}^{\varepsilon}$ behaves like $\overset{\sim}{\underset{-t}{u}}{}^{\varepsilon}$ and the differences between $\underset{-t}{v}^{\varepsilon}$ and $\overset{\sim}{\underset{-t}{u}}{}^{\varepsilon}$ result only from the effect of the scatterers which overlap the areas of self-intersection. Now if such an area is sufficiently small, say of the order of the support of a scatterer (i.e., of the order $\varepsilon^{4}$), then, since $\rho^{\varepsilon} \sim 1/\varepsilon^{4}$ ($\nu = 2$), the number of scatterers in this problematic area is roughly speaking finite; hence their total effect on the velocity of the particle will be of order $\varepsilon$.

It is easily seen that the area of self-intersection, at least of the

process $\overset{\sim}{\underset{-t}{y}}{}^{\varepsilon} = \int_{0}^{t} \overset{\sim}{\underset{-s}{u}}{}^{\varepsilon} ds + \underset{-o}{x}$, will be of order $\varepsilon^{4}$ if the self-intersections of

the limit process $\underset{-t}{x}$ are transversal, i.e., not tangential. For a tangential intersection there exist two time points $s < t$ such that

$$\underset{-}{x}(s) = \underset{-}{x}(t) \qquad \text{and} \qquad \underset{-}{v}(s) = \underset{-}{v}(t) \qquad \text{or} \qquad \underset{-}{v}(s) = -\underset{-}{v}(t).$$

If we can show that, with probability one, tangential intersections do not occur, we are almost done since then we can construct a good coupling of $\underset{-t}{v}^{\varepsilon}$ and the new process $\underset{-t}{u}^{\varepsilon}$ which is pieced together from copies of $\overset{\sim}{\underset{-t}{u}}{}^{\varepsilon}$ in the following way.

Set $\underset{-t}{u}^{\varepsilon} = \underset{-t}{v}^{\varepsilon} = \overset{\sim}{\underset{-t}{u}}{}^{\varepsilon}$ up to the "first self-intersection" of $\underset{-t}{x}^{\varepsilon}$ (i.e., the first time $\underset{-t}{x}^{\varepsilon}$ enters its past $2\varepsilon^{2}$-neighbourhood), which then is of course the first "self-intersection" of $\underset{-t}{y}^{\varepsilon}$. Then $u_{t}^{\varepsilon}$ continues to be like $\overset{\sim}{\underset{-t}{u}}{}^{\varepsilon}$ until $\underset{-t}{x}^{\varepsilon}$ leaves its past $2\varepsilon^{2}$-neighbourhood again, say at time $\tau$. Then set $\underset{-t}{u}^{\varepsilon} = \underset{-t}{v}^{\varepsilon} = \overset{\sim}{\underset{-t}{u}}{}^{\varepsilon}(\{\underset{-}{v}^{\varepsilon}\})$ until the next self-intersection and so on. Here $\overset{\sim}{\underset{-t}{u}}{}^{\varepsilon}(\{\underset{-\tau}{v}^{\varepsilon}\})$ denotes the process $\overset{\sim}{\underset{-t}{u}}{}^{\varepsilon}$ starting at time $\tau$ with the initial conditions $\underline{v}_\tau^\epsilon$ and the scatterer configuration which overlaps $\underline{x}_\tau^\epsilon$.

Roughly speaking, $\underline{u}_t^\epsilon$ is made up of pieces of $\underline{\tilde{u}}_t^\epsilon$ which converge in distribution to pieces of the limit process $\underline{v}_t$; moreover the pieces taken together even constitute the limit process. Hence $\underline{u}_t^\epsilon \to \underline{v}_t$, but according to the construction $\underline{u}_t^\epsilon$ is well coupled to $\underline{v}_t^\epsilon$. Thus $\underline{v}_t^\epsilon \to \underline{v}_t$.

## 4. PROBABILITY OF TANGENTIAL SELF-INTERSECTION OF $\underline{x}_t$ OCCURRING WITHIN $[0,T]$

I conclude this note with an (informal) proof of the statement: the probability of tangential self-intersection of $\underline{x}_t$ occurring in the interval $[0,T]$ is zero.

We shall employ the following facts:

(i) $(\underline{x}_t,\underline{v}_t)$ is Markovian;

(ii) the conditional probability distribution of the random variable $(\underline{x}(t), \underline{v}(t))$ given $(\underline{x}(s),\underline{v}(s))$ is absolutely continuous with bounded density for $t > s$;

(iii) $|\underline{v}_t| = v_o = 1$;

(iv) $\underline{v}_t$ is a Brownian motion and therefore Hölder continuous with exponent $\alpha < \frac{1}{2}$, i.e., for $t,s \in [0,T]$,

$$|v(t) - v(s)| \leqq K|t-s|^\alpha. \tag{11}$$

(i) is clear;   (ii) follows from the hypoellipticity of the operator

$$-\frac{\partial}{\partial t} + \underline{v} \cdot \frac{\partial}{\partial \underline{x}} + A\,(cf.\ (8));$$

(iii) is clear; and (iv) is a well-known fact about Brownian motion.

Now take a path $\underline{x}(s)$, $0 \leqq s \leqq T$, of $\underline{x}_t$. By (i), we can take any point $\underline{x}(\tau) = \underline{x}\ (\underline{v}(\tau) = \underline{v})$ on this path, such that $\underline{x}(s)$, $s < \tau$, and $\underline{x}(s)$, $s > \tau$, are independent given $(\underline{x},\underline{v})$. We are then close to posing the problem of a tangential self-intersection in the following way:  given two independent copies $\underline{x}_t^{(1)}$, $\underline{x}_t^{(2)}$ of $\underline{x}_t$, starting say at zero, what is

$$\text{Prob}\ (\underline{x}_t^{(1)} \text{ and } \underline{x}_t^{(2)} \text{ intersect each other tangentially in } [\delta,T])? \tag{12}$$

$\delta > 0$, afterwards we can let $\delta$ go to zero.

This we can reduce to the conditional probability given a curve $\underline{x}^{(1)}(s)$, $\delta \leq s \leq T$, and we need only show

$$\text{Prob } (\underline{x}_t^{(2)} \text{ intersects } \underline{x}^{(1)}(s), \; 0 \leq s \leq T, \text{ tangentially in}$$

$$[\delta,T] \,\big|\, \underline{x}_s^{(1)} = \underline{x}^{(1)}(s), \; s \in [\delta,T]) = 0 \tag{13}$$

since (12) is the average over all paths $\underline{x}^{(1)}(s)$ of (13).

Draw a tube around (the $C^1$-curve) $\underline{x}^1(s)$, $0 \leq s \leq T$, of thickness $2\epsilon$ and split $[0,T]$ into disjoint intervals $[t_i^\epsilon, t_i^\epsilon + \epsilon)$. There are $T/\epsilon$ many such intervals. At each $t_i^\epsilon$ draw a line orthogonal to the curve $\underline{x}^1(\cdot)$ which cuts the tube into disjoint segments (for $\epsilon$ small enough) $A_i^\epsilon$ with $|A_i^\epsilon| \sim \epsilon^2$ (recall that $|\underline{v}_t| = 1$), and there are $T/\epsilon$ many $A_i^\epsilon$'s.

For $\underline{x}_t^{(2)}$ intersection $\underline{x}^{(1)}(\cdot)$ we must have $\underline{x}^{(2)}(t_i^\epsilon) \in A_j^\epsilon$ for some i and j; otherwise, because of (iii), $\underline{x}_t^{(2)}$ will never intersect $x^{(1)}(\cdot)$. If $\underline{x}^{(2)}(t_i^\epsilon) \in A_j^\epsilon$ for some i and j then for a tangential intersection there must be some time $\overset{\sim}{\delta}$ in $I_\epsilon(i) = [t_i^\epsilon - \epsilon, \; t_i^\epsilon + \epsilon]$ at which $\underline{v}^{(1)}(\overset{\sim}{\delta}) = \pm\underline{v}^{(2)}(\overset{\sim}{\delta})$. By (11), $|\underline{v}^{(1)}(s) - \underline{v}^{(1)}(t_i^\epsilon)| \leq K(2\epsilon)^\alpha$ for $s \in I_\epsilon(i)$ and hence for a tangential collision there must be some $\overset{\sim}{\delta} \in [t_i^\epsilon - \epsilon, \; t_i^\epsilon + \epsilon]$ for which $\underline{v}^{(2)}(\overset{\sim}{\delta}) \in H_\epsilon(i) = [-K(2\epsilon)^\alpha \pm \underline{v}^{(1)}(t_i^\epsilon), \; \pm\underline{v}^{(1)}(t_i^\epsilon) + K(2\epsilon)^\alpha]$. Thus we have

$$\text{Prob } (\underline{x}_t^2 \text{ intersects } \underline{x}^1(s) \text{ tangentially during } [\delta,T] \,\big|\, \underline{x}^1(s); \, s \in [\delta,T]$$

$$\leq \text{Prob } (\cup_i \cup_j \{\underline{x}^{(2)}(t_i^\epsilon) \in A_j^\epsilon\} \cap \{\underline{v}^{(2)}(\overset{\sim}{\delta}) \in H_\epsilon(i) \text{ for some } \overset{\sim}{\delta} \in I_\epsilon(i)\})$$

$$\leq \sum_{i,j} \text{Prob}(\{\underline{x}^{(2)}(t_i^\epsilon) \in A_j^\epsilon\} \cap \{\underline{v}^{(2)}(\overset{\sim}{\delta}) \in H_\epsilon(i) \text{ for some } \overset{\sim}{\delta} \in I_\epsilon(i)\}$$

$$\times \{\sup_{\overset{\sim}{\delta} \in I_\epsilon(i)} |\underline{v}^{(2)}(\overset{\sim}{\delta}) - \underline{v}^{(2)}(t_i^\epsilon)| \leq \epsilon^{1/4}\})$$

$$+ \sum_{i,j} \text{Prob } \{\sup_{\overset{\sim}{\delta} \in I_\epsilon(i)} |\underline{v}^{(2)}(\overset{\sim}{\delta}) - \underline{v}^{(2)}(t_i^\epsilon)| > \epsilon^{1/4}\}.$$

By (ii) we have that the probability in the first sum is of the order of $|A_j^\epsilon| \sim \epsilon^2$ times $\epsilon^{1/4}$, since $\underline{v}^{(2)}(t_i^\epsilon)$ has to be in an interval around

$\underline{v}^{(1)}(t_i^\varepsilon)$ of length $2(K(2\varepsilon)^\alpha + \varepsilon^{1/4})$ and choosing $\alpha > \frac{1}{4}$ .

Since the sum consists of $(\frac{T}{\varepsilon})^2$ terms, this term goes like $\varepsilon^{1/4} \to 0$ as $\varepsilon \to 0$.

In the second sum we use again that $\underline{v}_t^{(2)}$ is a Brownian motion, so we have the reflection principle which gives us that

$$\text{Prob} \{\sup_{\delta \in I_\varepsilon(i)} |\underline{v}^{(2)}(\delta) - \underline{v}^{(2)}(t_i^\varepsilon)| > \varepsilon^{1/4}\}$$

$$\leq 4 \text{ Prob} \{|\underline{v}^{(2)}(t_i^\varepsilon - \varepsilon) - \underline{v}^{(2)}(t_i^\varepsilon)| > \varepsilon^{1/4}\}$$

$$\leq 4\varepsilon^{-n/4} E(|v^{(2)}(\varepsilon)|^n) \leq C\varepsilon^{n/4}$$

by Chebychev's inequality. Choosing $n = 10$ the sum vanishes like $\varepsilon^{1/2}$, upon which (13) follows.

REFERENCES


[1] Dell'Antonio, G.E.  The van Hove limit in classical and quantum mechanics in *Stochastic Processes in Quantum Theory and Statistical Physics*, Proceedings, Marseille (1981).

[2] Kesten, H. and Papanicolaou, G. *Commun. Math. Phys. 78* (1981).

[3] Dürr, D., Goldstein, S. and Lebowitz, J.L.  A limit theorem for stochastic acceleration in two dimensions, with an Appendix by B. Israel (in preparation).

[4] Billingsley, P. *Convergence of Probability Measures*, Wiley (1969).

[5] Dürr, D., Goldstein, S. and Lebowitz, J.L. *Commun. Math. Phys. 78*, 507 (1981) and *Z. Wahrscheinlichkeitstheorie 62*, 427 (1983).



Detlef Dürr
Fachbereich Mathematik
Ruhr-Universität Bochum
West Germany

M FUKUSHIMA & K KANEKO

# $(r, p)$-Capacities for general Markovian semigroups

## 1. INTRODUCTION

Consider the Gamma transform

$$V_r(x) = \frac{1}{\Gamma(r/2)} \int_0^\infty t^{\frac{r}{2} - 1} e^{-t} p_t(x)dt, \qquad r > 0,$$

of the transition density $p_t(x) = (2\pi t)^{-d/2} e^{-|x|^2/2t}$ of the d-dimensional standard Brownian motion. $V_r$ is then the Bessel convolution kernel $(\hat{V}_r(\xi) = (1 + \frac{|\xi|^2}{2})^{-r/2})$ and the space of Bessel potentials $V_r * f$ $L^p$-functions $f$ coincides with the Sobolev space $W_r^p$ when $r$ is an integer ([9]). This observation suggests a way of introducing an analogue $F_{r,p}$ of the Sobolev space and a relevant capacity $C_{r,p}$ for a much more general Markovian semigroup $\{T_t, t > 0\}$.

Already Meyers [7] has made a general approach to this kind of capacity starting with a lower semicontinuous kernel $k(x,y)$. But we wish to avoid the assumption that the operator $V_r$ admits such a smooth kernal $k$, because we are particularly interested in the application to cases where the underlying spaces are infinite dimensional ([4], [6], [10]).

Accordingly the proof of the continuity

$$A_n \uparrow \Rightarrow C_{r,p}(\cup_n A_n) = \sup_n C_{r,p}(A_n)$$

of the set function $C_{r,p}$ becomes more difficult in our setting. When the underlying space is locally compact, this continuity of the outer capacity implies the capacitability of the analytic set. Even in non-locally-compact cases such as the Wiener space case, this property is essentially important in carrying out the relevant analysis ([4], [10]). In section 3 we prove this property by assuming that the space $F_{r,p}$ is regular, namely, that it densely contains continuous functions. We can then appeal to Deny's principle [2] that the regularity of a function space implies the continuity of the capacity.

## 2. <u>MARKOVIAN SEMIGROUP AND $(r,p)$-CAPACITY</u>

Let X be a metric space, m be a measure on X and $T_t$, $t > 0$ be a strongly
continuous contraction semigroup of linear operators on $L^p = L^p(X,m)$, $p > 1$
being fixed. We suppose that $T_t$, $t > 0$, are *Markovian*:

$$f \in L^p, \quad 0 \leq f \leq 1 \quad m\text{-a.e} \Rightarrow 0 \leq T_t f \leq 1 \quad m\text{-a.e.} \tag{2.1}$$

We then let, for each $r > 0$,

$$V_r = \frac{1}{\Gamma(r/2)} \int_0^\infty t^{\frac{r}{2} - 1} e^{-t} T_t \, dt. \tag{2.2}$$

$V_r$ is Markovian. It is contractive:

$$\| V_r f \|_{L^p} \leq \| f \|_{L^p}, \quad f \in L^p. \tag{2.3}$$

Furthermore,

$$V_r V_{r'} = V_{r+r'}, \qquad r, r' > 0. \tag{2.4}$$

To see this, it suffices to express the left-hand side as a double integral
of two variables, say, t and s and changes them by $t = \xi \cos^2\theta$, $s = \xi \sin^2\theta$.
Suppose $V_r f = 0$, then $V_{r'} f = 0$ for all $r' \geq r$ by (2.4) and consequently
$T_t f = 0$, $t > 0$, which implies $f = 0$. Thus, each $V_r$ is injective on $L^p$ and
hence the space $(F_{r,p}, \| \ \|_{r,p})$ defined below gives us a Banach space:

$$\begin{cases} F_{r,p} = V_r(L^p) \\ \| u \|_{r,p} = \| f \|_{L^p} \quad \text{for} \quad u = V_r f, \quad f \in L^p. \end{cases} \tag{2.5}$$

The associated set function $C_{r,p}$ is then defined, for open set A, by

$$C_{r,p}(A) = \inf \{ \| u \|_{r,p}^p : u \in F_{r,p}, \ u \geq 1 \ m\text{-a.e. on } A \}$$

and, for any set A, by

$$C_{r,p}(A) = \inf\{ C_{r,p}(B) : A \subset B, \ B \text{ is open}\}.$$

We call $C_{r,p}$ the $(r,p)$-*capacity* for $\{T_t, \ t > 0\}$.

$\underline{\text{Theorem 1.}}$   (i)  $m(A) \leq C_{r,p}(A)$

(ii)  $r < r' \Rightarrow C_{r,p}(A) \leq C_{r',p}(A)$

(iii) $A \subset B \Rightarrow C_{r,p}(A) \leq C_{r,p}(B)$

(iv)  $C_{r,p}(\underset{n}{\cup} A_n) \leq \underset{n}{\Sigma} C_{r,p}(A_n)$.

(i) of Theorem 1 is immediate from (2.3).  (ii) follows from (2.3) and (2.4).
(iii) is trivial.  (iv) can be proved easily ([7]) but we give its proof by
using the next lemma for later convenience.

$\underline{\text{Lemma 1.}}$   For any open set A with finite $(r,p)$-capacity, there exists a
unique element $u_A$ in $F_{r,p}$ such that $u_A \geq 1$ m-a.e. on A and $\|u_A\|_{r,p}^p = C_{r,p}(A)$.
$u_A$ admits an expression $u_A = V_r f$ for some non-negative $f \in L^p$.

$\underline{\text{Proof.}}$   Since $(F_{r,p}, \| \ \|_{r,p})$ is isometric to $L^p(X;m)$, it is uniformly
convex:  for any $\varepsilon > 0$ and $M > 0$, there is a $\delta > 0$ such that
$\|u\|_{r,p} \leq M, \|v\|_{r,p} \leq M$ and $\|u-v\|_{r,p} > \varepsilon$ imply $\|u+v\|_{r,p} \leq 2M-\delta$.  From
this follows the uniqueness of $u_A$.  The existence follows from the Banach-
Saks theorem for $L^p$ space.  If $u_A = V_r f$, $f \in L^p$, then $V_r f \leq V_r f^+$,
$\|f^+\|_{L^p} \leq \|f\|_{L^p}$   and consequently $f = f^+$ by the uniqueness.

$\underline{\text{Proof of Theorem 1 (iv)}}$   It suffices to assume in the desired inequality
(iv) that $A_n$ are open and the right-hand side is finite.  Let $u_{A_n} = V_r f_n$,
$f_n \in L^p_+$ , and $f = \sup_n f_n$.  Since $f^p \leq \underset{n}{\Sigma} f_n^p$, we have $f \in L^p$ and
$\|V_r f\|_{r,p}^p \leq \underset{n}{\Sigma} C_{r,p}(A_n)$.  On the other hand, $C_{r,p}(\underset{n}{\cup} A_n) \leq \|V_r f\|_{r,p}^p$ because
$V_r f \geq 1$ m-a.e. on $\underset{n}{\cup} A_n$.

$\underline{\text{Remark 1.}}$   We see in the proof of Lemma 1 that, for an open set A,

$$C_{r,p}(A) = \inf\{\|f\|_{L^p}^p : V_r f \geq 1 \ \text{m-a.e. on A, } f \in L^p_+\}, \qquad (2.6)$$

where $L^p_+$ denotes the set of non-negative functions in $L^p$.  In the case of
the Bessel potentials on $R^d$ mentioned at the beginning of section 1, it
further holds that

$$C_{r,p}(A) = \inf \{ \| f \|_{L^p}^p : V_r f(x) \geq 1 \text{ for any } x \in A, f \in L^p_+ \}, \qquad (2.7)$$

because, for the Bessel potential $u = V_r f$, $f \in L^p_+$, we have $\lim_{r \, 0} M_r u(x) = u(x)$ at each $x \in R^d$, where $M_r u(x)$ denotes the volume average of $u$ on the ball of radius r, centred at x. In this case, we also know from [7] that the identity (2.7) holds for any set $A \subseteq R^d$.

<u>Remark 2.</u>   We have started with a strongly continuous contraction semigroup of Markovian linear operators on $L^p$ for a fixed $p \geq 1$. Suppose that we are first given such a semigroup on $L^1$. Then it uniquely determines a semi-group of the same properties on $L^p$ for every $p \geq 1$ ([5]) and hence there correspond (r,p)-capacities $C_{r,p}$ indexed by all $r > 0$ and all $p > 1$. In accordance with Malliavin [6], we say in this case that a set $A \subseteq X$ is *slim* if $C_{r,p}(A) = 0$ for all $r > 0$, $p > 1$.

In the case of the Bessel potentials on $R^d$ discussed in Remark 1, no non-empty slim set exists since each one point set of $R^d$ has a positive (r,p)-capacity whenever $rp > d$ ([1]). The situation may be quite different when the underlying space X is infinite-dimensional. In fact, there are many non-empty slim sets for the Ornstein-Uhlenbeck semigroup $\{T_t, t > 0\}$ on the Wiener space. In particular, each one-point set is slim ([10]).

<u>Remark 3.</u>   Suppose that we are first given a strongly continuous contraction semigroup $\{T_t, t > 0\}$ of Markovian linear operators on $L^2$. In addition, we assume that $T_t$ is symmetric and the measure m is $\sigma$-finite (this amounts to assuming that we are first given a Dirichlet space on $L^2$ ([3])). Then $\{T_t, t > 0\}$ uniquely decides a semigroup of the same properties on $L^1$ ([8]) and consequently on $L^p$ for every $p \geq 1$.

Denote by $L$ the infinitesimal generator of $\{T_t, t > 0\}$ on $L^2$. Substituting the spectral representation of $L$ into (2.2), we easily see that

$$V_r = (I-L)^{-r/2} \quad \text{on } L^2 \qquad (2.8)$$

$$F_{r,2} = \mathcal{D}((I-L)^{r/2}), \quad \| u \|_{r,2} = \| (I-L)^{r/2} u \|_{L^2}, \qquad (2.9)$$

and consequently $F_{1,2}$ coincides with the associated Dirichlet space on $L^2$ and $\| u \|_{1,2}$ is the 1-order Dirichlet norm. $C_{1,2}$ is exactly the same as the capacity Cap defined in [3] for this Dirichlet space. Hence the set of

(1,2)-capacity zero admits a probabilistic characterization related to the Markovian semigroup $\{T_t, t > 0\}$ ([3]), but no such interpretation seems to be possible for $(r,p)$-capacity when $(r,p) \neq (1,2)$.

In view of expression (2.9) and its formal extension

$$\|u\|_{r,p} = \|(I-L)^{r/2}u\|_{L^p} , \quad u \in F_{r,p}$$

we may well say that the space $F_{r,p}$ is a right analogue to the Sobolev space $W_{r,p}$ for the present general semigroup $\{T_t, t > 0\}$.

## 3. CONTINUITY OF CAPACITIES

As in section 2, we treat a strongly continuous contraction semigroup $\{T_t, t > 0\}$ of Markovian linear operators on $L^p = L^p(X;m)$. In this section, we assume that X is a separable metric space and $m(A) > 0$ for any non-empty open set A. We further assume that the space $F_{r,p}$ is regular:

$$F_{r,p} \cap C(X) \quad \text{is dense in} \quad (F_{r,p}, \| \ \|_{r,p}), \tag{3.1}$$

where $C(X)$ denotes the space of (not necessarily bounded) continuous functions on X.

We fix the index $(r,p)$. "Quasi-everywhere" or "q.e" will mean "except on a set of $(r,p)$-capacity zero". A function u on X is said to be *quasi-continuous* if, for any $\varepsilon > 0$, there exists an open set A with $C_{r,p}(A) < \varepsilon$ such that the restriction of u to X−A is continuous. Just as in the case of the Dirichlet space ([3]), we can show the following:

(a)  If u is quasi-continuous and $u \geq 0$ m-a.e. on an open set G, then $u \geq 0$ q.e. on G.

(b)  Each $u \in F_{r,p}$ admits a quasi-continuous modification (denoted by $\tilde{u}$) and

$$C_{r,p}(|\tilde{u}| > \lambda) \leq \frac{1}{\lambda^p} \|u\|_{r,p}^p , \qquad \lambda > 0. \tag{3.2}$$

(c)  If a sequence of quasi-continuous functions $u_n \in F_{r,p}$ converges to $u \in F_{r,p}$ in metric $\| \ \|_{r,p}$, then a subsequence of $u_n$ converges q.e. to a quasi-continuous modification of u.

<u>Lemma 2.</u>    For any set $B \subset X$ with finite $(r,p)$-capacity, there exists a unique element $e_B$ in the set $L_B = \{u \in F_{r,p} : \tilde{u} \geq 1 \ \text{q.e. on B}\}$ minimizing

the norm $\| \ \|_{r,p}$ .  $e_B$ is non-negative and

$$C_{r,p}(B) = \| e_B \|_{r,p}^{p} . \tag{3.3}$$

<u>Proof.</u>  The unique existence of $e_B$ and its non-negativity can be shown in the same way as in the proof of Lemma 1.  For any $\varepsilon > 0$, there exists an open set $A \supset B$ such that $C_{r,p}(B) > C_{r,p}(A) - \varepsilon$.  Since $u_A$ of Lemma 1 belongs to $L_B$ by (a), $C_{r,p}(A) = \| u_A \|_{r,p}^{p} \geq \| e_B \|_{r,p}^{p}$ and we get the inequality "$\geq$" in (3.3).

To prove the converse inequality, we adopt the method of Deny [2].  Take a quasi-continuous version $\tilde{e}_B$ of $e_B$.  For any $\varepsilon > 0$, choose an open set $A_\varepsilon$ such that $C_{r,p}(A_\varepsilon) < \varepsilon$, $\tilde{e}_B |_{X-A_\varepsilon}$ is continuous and $\tilde{e}_B \geq 1$ on $B \cap (X-A_\varepsilon)$.  Denote by $u_\varepsilon$ the function of Lemma 1 for the open set $A_\varepsilon$.  Now the set

$$G_\varepsilon = \{ x \in X-A_\varepsilon : \tilde{e}_B(x) > 1 - \varepsilon \} \cup A_\varepsilon$$

is open and $B \subset G_\varepsilon$.  Moreover, $e_B + u_\varepsilon \geq 1 - \varepsilon$ m-a.e. on $G_\varepsilon$.  Therefore $C_{r,p}(B) \leq C_{r,p}(G_\varepsilon) \leq (1-\varepsilon)^{-p} \| e_B + u_\varepsilon \|_{r,p}^{p} \leq (1-\varepsilon)^{-p} ( \| e_B \|_{r,p} + \varepsilon )^{p}$.  By letting $\varepsilon \downarrow 0$, we arrive at (3.3).

<u>Theorem 2.</u>   $A_n \uparrow \Rightarrow C_{r,p}(\underset{n}{\cup} A_n) = \underset{n}{\sup}\, C_{r,p}(A_n)$.

<u>Proof.</u>  We may assume that the right-hand side is finite.  We put $A = \underset{n}{\cup} A_n$.  Let $\dot{e}_n$ be the function of Lemma 2 for the set $A_n$.  Since $\| e_n \|_{r,p}^{p} = C_{r,p}(A_n)$ are bounded, a Cesaro mean $u_m$ of a subsequence of $e_n$ converges strongly to some $u \in F_{r,p}$ by virtue of the Banach-Saks theorem.  Take quasi-continuous modifications $\tilde{u}_m$ of $u_m$.  Then $\underset{m\to\infty}{\underline{\lim}}\, \tilde{u}_m(x) \geq 1$ q.e. on $A_n$ for each n and consequently q.e. on A.  By (c), a subsequence of $\tilde{u}_m$ converges q.e. to a quasi-continuous modification $\tilde{u}$ of u.  Hence $u \in L_A$ and

$$C_{r,p}(A) \leq \| u \|_{r,p}^{p} = \lim_{m\to\infty} \| u_m \|_{r,p}^{p} = \lim_{n\to\infty} \| e_n \|_{r,p}^{p} = \underset{n}{\sup}\, C_{r,p}(A_n).$$

<u>Remark 4.</u>   When $r = 1$ and $p = 2$, Theorem 2 holds without the regularity assumption (3.1) because $C_{1,2}$ is strongly subadditive, which follows from the characteristic property of the Dirichlet space that every normal contraction operates on $F_{1,2}$ ([3]).

46

## REFERENCES

[1] Adams, D. and Meyers, N.G. Bessel potentials, inclusion relations among classes of exceptional sets. *Indiana Univ. Math. J.*, 22 873-905 (1973).

[2] Deny, J. Theorie de la capacite dans les espaces fonctionnels. *Seminaire Brelot-Choquet-Deny, 9e annee,* Paris (1964/65).

[3] Fukushima, M. *Dirichlet Forms and Markov Processes.* Kodansha and North-Holland (1980).

[4] Fukushima, M. Basic properties of Brownian motion and a capacity on the Wiener space. *J. Math. Soc. Japan,* 36, 161-175 (1984).

[5] Garsia, A. *Topics in Almost Everywhere Convergence.* Chicago: Markham Pub. Co. (1970).

[6] Malliavin, P. Implicit functions in finite co-rank on the Wiener space. *Proc. Taniguchi Symp. on Stochastic Analysis, Kinokuniya* (1984).

[7] Meyers, N.G. A theory of capacities for potentials of functions in Lebesgue classes. *Math. Scand.* 26 255-292 (1970).

[8] Silverstein, M. *Symmetric Markov Processes.* Lecture Notes in Math. vol. 426, Springer (1974).

[9] Stein, E.M. Singular integrals and differentiability properties of functions. Princeton Univ. Press (1970).

[10] Takeda, M. (r,p)-capacities on the Wiener space and properties of Brownian motion. *Z. Wahrscheinlichkeitsth. verw. Gebiete,* 68 149-162 (1984).



M. Fukushima
College of General Education
Osaka University, Toyonaka, Osaka

H. Kaneko
Department of Mathematics
Osaka University, Toyonaka, Osaka

S KUSUOKA

# The path property of Edward's model for long polymer chains in three dimensions

## 1. INTRODUCTION

Edward's model for long polymer chains in d dimensions is a family of
probability measures $\nu(g)$, $g \geq 0$, which we call polymer measures, on
$C_o([0,1] \to R^d) = \{w \in C([0,1] \to R^d) \; ; \; w(0) = 0\}$ formally given by

$$\nu(g)(dw) = L^{-1} \exp(-g \int_0^1 \int_0^1 \delta(w(s)-w(t)) \; ds \; dt) \; \mu(dw), \qquad (1.1)$$

where $\delta$ is the Dirac function and $\mu$ is the d dimensional Wiener measure on
$C_o([0,1] \to R^d)$.

The term on the right-hand side of (1.1) has rigorous meaning for $d = 1$,
but not for $d \geq 2$. For $d = 2$ and $d = 3$, Varadhan in the Appendix to [2]
and Westwater [3], [4] respectively showed the existence of non-trivial
polymer measures. In the present paper, we treat polymer measures in three
dimensions which Westwater [3], [4] constructed, and show the following.

Theorem 1.1. Let $\nu(g)$, $g \geq 0$, be polymer measures in three dimensions.
For each $g \geq 0$, there exist measurable functions $B^g:[0,1] \times C_o([0,1] \to R^3) \to R^3$
and $R^g:[0,1] \times C_o([0,1] \to R^3) \to R^3$ such that

(1) $B^g(\cdot,w):[0,1] \to R^3$ and $R^g(\cdot,w):[0,1] \to R^3$ are continuous for $\nu(g)$-a.e.w,

(2) $w(t) = B^g(t,w) + R^g(t,w)$ for any $(t,w) \in [0,1] \times C_o([0,1] \to R^3)$,

(3) $B^g(t,\cdot):C_o([0,1] \to R^3) \to R^3$ and $R^g(t,\cdot):C_o([0,1] \to R^3) \to R^3$ are

   $F(t)$-measurable, $0 \leq t \leq 1$, where $F(t)$ is a $\sigma$-algebra over $C_o([0,1] \to R^3)$
   generated by $\{w(s); 0 \leq s \leq t\}$,

(4) $\{B^g(t,w); 0 \leq t \leq 1\}$ is a three-dimensional $F(t)$-Brownian motion, under
   $\nu(g)(dw)$, and

(5) for any $\alpha > \dfrac{8}{5}$,

$$\sum_{k=1}^{2^n} |R^g(k \cdot 2^{-n},w) - R^g((k-1)2^{-n},w)|^\alpha \to 0, \quad n \to \infty,$$

   for $\nu(g)$-a.e.w.

Thus $\{w(t);\ 0 \le t \le 1\}$ under $\nu(g)(dw)$ is a Dirichlet process in the sense of Föllmer [1], and its principal part is still a Brownian motion.

Since polymer measures in one or two dimensions are absolutely continuous relative to the Wiener measures in one or two dimensions, by virtue of the Cameron-Martin-Maruyama-Girsanov formula we see that Theorem 1.1 holds and

$$\int_0^1 |(d/dt)R^g(t,w)|^2\ dt < \infty \quad \text{for } \nu(g)\text{-a.e.w in the case of one or two}$$

dimensions. However, as Westwater [4] has shown, polymer measures in three dimensions are singular relative to the Wiener measure, and so we have

$$\int_0^1 |(d/dt)R^g(t,w)|^2\ dt = \infty \quad \text{for } \nu(g)\text{-a.e.w.}$$

It is a big problem whether there exists a non-trivial polymer measure in four dimensions. If it exists, it is unlikely that the principal part of its paths is a Brownian motion because of the renormalization order.

2. <u>WESTWATER'S RESULTS AND SOME OBSERVATIONS ON THEM</u>

In this section, we recall results in Westwater [3], [4], and make some observations on them.

Let $U_n = \bigcup_{k=1}^{2^n} [(k-1)2^{-n},\ k\cdot 2^{-n}]^2 \subset [0,1]^2$. We denote $C_o([0,1] \to R^3)$ by B and the three-dimensional Wiener measure by $\mu$. For each $g \ge 0$ and $n \ge 0$, we define a probability measure $\nu_n(g)$ on B by

$$\nu_n(g)(dw) = L_n(g)^{-1} \exp(-g \int_{(0,1)^2-U_n} \delta(w(s)-w(t))\ ds\ dt)\ \mu(dw),$$

where $L_n(g) = E_\mu\left[\exp(-g \int_{(0,1)^2-U_n} \delta(w(s)-w(t))\ ds\ dt)\right]$.

Let $G(m)$, $m \ge 0$, denote the $\sigma$-algebra over B generated by $w(k\cdot 2^{-m})$, $0 \le k \le 2^m$, and let $f_{n,m}(g)(w) = (d\nu_n(g)/d\mu)|_{G(m)}(w)$. Then we have the following by Westwater [3], [4].

<u>Theorem 2.1.</u>   There exists a $G(m)$-martingale $\{f_m(g)\}_{m=1}^\infty$ under $\mu(dw)$ for each $g \ge 0$ such that

$$E_\mu\left[|f_m(g) - f_{n,m}(g)|^p\right] \to 0 \quad \text{as } n \to \infty \quad \text{for any } m \ge 0 \quad \text{and } p > 1.$$

Furthermore, for any $\rho > 1$, $p > 1$ and $g_o > 0$, there exists a constant $K(\rho,p,g_o)$ such that

$$E_\mu\left[|f_m(g)|^p\right] \le K(\rho,p,g_o)\cdot\rho^m \quad \text{for any } m \ge 0 \text{ and } 0 \le g \le g_o.$$

__Proposition 2.2.__  There exists a probability measure $\nu(g)$ on $B$ for each $g \ge 0$ such that $d\nu(g)\big|_{G(m)} = f_m(g)\,d\mu\big|_{G(m)}$, $m \ge 0$.


__Proof.__  By Kolmogorov's extrension theorem, there exists a probability measure $\bar\nu(g)$ defined in $\sum\limits_{m=1}^{\infty} G(m)$ such that $d\bar\nu(g)\big|_{G(m)} = f_m(g)\,d\mu\big|_{G(m)}$.

Let $0 \le s < t \le 1$, $s,t \in \{k\cdot 2^{-n};\ n \ge 1,\ 0 \le k \le 2^n\}$.  Let

$N = \min\{n \ge 0;\ t - s \ge 2^{-n}\}$.  Then there exists an integer $\ell$ such that $(\ell-2)2^{-N} < s \le \ell\cdot 2^{-N} < t < (\ell+2)2^{-N}$.  Therefore $s$ and $t$ are expressed as follows:

$$s = \ell\cdot 2^{-N} - \sum_{k=N}^{N_o} a_k\cdot 2^{-k}, \qquad a_k = 0 \text{ or } 1,$$

and

$$t = \ell\cdot 2^{-N} + \sum_{k=N}^{N_o} b_k\cdot 2^{-k}, \qquad b_k = 0 \text{ or } 1$$

for some sufficiently large integer $N_o$.  Let $s_m = \ell\cdot 2^{-N} - \sum\limits_{k=N}^{m} a_k\cdot 2^{-k}$ and $t_m = \ell\cdot 2^{-N} + \sum\limits_{k=N}^{m} b_k\cdot 2^{-k}$, $m \ge N-1$.  It is easy to see that

$$E_{\bar\nu(g)}\left[|\,w(t_{m+1}) - w(t_m)|^4\right]$$

$$\le E_\mu\left[|w(t_{m+1}) - w(t_m)|^4\, f_m(g)\right]$$

$$\le E_\mu\left[|w(t_{m+1}) - w(t_m)|^8\right]^{1/2} E_\mu\left[f_m(g)^2\right]^{1/2}$$

Thus we have, from Theorem 2.1, for any $m \ge 0$,

$$E_{\bar\nu(g)}\left[|w(t_{m+1}) - w(t_m)|^4\right]^{1/4} \le C\cdot 2^{-(2-\alpha)m/4}$$

where $\alpha$ and $C$ are constants independent of $m$ with $0 < \alpha < 1$.

Similarly we have, for any $m \ge 0$,

$$E_{\bar{\nu}(g)}\left[\left|w(s_{m+1}) - w(s_m)\right|^4\right]^{1/4} \leq C\cdot 2^{-(2-\alpha)m/4}.$$

Since $s_{N-1} = t_{N-1}$, we get

$$E_{\bar{\nu}(g)}\left[\left|w(t) - w(s)\right|^4\right]^{1/4} \leq C'\cdot 2^{-(2-\alpha)N/4}$$

$$\leq C'\cdot |t-s|^{(2-\alpha)/4}.$$

Therefore, noting that $2 - \alpha > 1$, we see by virtue of Kolmogorov's theorem that there exists a probability measure $\nu(g)$ on B such that $\nu(g)\big|\sum_{m=1}^{\infty} G(m) = \bar{\nu}(g)$. This completes the proof.

<u>Lemma 2.3.</u>  Let $\phi: R^3 \to [0,\infty)$ be a bounded continuous function with $\int_{R^3} \phi(x)\, dx = 1$.  Let $\nu^\varepsilon(g)$, $g \geq 0$ and $0 < \varepsilon \leq 1$, be a probability measure on B given by

$$\nu^\varepsilon(g)(dw) = L^\varepsilon(g)^{-1} \exp\left(-g\int_0^1\int_0^1 \phi_\varepsilon(w(s)-w(t))\, ds\, dt\right) \mu(dw),$$

where $\phi_\varepsilon(x) = \varepsilon^{-3}\phi(\varepsilon^{-1}x)$, $x \in R^3$, and $L^\varepsilon(g)$ is the normalizing constant. Then $\nu^\varepsilon(g) \to \nu(g)$ as $\varepsilon \downarrow 0$ with respect to the Prohorov topology of the space of probability measures on B.

<u>Proof.</u>  We give only a sketch of the proof.  Let $\nu_n^\varepsilon(g)$, $n \geq 1$, $0 < \varepsilon \leq 1$ and $g \geq 0$, be probability measures on B given by

$$\nu_n^\varepsilon(g)(dw)$$

$$= L_n^\varepsilon(g)^{-1} \exp\left(-g\int_{(0,1)^2-U_n} \phi_\varepsilon(w(s)-w(t))\, ds\, dt\right) \mu(dw),$$

and let

$$f_m^\varepsilon(g)(w) = \frac{d\nu^\varepsilon(g)}{d\mu}\bigg|_{G(m)}(w)$$

and

$$f_{n,m}^\varepsilon(g)(w) = \frac{d\nu_n(g)}{d\mu}\bigg|_{G(m)}(w).$$

Then, by the same argument of Westwater [3], we see that, for any $\rho > 1$ and $p > 2$, and for any $m \geq 0$ and $0 \leq g \leq g_o$ there exist constants $g_o > 0$ and

$K > 0$ such that,

$$\sup_{0 < \varepsilon \leq 1} E_\mu \left[ |f_m^\varepsilon(g) - f_{n,m}^\varepsilon(g)|^p \right] \to 0, \quad n \to \infty, \tag{2.1}$$

and

$$\sup_{0 < \varepsilon \leq 1} E_\mu \left[ |f_m^\varepsilon(g)|^p \right] \leq K \cdot \rho^m. \tag{2.2}$$

It is easy to see that $E_\mu \left[ |f_{n,m}(g) - f_{n,m}^\varepsilon(g)| \right] \to 0$, $\varepsilon \downarrow 0$. Thus it follows from Theorem 2.1 and the proof of Proposition 2.2 that, for any $0 \leq g \leq g_o$,

$$\nu^\varepsilon(g) \to \nu(g) \quad \text{as} \quad \varepsilon \downarrow 0. \tag{2.3}$$

Then a similar argument of Westwater [4] implies that (2.3) holds for any $g \geq 0$. This completes the proof.

<u>Remark 2.4.</u>  By using the argument in Westwater [4], we can improve (2.2) to the following:  for any $\rho > 1$, $p > 1$ and $g_o \geq 0$, there exists a constant $K'(\rho, p, g_o)$ such that

$$\sup_{0 < \varepsilon \leq 1} E_\mu \left[ |f_m^\varepsilon(g)|^p \right] \leq K'(\rho, p, g_o) \cdot \rho^m$$

for any $m \geq 0$ and $0 \leq g \leq g_o$.

For $w_1 \in C_o([0, T_1] \to R^3)$ and $w_2 \in C_o([0, T_2] \to R^3)$, we define $w_1 * w_2 \in C_o([0, T_1 + T_2] \to R^3)$ by

$$w_1 * w_2 = \begin{cases} w_1(t) & \text{for } 0 \leq t \leq T_1 \\[2mm] w_1(T_1) + w_2(t - T_2) & \text{for } T_1 \leq t \leq T_1 + T_2. \end{cases}$$

<u>Lemma 2.5.</u>  For any $0 < \tau < 1$, we have

$$E_{\nu(g)} \left[ \Phi \right]$$

$$= L_{\tau, g}^{-1} \int_{B \times B} \Phi \left( (\tau^{1/2} w_1(\cdot / \tau)) * ((1 - \tau)^{1/2} w_2(\cdot / (1 - \tau))) \right)$$

$$\times \exp \left( -2g \cdot \tau (1 - \tau) \int_0^1 \int_0^1 \delta(\tau^{1/2}(w_1(1) - w_1(s)) + (1 - \tau)^{1/2} w_2(t)) \, ds \, dt \right)$$

$$\times \nu(\tau^{1/2} g)(dw_1) \times \nu((1 - \tau)^{1/2} g)(dw_2)$$

for any bounded continuous function $\Phi : B \to R$.

$\underline{\text{Proof.}}$   Lemma 2.5 is formally true as follows.  Observe that

$$E_{\nu(g)}\left[\Phi\right]$$

$$= L^{-1} \int_B \Phi(w) \exp(-g \int_0^1 \int_0^1 \delta(w(s)-w(t)) \ ds \ dt) \ \mu(dw).$$

Then note that

$$\int_0^1 \int_0^1 \delta(w(s)-w(t)) \ ds \ dt$$

$$= \int_0^\tau \int_0^\tau \delta(w(s)-w(t)) \ ds \ dt + \int_\tau^1 \int_\tau^1 \delta(w(s)-w(t)) \ ds \ dt$$

$$+ 2 \int_0^\tau \int_\tau^1 \delta(w(s)-w(t)) \ ds \ dt,$$

that $\{w_1(t) = \tau^{-1/2}w(t\cdot\tau); \ 0 \leq t \leq 1\}$ and

$$\{w_2(t) = (1-\tau)^{-1/2}(w(\tau+t(1-\tau))-w(\tau)); \ 0 \leq t \leq 1\}$$

are independent Wiener processes under $\mu(dw)$, and that $\delta(ax) = a^{-3}\delta(x)$, $a > 0$ and $x \in R^3$.  Thus we have

$$E_{\nu(g)}\left[\Phi\right] = L^{-1} \int_{B\times B} \Phi((\tau^{1/2}w_1(\cdot/\tau))*((1-\tau)^{1/2}w_2(\cdot/(1-\tau)))\ )$$

$$\times \exp(-2g\cdot\tau(1-\tau)\int_0^1\int_0^1 \delta(\tau^{1/2}(w_1(1)-w_1(s))+(1-\tau)^{1/2}w_2(t)) \ ds \ dt$$

$$- g\cdot\tau^{1/2} \int_0^1 \int_0^1 \delta(w_1(s)-w_1(t)) \ ds \ dt$$

$$- g\cdot(1-\tau)^{1/2} \int_0^1 \int_0^1 \delta(w_1(s)-w_1(t)) \ ds \ dt) \ \mu(dw_1) \times \mu(dw_2).$$

This implies Lemma 2.5.

Following this idea we can prove Lemma 2.5 rigorously by using Theorem 2.1,  Lemma 2.3, Remark 2.4 and the argument in Westwater $\left[4\right]$.

$\underline{\text{Remark 2.6.}}$   The probability law of $-w(\cdot)$ under $\nu(g)(dw)$ is equal to $\nu(g)$.

## 3. BASIC ESTIMATES

Let $Y_i(n,k)$, $i = 1,2,3$, $n \geq 1$ and $1 \leq k \leq 2^n$, be non-negative measurable functions on $B$ given by

$$Y_1(n,k;w) = \int_0^{(k-1)2^{-n}} \int_{(k-1)2^{-n}}^{k \cdot 2^{-n}} \delta(w(s)-w(t)) \, ds \, dt,$$

$$Y_2(n,k;w) = \int_{(k-1)2^{-n}}^{k \cdot 2^{-n}} \int_{k \cdot 2^{-n}}^1 \delta(w(s)-w(t)) \, ds \, dt,$$

and

$$Y_3(n,k;w) = \int_0^{(k-1)2^{-n}} \int_{k \cdot 2^{-n}}^1 \delta(w(s)-w(t)) \, ds \, dt,$$

Let $T(n,k):B \to B$, $n \geq 1$ and $1 \leq k \leq 2^n$, be measurable maps given by

$$(T(n,k)w)(t) = \begin{cases} w(t), & 0 \leq t \leq (k-1)2^{-n}, \\[2mm] -w(t) + 2 \cdot w((k-1)2^{-n}), & (k-1)2^{-n} \leq t \leq k \cdot 2^{-n}, \\[2mm] w(t) - 2 \cdot w(k \cdot 2^{-n}) + 2 \cdot w((k-1)2^{-n}), & k \cdot 2^{-n} \leq t \leq 1. \end{cases}$$

Then $\{(T(n,k)w)(t); \ 0 \leq t \leq 1\}$ is a Wiener process under $\mu(dw)$.

Proposition 3.1. For any $\lambda_1$, $\lambda_2 > 0$ with $\lambda_1 + \lambda_2 < \frac{1}{4}$, there exists a constant $C_{\lambda_1,\lambda_2}$ such that

$$E_\mu\left[ |Y_i(n,k;w) - E_\mu[Y_i(n,k;w)|G(m)]|^2 \right]$$

$$\leq C_{\lambda_1,\lambda_2} 2^{-\lambda_1 m} 2^{-2\lambda_2 n}, \quad i = 1,2, \tag{1}$$

$$E_\mu\left[ |Y_i(n,k;T(n,k)w) - E_\mu[Y_i(n,k;T(n,k)w)|G(m)]|^2 \right]$$

$$\leq C_{\lambda_1,\lambda_2} 2^{-\lambda_1 m} 2^{-2\lambda_2 n}, \quad i = 1,2, \tag{2}$$

for any $m \geq 0$, $n \geq 1$ and $1 \leq k \leq 2^n$.

Proof. By the same argument in Westwater [3] Section 3.1, we have

$$E_\mu\left[ |Y_1(n,k;w) - E_\mu[Y_1(n,k;w)|G(m)]|^2 \right]$$

54

$$\leq (2\pi)^{-3} \int_K \det A(\underline{s},\underline{t})^{-1/2} \, ,$$

where $\underline{s} = (s_1, s_2)$, $\underline{t} = (t_1, t_2)$,

$$K = \{(s_1, s_2, t_1, t_2) \in (0, (k-1)2^{-n})^2 \times ((k-1)2^{-n}, k \cdot 2^{-n})^2;$$
$$[2^m s_1] = [2^m s_2] \quad \text{or} \quad [2^m t_1] = [2^m t_2]\}$$

and

$$A(\underline{s},\underline{t}) = \begin{bmatrix} A_o(\underline{s},\underline{t}) & & 0 \\ & A_o(\underline{s},\underline{t}) & \\ 0 & & A_o(\underline{s},\underline{t}) \end{bmatrix},$$

$$A_o(\underline{s},\underline{t}) = \begin{bmatrix} |t_1 - s_1| & t_1 \wedge t_2 - s_1 \, s_2 \\ t_1 \wedge t_2 - s_1 \, s_2 & |t_2 - s_2| \end{bmatrix}.$$

Thus we get

$$E_\mu \left[ |Y_1(n,k;w) - E_\mu [Y_1(n,k;w)|G(m)]|^2 \right]$$

$$\leq (2\pi)^{-3} |K|^{1/p} \left\{ \int_{(-1,0)^2 \times (0,1)^2} d\underline{s} \; d\underline{t} \; \det A_o(\underline{s},\underline{t})^{-3q/2} \right\}^{1/q},$$

for $p, q > 1$ and $(1/p) + (1/q) = 1$.

Observe that $\displaystyle\int_{(-1,0)^2 \times (0,1)^2} d\underline{s} \; d\underline{t} \; \det A_o(\underline{s},\underline{t})^{-\alpha} < \infty$ if $0 < \alpha < 2$,

and that $|K| \leq 2 \cdot 2^{-m} \wedge 2^{-2n}$. Then we have our assertion (1) for $i = 1$. The case where $i = 2$ is similar.

It is obvious that

$$E_\mu \left[ |Y_1(n,k;T(n,k)w) - E_\mu [Y_1(n,k;T(n,k)w)|G(m)]|^2 \right]$$

$$\leq E_\mu \left[ |Y_1(n,k)|^2 \right]$$

$$= (2\pi)^{-3} \int_{(0,(k-1)2^{-n})^2 \times ((k-1)2^{-n}, k \cdot 2^{-n})^2} d\underline{s} \; d\underline{t} \; \det A_o(s,t)^{-3/2}$$

$$\leq (2\pi)^{-3} \, 2^{-2n/p} \left\{ \int_{(-1,0)^2 \times (0,1)^2} d\underline{s} \; d\underline{t} \; \det A_o(\underline{s},\underline{t})^{-3q/2} \right\}^{1/q},$$

for $p, q > 1$, $(1/p) + (1/q) = 1$. On the other hand, we see that if $m \geq n$,

$$E_\mu \left[ |Y_1(n,k;T(n,k)w) - E_\mu \left[ Y_1(n,k;T(n,k)w) | G(m) \right] |^2 \right]$$

$$= E_\mu \left[ |Y_1(n,k;w) - E_\mu \left[ Y_1(n,k;w) | G(m) \right] |^2 \right] .$$

Therefore, by the same argument used for assertion (1), we have our assertion (2). This completes the proof.

By an argument similar to the proof of Proposition 3.1, we have the following.

Proposition 3.2. For any $\alpha < \frac{1}{4}$, there exists a constant $C_\alpha$ such that

$$E_\mu \left[ |Y_3(n,k;w) - E_\mu \left[ Y_3(n,k;w) | G(m) \right] |^2 \right] \leq C_\alpha 2^{-\alpha m} \tag{1}$$

for any $m \geq 0$, $n \geq 1$ and $1 \leq k \leq 2^n$, and

$$E_\mu \left[ |Y_3(n,k;T(n,k)w) - E_\mu \left[ Y_3(n,k;T(n,k)w | G(m) \right] |^2 \right] \leq C_\alpha 2^{-\alpha m} \tag{2}$$

for any $m \geq n \geq 1$ and $1 \leq k \leq 2^n$.

Let $Y_0(n,k;w) = Y_3(n,k;w) - Y_3(n,k;T(n,k)w)$. Then we have the following.

Proposition 3.3. For any $\alpha < \frac{1}{2}$, there exists a constant $C_\alpha$ such that
$E_\mu \left[ |Y_0(n,k;w)|^2 \right] \leq C_\alpha 2^{-\alpha n}$ for any $n \geq 1$ and $1 \leq k \leq 2^n$.

Proof. Observe that

$$E_\mu \left[ |Y_0(n,k;w)|^2 \right]$$

$$= (2\pi \cdot 2^{-n})^{-3/2} \int_{R^3} dx \; \{ \exp(- \frac{|x|^2}{2 \cdot 2^{-n}})$$

$$\times E_\mu \left[ | \int_0^{(k-1)2^{-n}} \int_{(k-1)2^{-n}}^{1-2^{-n}} \{ \delta(w(s)-w(t) + x) - \delta(w(s)-w(t)-x) \} \; ds \; dt |^2 \right] \}$$

$$= (2\pi \cdot 2^{-n})^{-3/2} \int_{R^3} dx \; \{ \exp(- \frac{|x|^2}{2 \cdot 2^{-n}})$$

$$\times (2\pi)^{-3} \int_{(0,(k-1)2^{-n})^2 \times ((k-1)2^{-n},1-2^{-n})^2} ds \; dt \; \det A(\underline{s},\underline{t})^{-1/2}$$

$$\times \{ 2 \exp(- \frac{1}{2} \underline{x} \cdot A(s,t)^{-1} \cdot {}^t\underline{x}) - 2 \exp(- \frac{1}{2} \overset{\sim}{\underline{x}} \cdot A(s,t)^{-1} \cdot {}^t\overset{\sim}{\underline{x}}) \} \} ,$$

where $\underline{x} = (x_1,x_1,x_2,x_2,x_3,x_3)$ and $\underline{\widetilde{x}} = (x_1,-x_1,x_2,-x_2,x_3,-x_3)$. Note that

$$\left| \exp(-\tfrac{1}{2}\underline{x}\cdot A(s,t)^{-1}\cdot{}^t\underline{x}) - \exp(-\tfrac{1}{2}\underline{\widetilde{x}}\cdot A(s,t)^{-1}\cdot{}^t\underline{\widetilde{x}}) \right.$$

$$\left. \leq \min\{1,\ 18|x|^2\ \|A(\underline{s},\underline{t})^{-1}\|_{\text{operator}}\} \right.,$$

and

$$\|A(\underline{s},\underline{t})^{-1}\|_{\text{operator}} \leq 2\cdot\det A_o(\underline{s},\underline{t})^{-1}.$$

Thus we see that

$$E_\mu\left[\left| \int_0^{(k-1)2^{-n}} \int_{(k-1)2^{-n}}^{1-2^{-n}} \{\delta(w(s)-w(t)+x)-\delta(w(s)-w(t)-x)\}ds\ dt \right|^2\right]$$

$$\leq C'\ \left(\int_{(-1,0)^2\times(0,1)^2} \det A_o(\underline{s},\underline{t})^{-3/2-\alpha}d\underline{s}\ d\underline{t}\right)\ |x|^{2\alpha},$$

$$0 < \alpha < \tfrac{1}{2}\ .$$

Therefore we have

$$E_\mu\left[|Y_o(n,k;w)|^2\right]$$

$$\leq C'_\alpha(2\pi\cdot2^{-n})^{-3/2} \int_{R^3} |x|^{2\alpha}\ \exp(-\frac{|x|^2}{2\cdot2^{-n}})\ dx$$

$$= C_\alpha 2^{-\alpha n}.$$

This completes the proof.

From Propositions 3.2 and 3.3 we have the following.

<u>Proposition 3.4.</u>   For any $\lambda_1$, $\lambda_2 > 0$ with $\lambda_1 + \lambda_2 < \tfrac{1}{4}$ , there exists a constant $C_{\lambda_1,\lambda_2}$ such that

$$E_\mu\left[|Y_o(n,k;w) - E_\mu[Y_o(n,k;w)|\mathcal{C}(m)]|^2\right] \leq C_{\lambda_1,\lambda_2}\ 2^{-\lambda_1 m}\ 2^{-2\lambda_2 n}$$

for any $m \geq 0$, $n \geq 1$ and $1 \leq k \leq 2^n$.

Let $Y_{-1}(n,k;w) = Y_1(n,k;T(n,k)w)$ and $Y_{-2}(n,k;w) = Y_2(n,k;T(n,k)w)$.   Then we have the following.

__Lemma 3.5.__  For any $g \geq 0$, $0 < \alpha < \frac{1}{4}$ and $1 < r < 2$, there exists a constant C such that

$$E_{\nu(g)}\left[|Y_i(n,k;w)|^r\right]^{1/r} \leq C\, 2^{-\alpha n}, \quad i = -2, -1, 0, 1, 2,$$

for any $n \geq 1$ and $1 \leq k \leq 2^n$.

__Proof.__  Let $G(-1) = \{B, \phi\}$.  Note that

$$Y_i(n,k)$$

$$= E_\mu\left[Y_i(n,k)\right] + \sum_{m=0}^{\infty} \{E_\mu\left[Y_i(n,k)\,|\,G(m)\right] - E_\mu\left[Y_i(n,k)\,|\,G(m-1)\right]\}.$$

Then we see that

$$E_{\nu(g)}\left[|Y_i(n,k)|^r\right]^{1/r}$$

$$\leq E_\mu\left[Y_i(n,k)\right] + \sum_{m=0}^{\infty} E_{\nu(g)}\left[\left|E_\mu\left[Y_i(n,k)\,|\,G(m)\right] - E_\mu\left[Y_i(n,k)\,|\,G(m-1)\right]\right|^r\right]^{1/r}$$

$$\leq E_\mu\left[Y_i(n,k)^2\right]^{1/2} + \sum_{m=0}^{\infty} E_\mu\left[\left|E_\mu\left[Y_i(n,k)\,|\,G(m)\right] - E_\mu\left[Y_i(n,k)\,|\,G(m-1)\right]\right|^{pr}\right]^{1/(pr)}$$

$$\times E_\mu\left[f_m(g)^q\right]^{1/(qr)}, \quad \frac{1}{p} + \frac{1}{q} = 1, \quad p,q > 1.$$

Take $p = \frac{2}{r}$.  Then our assertion follows from Theorem 2.1 and Propositions 3.1 and 3.4.

## 4.  THE PROOF OF THEOREM 1.1

We fix $g \geq 0$ throughout.  First we observe the following.

__Proposition 4.1.__  For any $0 \leq t \leq 1$,

$$\sum_{k=1}^{[2^n t]} (w^i(k \cdot 2^{-n}) - w^i((k-1)2^{-n})) \cdot (w^j(k\, 2^{-n}) - w^j((k-1)2^{-n}))$$

$$\to \delta_{i,j}\, t, \quad i,j = 1,2,3,$$

for $\nu(g)$-a.e.w.  Here $\delta_{i,j}$ is Kronecker's delta.

__Proof.__  Note that

$$E_\mu\left[\,|\sum_{k=1}^{[2^n t]} \{(w^i(k \cdot 2^{-n}) - w^i((k-1)2^{-n})) \times \right.$$

$$\times \ (w^j(k \cdot 2^{-n}) - w^j((k-1)2^{-n})) - 2^{-n}\}|^2\big] \leq 3 \cdot 2^{-n}.$$

Thus, by Theorem 2.1, we see that

$$\sum_{n=1}^{\infty} E_{\nu(g)}\Big[\big|\sum_{k=1}^{[2^n t]} (w^i(k \cdot 2^{-n}) - w^i((k-1)2^{-n}))$$

$$\times \ (w^j(k \cdot 2^{-n}) - w^j((k-1)2^{-n})) - 2^{-n}[2^n t]\big|^2\Big] < \infty.$$

This proves our assertion.

Now let

$$N(n,k;w) = w(k \cdot 2^{-n}) - E_{\nu(g)}\big[w(k \cdot 2^{-n})\,|\,F((k-1)2^{-n})\big], \qquad (4.1)$$

and

$$R(n,k;w) = E_{\nu(g)}\big[w(k \cdot 2^{-n}) - w((k-1)2^{-n})\,|\,F((k-1)2^{-n})\big], \qquad (4.2)$$

for each $n \geq 1$ and $1 \leq k \leq 2^n$.

Then it is obvious that

$$N(n,k;w) + R(n,k;w) = w(k \cdot 2^{-n}) - w((k-1)2^{-n}). \qquad (4.3)$$

Proposition 4.2. For any $p > 1$ and $\varepsilon > 0$, there exists a constant $C_{p,\varepsilon}$ such that

$$E_{\nu(g)}\big[|R(n,k)|^p\big] \leq C_{p,\varepsilon}\, 2^{-p(1-\varepsilon)n/2}$$

for any $n \geq 1$ and $1 \leq k \leq 2^n$.

Proof. Our assertion follows from Theorem 2.1 and the facts that

$$E_{\nu(g)}\big[|R(n,k)|^p\big] \leq E_{\nu(g)}\big[|w(k \cdot 2^{-n}) - w((k-1)2^{-n})|^p\big],$$

and that

$$E_{\mu}\big[|w(k \cdot 2^{-n}) - w((k-1)2^{-n})|^p\big] = C_p \cdot 2^{-pn/2}.$$

Lemma 4.3. For any $\alpha < \dfrac{3}{4}$, there exists a constant $C_{1,\alpha}$ such that, for any $n \geq 1$ and $1 \leq k \leq 2^n$,

$$E_{\nu(g)}\big[|R(n,k)|\big] \leq C_{1,\alpha} \cdot 2^{-\alpha n}.$$

<u>Proof.</u>    Let $\overline{F}(t)$ denote the $\sigma$-algebra over B generated by

$\{w(1) - w(s);\ t \le s \le 1\}$.  Let $\tau_1 = (k-1)2^{-n}$, $\tau_2 = k \cdot 2^{-n}$ and $\Delta\tau = 2^{-n}$.  Then we have

$$E_{\nu(g)}\left[|R(n,k)|\right] \le E_{\nu(g)}\left[|E_{n(g)}\left[w(\tau_1) - w(\tau_2)|F(\tau_1)\ \overline{F}(\tau_2)\right]|\right].$$

Let $\underline{w} = (w_1, w_2, w_3) \in B^3$,

$$Z_o(\underline{w}) = 2g(1-\tau_2)\tau_1 \int_0^1 \int_0^1 \delta((1-\tau_2)^{1/2}w_3(t) + \Delta\tau^{1/2}w_2(1) + \tau_1^{1/2}(w_1(1)-w_1(s))\ )\ ds\ dt,$$

$$Z_1(\underline{w}) = 2g\tau_1\Delta\tau \int_0^1 \int_0^1 \delta(\Delta\tau^{1/2}w_2(t) + \tau_1^{1/2}(w_1(1)-w_1(s))\ )\ ds\ dt,$$

$$Z_2(\underline{w}) = 2g(1-\tau_2)\Delta\tau \int_0^1 \int_0^1 \delta((1-\tau_2)^{1/2}w_3(t) + \Delta\tau^{1/2}(w_2(1)-w_2(s))\ )\ ds\ dt,$$

and

$$Z(\underline{w}) = Z_o(\underline{w}) + Z_1(\underline{w}) + Z_2(\underline{w}).$$

Then we obtain from Lemma 2.5

$$E_{\nu(g)}\left[\Phi\right] = L^{-1}\ \overset{\sim}{E}\left[\exp(-Z(\underline{w}))\right.$$

$$\left. \times\ \Phi((\tau_1^{1/2}\ w_1(\cdot/\tau_1))*(\Delta\tau^{1/2}w_2(\cdot/\Delta\tau))*((1-\tau_2)^{1/2}w_3(\cdot/(1-\tau_2)))\ )\right] \tag{4.5}$$

for any bounded measurable function $\Phi: B \to R$,  where $\overset{\sim}{E}[\cdot]$ denotes

$$\int_{B^3} \cdot \nu_1(dw_1) \times \nu_2(dw_2) \times \nu_3(dw_3),$$

$$\nu_1(dw) = \nu(\tau_1^{1/2}g)(dw),\quad \nu_2(dw) = \nu(\Delta\tau^{1/2}g)(dw),$$

$$\nu_3(dw) = \nu((1-\tau_2)^{1/2}g)(dw)\quad \text{and}\quad L = \overset{\sim}{E}\left[\exp(-Z(\underline{w})\ )\right].$$

Then it is easy to see that

$$E_{\nu(g)}\left[|E_{\nu(g)}\left[w(\tau_2) - w(\tau_1)|F(\tau_1)\ \overline{F}(\tau_2)\right]|\ \right]$$

$$\tag{4.6}$$

$$= L^{-1} \int_{B\times B} \nu_1(dw_1) \times \nu_3(dw_3)|\int_B \Delta\tau^{1/2}w_2(1)\ e^{-Z(\underline{w})}\ \nu_2(dw_2)|.$$

Let $Z'(\underline{w}) = Z(w_1, -w_2, w_3)$. Noting that $\left|(e^{-x} - e^{-y}/e^{-x} + e^{-y})\right| \leq |x-y|$ for any $x, y \geq 0$, we see from Remark 2.6 that

$$\left| \int_B w_2(1)\, e^{-Z(\underline{w})}\, \nu_2(dw_2) \right|$$

$$= \frac{1}{2} \left| \int_B w_2(1)\, (e^{-Z(\underline{w})} - e^{-Z'(\underline{w})})\, \nu_2(dw_2) \right|$$

$$\leq \frac{1}{2} \int_B |w_2(1)| \cdot |Z(\underline{w}) - Z'(\underline{w})|\, (e^{-Z(\underline{w})} + e^{-Z'(\underline{w})})\, \nu_2(dw_2)$$

$$= \int_B |w_2(1)| \cdot |Z(\underline{w}) - Z'(\underline{w})| \cdot e^{-Z(\underline{w})}\, \nu_2(dw_2).$$

Therefore we have

$$E_{\nu(g)}\big[|R(n,k;w)|\big]$$

$$\leq L^{-1}\, \overset{\sim}{E}\Big[\Delta\tau^{1/2}|w_2(1)| \cdot \{|Z_2'(\underline{w})| + |Z_1'(\underline{w})| + |Z_o(\underline{w}) - Z_o'(\underline{w})| \tag{4.7}$$

$$+ |Z_1(\underline{w})| + |Z_2(\underline{w})| \quad \exp(-Z(\underline{w}))\Big]$$

$$= 2g\, E_{\nu(g)}\Big[|w(k\cdot 2^{-n}) - w((k-1)2^{-n})| \cdot \sum_{i=-2}^{2} |Y_i(n,k;w)|\Big],$$

where $Z_i'(\underline{w}) = Z_i(w_1, -w_2, w_3)$, $i = 0,1,2$, and $Y_i'$'s are as in Section 3. Thus by Hörder's inequality we have

$$E_{\nu(g)}\big[|R(n,k;w)|\big]$$

$$\leq 2g\, E_{\nu(g)}\Big[|w(k\cdot 2^{-n}) - w((k-1)2^{-n})|^{(2-\varepsilon)/(1-\varepsilon)}\Big]^{(1-\varepsilon)/(2-\varepsilon)}$$

$$\times \sum_{i=-2}^{2} E_{\nu(g)}\Big[|Y_i(n,k;w)|^{2-\varepsilon}\Big]^{1/(2-\varepsilon)}.$$

Therefore Theorem 2.1 and Lemma 3.5 lead us to our assertion.

<u>Lemma 4.4.</u>  For any $\alpha > 3/2$ and $\varepsilon > 0$ with $\varepsilon < (\alpha/2) - (3/4)$, there exists a constant $C(\alpha,\varepsilon)$ such that

$$E_{\nu(g)}\big[|R(n,k;w)|^{\alpha}\big] \leq C(\alpha,\varepsilon) \cdot 2^{-(1+\varepsilon)n}$$

for any $n \geq 1$ and $1 \leq k \leq 2^n$.

$\underline{\text{Proof.}}$   By Hölder's inequality, we have for any $r > \alpha$

$$E_{\nu(g)}\left[|R(n,k;w)|^{\alpha}\right]$$

$$\leq E_{\nu(g)}\left[|R(n,k;w)|^{r}\right]^{(\alpha-1)/(r-1)} \cdot E_{\nu(g)}\left[|R(n,k;w)|\right]^{(r-\alpha)/(r-1)}.$$

Then Proposition 4.2 and Lemma 4.4 lead us to our assertion.

$$\text{Now let } M_n(w) = \sum_{k=1}^{2^n} N(n,k;w), \quad n \geq 1. \quad \text{Then we have}$$

$\underline{\text{Lemma 4.5.}}$   For any $n \geq 1$ and $0 < \varepsilon < \dfrac{1}{4}$ ,

$$E_{\nu(g)}\left[|M_{n+1}(w) - M_n(w)|^{2}\right] \leq C(2,\varepsilon)\cdot 2^{-\varepsilon n},$$

where $C(2,\varepsilon)$ is a constant in Lemma 4.4.

$\underline{\text{Proof.}}$   It is obvious that

$$E_{\nu(g)}\left[|M_{n+1}(w) - M_n(w)|^{2}\right]$$

$$= \sum_{k=1}^{2^n} E_{\nu(g)}\left[(N(n+1,2k) + N(n+1,2k-1) - N(n,k))^{2}\right],$$

and that

$$N(n+1,2k) + N(n+1,2k-1) - N(n,k)$$

$$= -R(n+1,2k) + E_{\nu(g)}\left[R(n+1,2k)\mid F((k-1)2^{-n})\right].$$

Therefore we have

$$E_{\nu(g)}\left[|M_{n+1}(w) - M_n(w)|^{2}\right]$$

$$\leq \sum_{k=1}^{2^n} E_{\nu(g)}\left[R(n+1,2k)^{2}\right]$$

$$\leq C(2,\varepsilon)\cdot 2^{-\varepsilon n}.$$

This completes the proof.

From Lemma 4.5, $\{M_n(w)\}_{n=1}^{\infty}$ is a Cauchy sequence in $L^{2}(B,d\nu(g))$.   Let

$$M(w) = \lim_{n \to \infty} M_n(w) \text{ and } M(t;w) = E_{\nu(g)}\left[M \mid F(t)\right], \quad 0 \le t \le 1.$$

Then it is easy to see that

$$M(k \cdot 2^{-m}) = \lim_{n \to \infty} \sum_{j=1}^{k \cdot 2^{n-m}} N(n,j). \tag{4.8}$$

Let $R(t,w) = w(t) - M(t;w), \quad 0 \le t \le 1.$

<u>Lemma 4.6.</u>   For any $\alpha > 8/5$, there exist constants $C > 0$ and $\varepsilon > 0$ such that

$$E_{\nu(g)}\left[\left|R(k \cdot 2^{-m}) - R((k-1)2^{-m})\right|^{\alpha}\right] \le C \cdot 2^{-(1+\varepsilon)m}$$

for any $m \ge 1$ and $1 \le k \le 2^m$.

<u>Proof.</u>   Let $\Delta M_n = \sum_{j=(k-1)2^{n-m}+1}^{k \cdot 2^{n-m}} N(n,j), \; n \ge m.$  Then, by the same argument as the proof of Lemma 4.5, we have

$$E_{\nu(g)}\left[\left|\Delta M_{n+1} - \Delta M_n\right|^2\right] \le C(2,\varepsilon) \cdot 2^{-m} \cdot 2^{-\varepsilon n}, \quad 0 < \varepsilon < \frac{1}{4}. \tag{4.9}$$

Observe that

$$R(k \cdot 2^{-m}) - R((k-1)2^{-m})$$

$$= w(k \cdot 2^{-m}) - w((k-1)2^{-m}) - (M(k \cdot 2^{-m}) - M((k-1)2^{-m}))$$

$$= R(m,k) - \sum_{n=m}^{\infty} (\Delta M_{n+1} - \Delta M_n).$$

Therefore we have

$$E_{\nu(g)}\left[\left|R(k \cdot 2^{-m}) - R((k-1)2^{-m})\right|^2\right]^{1/2}$$

$$\le C(2,\varepsilon)^{1/2} \left\{2^{-(1+\varepsilon)m/2} + \sum_{n=m}^{\infty} 2^{-m/2} \cdot 2^{-\varepsilon n/2}\right\}$$

$$= C_{\varepsilon} \cdot 2^{-(1+\varepsilon)m/2}, \quad 0 < \varepsilon < \frac{1}{4}.$$

Thus for $0 < \alpha < 2$, we have

$$E_{\nu(g)}\left[\,|R(k \cdot 2^{-m}) - R((k-1)2^{-m})|^{\alpha}\,\right]$$

$$\leq C_{\varepsilon}^{\alpha} \cdot 2^{-\alpha(1+\varepsilon)m/2}.$$

This completes the proof.

Now we are ready to prove Theorem 1.1.  Lemma 4.6 shows that, for any $\alpha > 8/5$,

$$\sum_{k=1}^{2^n} |R(k \cdot 2^{-n};w) - R((k-1)2^{-n};w)|^{\alpha} \to 0 \quad \text{as} \quad n \to \infty \qquad (4.10)$$

for $\nu(g)$-a.e.w.  Since $M(t)$ is an $F(t)$-martingale, by virtue of Doob's theorem, $M(\cdot;w)$ has a right limit and a left limit at each point $t \in [0,1]$ for $\nu(g)$-a.e.w, and so does $R(\cdot;w)$.  But (4.10) shows that $R(t-0;w) = R(t+0;w)$ at each $t \in [0,1]$ for $\nu(g)$-a.e.w, and so $R(\cdot;w)$ has a continuous version.  Thus we may assume that $R(t;w)$ and $M(t;w)$ are continuous in t for $\nu(g)$-a.e.w.

Proposition 4.1 and (4.10) imply that

$$\sum_{k=1}^{[2^n t]} (M^i(k \cdot 2^{-n}) - M^i((k-1)2^{-n})) \cdot (M^j(k \cdot 2^{-n}) - M^j((k-1)2^{-n}))$$

$$\to \delta_{i,j}t, \quad i,j = 1,2,3, \quad \text{as} \quad n \to \infty \text{ for } \nu(g)\text{-a.e.w.}$$

Thus $\{M(t);\ 0 \leq t \leq 1\}$ is a continuous $F(t)$-martingale with $<M^i,M^j>(t) = \delta_{i,j}t$, $i,j = 1,2,3$.  Therefore $\{M(t);\ 0 \leq t \leq 1\}$ is a Wiener process.

This completes the proof.

REFERENCES


[1]  Föllmer, H.   Dirichlet processes, *Proceedings of LMS Durham Symposium 1980*, ed. D. Williams, Springer Lecture Notes in Math. vol. 851 476–478.

[2]  Symanzik, K.  Euclidean quantum field theory, *'Local Quantum Theory'*, *Proceedings of the International School of Physics 'Enrico Fermi'* vol.45, ed. R. Jost. New York: Academic Press (1969).

[3]  Westwater, J.  On Edward's model for long polymer chains, *Comm. Math. Phys.* 72 131-174, (1980).

[4]  Westwater, J.  On Edward's model for polymer chains III. Borel summability,  *Comm. Math. Phys.* 84 459-470, (1982).

Shigeo Kusuoka
Department of Mathematics
Faculty of Science
University of Tokyo
Tokyo, Japan.

S KUSUOKA
# Asymptotics of polymer measures in one dimension

## 1. INTRODUCTION

In this report, we consider probability measures $\nu_{T,b}$, $T > 0$ and $b \in R$, on $C([0,T] \to R)$ given by

$$\nu_{T,b}(dw) = Z_T(b)^{-1} \exp(b \cdot w(T) - \int_0^T \int_0^T \delta(w(t)-w(s))\, dt\, ds)\, \mu_T(dw),$$

where $\mu_T$ is the Wiener measure on $C([0,T] \to R)$ and

$$Z_T(b) = E_{\mu_T}\left[\exp(b \cdot w(T) - \int_0^T \int_0^T \delta(w(t)-w(s))\, dt\, ds)\right].$$

Then $\nu_{T,0}$ is Edward's model of long polymer chains in one dimension.

We study the asymptotic behaviour of $E_{\nu_{T,b}}\left[|w(T)|^\alpha\right]$, $\alpha > 0$, as $T \to \infty$. Our main results are the following.

(1) $f(b) = \lim\limits_{T\to\infty} \frac{1}{T} \log Z_T(b)$ exists for each $b \in R$, and $f : R \to R$ is convex (Lemma 3.1).

(2) If f is differentiable at b, then

$$\lim_{T\to\infty} E_{\nu_{T,b}}\left[(\frac{|w(T)|}{T})^\alpha\right] = |f'(b)|^\alpha, \quad \alpha > 0 \quad \text{(Theorem 3.9)}.$$

We also give a remark on the asymptotic behaviour of $E_{\nu_{T,0}}\left[|w(T)|^\alpha\right]$.

To show these results, we give an abstract version of Schilder's result [5] in Section 2.

Remark. The author has learned that J. Westwater showed that the probability law of $w(T\cdot)/T$ on $C([0,1] \to R)$ under $\nu_{T,0}(dw)$ converges to $\frac{1}{2}(\nu_+ + \nu_-)$ as $T \to \infty$, where $\nu_+$ and $\nu_-$ are probability measures on $C([0,1] \to R)$ concentrated at functions $g_+$ and $g_-$ respectively, $g_+(t) = (\frac{d^+}{db} f)(0) \cdot t$ and $g_-(t) = -(\frac{d^+}{db})f(0) \cdot t$, $0 \le t \le 1$. His result will appear in [6].

## 2. ABSTRACT SCHILDER-TYPE THEORY

Let $(\mu,H,B)$ be an abstract Wiener space, i.e., B is a separable real Banach space, H is a separable real Hilbert space which is continuously and densely embedded in B, and $\mu$ is a Gaussian measure on B such that

$$E_\mu\left[\exp(\sqrt{-1}\ _B\langle w,u\rangle_{B^*})\right] = \exp(-\|u\|_H^2\ /2), \quad u \in B^* \subset H, \tag{2.1}$$

where we identify the dual space $H^*$ of H with the original space H through the inner product of H, and $E_\mu$ denotes the expectation with respect to $\mu(dw)$. It is easy to see that $B^*$ is a dense subspace of H and $(u,v)_H = {}_B\langle u,v\rangle_{B^*}$ for any $u \in H \subset B$ and $v \in B^* \subset H$.

Let $F_\lambda : B \to R$, $0 < \lambda \leq 1$, be measurable functions. Throughout this section we impose the following assumption on $F_\lambda$.

(A-1)  There exist constants $C > 0$, $p_0 > 1$, and a measurable function $g$ on B such that

$$\lambda^{-2}F_\lambda(w) \leq g(w) + \lambda^{-2}C \quad \text{for any } w \in B \text{ and } 0 < \lambda \leq 1, \text{ and} \tag{2.2}$$

$$E_\mu\left[\exp(p_0 g(w))\right] < \infty. \tag{2.3}$$

Let $Z_\lambda = E_\mu\left[\exp(\lambda^{-2}F_\lambda(w))\right]$. Then, from assumption (A-1), we see that $\varlimsup_{\lambda\to 0} \lambda^2\log Z_\lambda \leq C$.

**Definition 2.1.**  We define $I : B \to [-\infty,\infty)$ and $\tilde{I} : B \to [-\infty,\infty)$ by

$$I(u) = \lim_{\varepsilon\to 0} \varlimsup_{\lambda\to 0} \lambda^2\log E_\mu\left[\exp(\lambda^{-2}F_\lambda(w)), \|\lambda w-u\|_B < \varepsilon\right], \tag{2.4}$$

and

$$\tilde{I}(u) = \lim_{\varepsilon\to 0} \varliminf_{\lambda\to 0} \lambda^2\log E_\mu\left[\exp(\lambda^{-2}F_\lambda(w)), \|\lambda w-u\|_B < \varepsilon\right], \tag{2.5}$$

for any $u \in B$.

**Remark 2.2.**  It is obvious that $I(u) \leq \varlimsup_{\lambda\to 0} \lambda^2\log Z_\lambda \leq C$.

**Proposition 2.3.**  (1) $\tilde{I}(u) \leq I(u)$ for any $u \in B$.

(2)  $I : B \to [-\infty,\infty)$ and $\tilde{I} : B \to [-\infty,\infty)$ are upper semi-continuous.

Proof. Assertion (1) is obvious. Suppose that $u_n \to u$ in $B$. Then we see that

$$\lambda^2 \log E_\mu \left[ \exp(\lambda^{-2} F_\lambda(w)), \quad \| \lambda w - u \|_B < \varepsilon \right]$$

$$\geqq \lambda^2 \log E_\mu \left[ \exp(\lambda^{-2} F_\lambda(w)), \quad \| \lambda w - u_n \|_B < \varepsilon - \| u - u_n \|_B \right].$$

Thus we have, if $\| u - u_n \|_B < \varepsilon$,

$$\varlimsup_{\lambda \to 0} \lambda^2 \log E_\mu \left[ \exp(\lambda^{-2} F_\lambda(w)), \quad \| \lambda w - u \|_B < \varepsilon \right] \geqq I(u_n),$$

which shows that $\varlimsup_{n \to \infty} I(u_n) \leqq I(u)$. This proves the upper semi-continuity of $I$. The proof for $\tilde{I}$ is similar.

Proposition 2.4. $I(u) = -\infty$ for any $u \in B-H$.

Proof. Let $p_1 > 1$ be such that $\dfrac{1}{p_0} + \dfrac{1}{p_1} = 1$. Then we have

$$\lambda^2 \log E_\mu \left[ \exp(\lambda^{-2} F_\lambda(w)), \quad \| \lambda w - u \|_B < \varepsilon \right]$$

$$\leqq \lambda^2 \log E_\mu \left[ \exp(g(w) + \lambda^{-2} C), \quad \| \lambda w - u \|_B < \varepsilon \right]$$

$$\leqq C + \frac{\lambda^2}{p_0} \log E_\mu \left[ \exp(p_0 g(w)) \right] + \frac{\lambda^2}{p_1} \log \mu( \| \lambda w - u \|_B < \varepsilon ).$$

Therefore we see that

$$I(u) \leqq C + \frac{1}{p_1} \lim_{\varepsilon \to 0} \varlimsup_{\lambda \to 0} \lambda^2 \log \mu( \| \lambda w - u \|_B < \varepsilon ). \tag{2.6}$$

Let $\{v_n\}_{n=1}^\infty$ be a dense subset of $B$. Then, by virtue of the Hahn-Banach theorem, we can take $\{x_n\}_{n=1}^\infty \subset B^*$ such that $\| x_n \|_{B^*} = 1$ and

$$_B\langle v_n, x_n \rangle_{B^*} = \| v_n \|_B .$$ Then we have

$$\| v \|_B = \sup_n {}_B\langle v, x_n \rangle_{B^*} \quad \text{for any } v \in B. \tag{2.7}$$

Therefore $\{x_n\}_{n=1}^\infty$ is dense in $H$. By using Schmidt's orthogonalization, we can get a complete orthonormal base $\{e_n\}_{n=1}^\infty \subset B^*$ of $H$ such that $Re_1 + \ldots + Re_n \supset Rx_1 + \ldots + Rx_n .$

Let $u_n = \sum_{k=1}^{n} {}_B\langle u, e_k \rangle_{B^*} e_k$. Then we see that

$${}_B\langle u - u_n, e_k \rangle_{B^*} = 0, \quad k = 1,\ldots,n. \tag{2.8}$$

By virtue of Cameron-Martin's formula and (2.8), we have

$$\lambda^2 \log \mu(\| \lambda w - u \|_B < \varepsilon)$$

$$= \lambda^2 \log E_\mu\left[\exp(- \lambda^{-2} {}_B\langle \lambda w, u_n \rangle_{B^*} - \tfrac{1}{2} \lambda^{-2} | u_n \|_H^2) , \| \lambda w - (u - u_n) \|_B < \varepsilon\right]$$

$$\leq -\varepsilon \| u_n \|_{B^*} - \tfrac{1}{2} \| u_n \|_H^2 .$$

Therefore we get from (2.6),

$$I(u) \leq C - \frac{1}{2p_1} \| u_n \|_H^2 , \quad n = 1,2,\ldots . \tag{2.9}$$

Thus it is sufficient to prove that $\lim_{n\to\infty} \| u_n \|_H = \infty$ if $u \in B-H$. Suppose that $\lim_{n\to\infty} \| u_n \|_H < \infty$. Since $\lim_{n\to\infty} (u_n, e_k)_H = {}_B\langle u, e_k \rangle_{B^*}$, $k = 1,2,\ldots$, there exists a $u' \in H$ such that $u_n \to u'$ weakly in H. Then we have ${}_B\langle u - u', e_n \rangle_{B^*} = 0$, $k = 1,2,\ldots$ . Therefore, by (2.7) and the assumption for $\{e_n\}_{n=1}^{\infty}$, we have

$$\| u - u' \|_B = \sup_n {}_B\langle u - u', x_n \rangle_{B^*} = 0.$$

Therefore we get $u = u' \in H$, which contradicts the assumption that $u \in B-H$. This completes the proof.

Definition 2.5. We say that a measurable function $q : B \to [0,\infty]$ is a measurable semi-norm, if

(1) $\mu(\{w \in B; q(w) < \infty\}) = 1$,

(2) $q(0) = 0$, $q(aw) = a q(w)$ for any $w \in B$ and $a > 0$, and

(3) $q(w_1 + w_2) \leq q(w_1) + q(w_2)$ for any $w_1, w_2 \in B$.

Proposition 2.6. There exists a compact subset K of B such that

$$\overline{\lim_{\lambda\to 0}} \lambda^2 \log E_\mu\left[\exp(\lambda^{-2} F_\lambda(w)), \lambda w \in B-K\right] \leq \overline{\lim_{\lambda\to 0}} \lambda^2 \log Z_\lambda - 1.$$

Proof.  If $\overline{\lim_{\lambda \to 0}} \; \lambda^2 \log Z_\lambda = -\infty$, the assertion is obvious.  Thus we assume that $\overline{\lim_{\lambda \to 0}} \; \lambda^2 \log Z_\lambda > -\infty$.  From Kuo [3, Lemma 4.5] and the Landau-Shepp-Fernique theorem (see Kuo [3, Theorem 3.1]), we see that there exist a constant $\alpha > 0$ and a measurable semi-norm $q$ such that $\{w \in B; q(w) \leq N\}$ is compact in $B$ for any $N \geq 0$ and $E_\mu[\exp(\alpha q(w)^2)] < \infty$.  Let $p_1 > 1$ be such that $(1/p_0) + (1/p_1) = 1$.  Then we see that

$$\lambda^2 \log E_\mu\left[\exp(\lambda^{-2} F_\lambda(w)), \; q(\lambda w) > N\right]$$

$$\leq C + \frac{\lambda^2}{p_0} \log E_\mu\left[\exp(p_0 g(w))\right] + \frac{\lambda^2}{p_1} \log \mu(q(\lambda w) > N)$$

$$\leq C + \frac{\lambda^2}{p_0} \log E_\mu\left[\exp(p_0 g(w))\right] - \frac{\alpha}{p_1} N^2 + \frac{\lambda^2}{p_1} \log E_\mu\left[\exp(\alpha q(w)^2)\right].$$

Take sufficiently large $N$ such that $C - \frac{1}{p_1} N^2 \leq \overline{\lim_{\lambda \to 0}} \; \lambda^2 \log Z_\lambda - 1$ and let $K = \{w \in B; q(w) \leq N\}$.  Then we have

$$\overline{\lim_{\lambda \to 0}} \; \lambda^2 \log E_\mu\left[\exp(\lambda^{-2} F_\lambda(w)), \; \lambda w \in B - \bar{K}\right] \leq \overline{\lim_{\lambda \to 0}} \; \lambda^2 \log Z_\lambda - 1.$$

This completes the proof.

Proposition 2.7.   (1) $\overline{\lim_{\lambda \to 0}} \; \lambda^2 \log Z_\lambda = \max\{I(u); \; u \in B\}$.

(2) $\underline{\lim_{\lambda \to 0}} \; \lambda^2 \log Z_\lambda \geq \sup\{\overset{\vee}{I}(u); \; u \in B\}$.


Proof.   The inequality $\geq$ is obvious.  Thus it is sufficient to show that there exists a $u \in B$ such that $\overline{\lim_{\lambda \to 0}} \; \lambda^2 \log Z_\lambda \leq I(u)$.  Suppose that $I(u) < \overline{\lim_{\lambda \to 0}} \; \lambda^2 \log Z_\lambda$ for any $u \in B$.  Then we see that there exists an open neighbourhood $U(u)$ of $u$ for each $u \in B$ such that

$$\overline{\lim_{\lambda \to 0}} \; \lambda^2 \log E_\mu\left[\exp(\lambda^{-2} F_\lambda(w)), \; \lambda w \in U(u)\right] < \overline{\lim_{\lambda \to 0}} \; \lambda^2 \log Z_\lambda.$$

Take a compact set $K$ in $B$ as in Proposition 2.6.  Then there exists $u_1, \ldots, u_n \in B$ such that $K \subset \bigcup_{k=1}^{n} U(u_k)$.  Therefore we see that

$$\overline{\lim_{\lambda \to 0}} \; \lambda^2 \log Z_\lambda$$

$$= \max\{ \overline{\lim_{\lambda \to 0}} \; \lambda^2 \log E_\mu \left[ \exp(\lambda^{-2} F_\lambda(w)), \; \lambda w \in 3\text{-}K \right],$$

$$\overline{\lim_{\lambda \to 0}} \; \lambda^2 \log E_\mu \left[ \exp(\lambda^{-2} F_\lambda(w)), \; \lambda w \in J(u_k) \right], \quad k = 1,\ldots,n \}$$

$$< \overline{\lim_{\lambda \to 0}} \; \lambda^2 \log Z_\lambda ,$$

which is contradiction.  This completes the proof.

<u>Proposition 2.8.</u>   For any $h \in H$, we have

(1)   $I(h) \; \leqq \; -\dfrac{1}{2} \, \|h\|_H^2 + \lim_{\varepsilon \to 0} \overline{\lim_{\lambda \to 0}} \; \operatorname*{ess.\,sup}_{\|\lambda w\|_B < \varepsilon} \; F_\lambda(w + \lambda^{-1} h), \quad$ and

(2)   $\overset{\sim}{I}(h) \; \geqq \; -\dfrac{1}{2} \, \|h\|_H^2 + \lim_{\varepsilon \to 0} \overline{\lim_{\lambda \to 0}} \; E_\mu \left[ F_\lambda(w + \lambda^{-1} h) \mid \; \|\lambda w\|_B < \varepsilon \right].$

<u>Proof.</u>   Since $B^*$ is dense in $H$, there exists $\{u_n\}_{n=1}^{\infty} \subset B^*$ such that $u_n \to h$ in $H$.  Then by virtue of Cameron-Martin's formula, we have

$$\lambda^2 \log E_\mu \left[ \exp(\lambda^{-2} F_\lambda(w)), \quad \|\lambda w - h\|_B < \varepsilon \right]$$

$$= \lambda^2 \log E_\mu \left[ \exp(\lambda^{-2} F_\lambda(w + \lambda^{-1} u_n) - \lambda^{-2} {}_B\langle \lambda w, u_n \rangle_{B^*} - \frac{1}{2} \lambda^{-2} \|u_n\|_H^2 ), \right.$$

$$\left. \|\lambda w - (h - u_n)\|_B < \varepsilon \right]$$

$$\leqq \varepsilon |u_n|_{B^*} - (h - u_n, u_n)_H - \frac{1}{2} \|u_n\|_H^2 + \operatorname*{ess.\,sup}_{\|\lambda w\|_B < \varepsilon + 2\|h - u_n\|_B} F_\lambda(w + \lambda^{-1} h).$$

Letting $\varepsilon \to 0$ and $n \to \infty$, we have our assertion (1).

On the other hand, by virtue of Cameron-Martin's formula and Jensen's inequality, we have

$$\lambda^2 \log E_\mu \left[ \exp(\lambda^{-2} F_\lambda(w)), \quad \|\lambda w - h\|_B < \varepsilon \right]$$

$$= \lambda^2 \log E_\mu \left[ \exp(\lambda^{-2} F_\lambda(w + \lambda^{-1} h) - \lambda^{-1}(w, h)\overset{\sim}{_H} - \frac{1}{2} \lambda^{-2} \|h\|_H^2 ), \; \|\lambda w\|_B < \varepsilon \right]$$

$$\geqq -\frac{1}{2} \|h\|_H^2 + E_\mu \left[ F_\lambda(w + \lambda^{-1} h) \mid \; \|\lambda w\|_B < \varepsilon \right] + \lambda^2 \log \mu(\|\lambda w\|_B < \varepsilon).$$

71

Here $(w,h)_H^{\sim}$ denotes an abstract Wiener integral, i.e. the stochastic extension of $(\cdot,h)_H$. Letting $\lambda \downarrow 0$ and $\varepsilon \downarrow 0$, we have our assertion (2). This completes the proof.

<u>Definition 2.9.</u>  We define probability measures $R_\lambda$ on B, $0 < \lambda \leq 1$, by $R_\lambda(A) = Z_\lambda^{-1} E_\mu \left[ \exp(\lambda^{-2} F_\lambda(w)), \quad \lambda w \in A \right]$ for any Borel set A in B.
   We will often use the following assumption on $Z_\lambda$:

(A-2) $-\infty < \varliminf_{\lambda \to 0} \lambda^2 \log Z_\lambda = \varlimsup_{\lambda \to 0} \lambda^2 \log Z_\lambda .$

<u>Lemma 2.10.</u>  Suppose that assumptions (A-1) and (A-2) hold.  Then

(1) there exists a compact set K in B such that $R_\lambda(B-K) \to 0$, $\lambda \downarrow 0$, and

(2) for any $u \in B$ with $I(u) < \lim\limits_{\lambda \to 0} \lambda^2 \log Z_\lambda$, there exists an $\varepsilon > 0$ such that $R_\lambda(\| w-u \|_B < \varepsilon ) \to 0$ as $\lambda \downarrow 0$.

   The proof of Lemma 2.10 is obvious from Proposition 2.6 and Definition 2.1.

<u>Theorem 2.11.</u>  Suppose that assumptions (A-1) and (A-2) hold.  Suppose furthermore that there exists a unique $h \in H$ such that $I(h) = \lim\limits_{\lambda \to 0} \lambda^2 \log Z_\lambda$. Then $E_{R_\lambda} \left[ k(w) \right] \to k(h)$ as $\lambda \downarrow 0$ for any continuous function k on B satisfying $N^{-2} \sup\limits_{\| w \|_B < N} \log(|k(w)|+1) \to 0$ as $N \to \infty$.

<u>Proof.</u>  If we show that

$$\varlimsup_{\lambda \to 0} E_{R_\lambda} \left[ |k(w)|, \quad \| w \|_B > N \right] \to 0, \quad N \to \infty, \tag{2.10}$$

then we have our assertion from Lemma 2.10.  Let $p_1 > 1$ such that $(1/p_0) + (1/p_1) = 1$.  Then it is easy to see that

$$\lambda^2 \log E_{R_\lambda} \left[ |k(w)|, \quad \| w \|_B > N \right]$$

$$\leqq -\lambda^2 \log Z_\lambda + \lambda^2 \log E_\mu \left[ |k(w)| \exp(g(w) + \lambda^{-2} C), \quad \| \lambda w \|_B > N \right]$$

$$\leq -\lambda^2 \log Z_\lambda + C + \frac{\lambda^2}{p_0} \log E_\mu \left[ \exp(p_0 g(w)) \right]$$

$$+ \frac{\lambda^2}{2p_1} \log E_\mu \left[ |k(w)|^{2p_1} \right] + \frac{\lambda^2}{2p_1} \log \mu( \| \lambda w \|_B > N).$$

From the assumption for k and the Landau–Shepp–Fernique theorem, we see that $E_\mu \left[ |k(w)|^{2p_1} \right] < \infty$. Then, by a similar argument to the proof of Proposition 2.6, we obtain (2.10). This completes the proof.

Remark 2.12.   Let $F : B \to R$ be a continuous function, let $F_\lambda(w) = F(\lambda w)$, $0 < \lambda \leq 1$ and $w \in B$, and suppose that assumption (A-1) holds for these $F_\lambda$'s.  Then, from Proposition 2.8, we see that

$$I(h) = \tilde{I}(h) = -\frac{1}{2} \| h \|_H^2 + F(h) \quad \text{for any } h \in H,$$

and that the assumption (A-2) holds and $\lim_{\lambda \to 0} \lambda^2 \log Z_\lambda = \sup_{h \in H} I(h)$.  In this case, Theorem 2.11 is nothing but what Schilder [5] and Kallianpur–Oodaira [2] have obtained.

3.  ASYMPTOTICS OF POLYMER MEASURES $\nu_{T,b}$
$-\!-\!-\!-\!-\!-\!-\!-\!-\!-\!-\!-\!-\!-\!-\!-\!-\!-\!-\!-\!-\!-\!-\!-\!-\!-\!-\!-\!-\!-\!-$

Let $g_\varepsilon : R \to [0,\infty)$, $\varepsilon > 0$, be functions given by

$$g_\varepsilon(x) = (\frac{1}{2\pi\varepsilon})^{1/2} \exp(-\frac{x^2}{2\varepsilon}), \quad x \in R,$$

and let $X_\varepsilon(T,a,w) = \int_0^T \int_0^T g_\varepsilon(a(w(t)-w(s))) \, dt \, ds$ for $T > 0$, $a > 0$ and

$w \in C([0,T] \to R)$.  Observing

$$X_\varepsilon(T,a,w) = \int_R a \, dx \, ( \int_0^T \frac{1}{a} g_{\varepsilon/2a^2}(w(t) - x) \, dt)^2, \tag{3.1}$$

we see that

$$X_\varepsilon(T,a,w) \to \frac{1}{a} \int_R L(T,x;w)^2 dx, \quad \varepsilon \downarrow 0, \quad \text{for } \mu_T\text{-a.e.w.} \tag{3.2}$$

Here $L(T,x;w)$ denotes the local time of the path w at the point x from time 0 till time T.  We take $\lim_{\varepsilon \downarrow 0} X_\varepsilon(T,a,w)$ as the definition of

$$\int_0^T \int_0^T \delta(a(w(t)-w(s))) \, dt \, ds, \quad T > 0 \text{ and } a > 0. \quad \text{Let}$$

$$\nu_{T,b}(dw) = Z_T(b)^{-1} \exp(b \cdot w(T) - \int_0^T \int_0^T \delta(w(t)-w(s)) \, dt \, ds) \, \mu_T(dw)$$

as given in the Introduction.

In this section we use the following notation:  B is a Banach space given by $B = \{w \in C([0,1] \to R) \; ; \; w(0) = 0\}$ with a norm given by $\|w\|_B = \max\{|w(t)| \; ; \; 0 \leq t \leq 1\}$;  H is a Hilbert space given by $H = \{h \in C([0,1] \to R) \; ; \; h(0) = 0,\; h(t)$ is absolutely continuous in $t \in [0,1]$ and $\int_0^1 |\frac{d}{dt} h(t)|^2 \, dt < \infty\}$ with an inner product given by $(h_1, h_2)_H = \int_0^1 \frac{d}{dt} h_1(t) \cdot \frac{d}{dt} h_2(t) \, dt$ ; and $\mu$ is the standard Wiener measure on B.  Then $(\mu, H, B)$ is an abstract Wiener space.

It is easy to see that, for any bounded measurable function k defined on B, we have

$$\int_{C([0,T] \to R)} k(\frac{w(T \cdot)}{T}) \, \nu_{T,b}(dw)$$

$$= Z_T(b)^{-1} \int_{C([0,T] \to R)} k(\frac{w(T \cdot)}{T})$$

$$\times \exp(b \cdot w(T) - T^2 \int_0^1 \int_0^1 \delta(w(Tt)-w(Ts)) \, dt \, ds \,) \, \mu_T(dw)$$

$$= Z_T(b)^{-1} E_\mu \Big[ k(T^{-1/2} w)$$

$$\times \exp(T\{b \cdot T^{-1/2} w(1) - \int_0^1 \int_0^1 \delta(T^{-1/2}(w(t)-w(s)) \,) \, dt \, ds\}) \Big].$$

Let $\lambda = T^{-1/2}$, $F_\lambda^{(b)}(w) = b \cdot \lambda \cdot w(1) - \int_0^1 \int_0^1 \delta(\lambda(w(t)-w(s)) \,) \, dt \, ds$, and $Z_\lambda^{(b)} = E_\mu \Big[ \exp(\lambda^{-2} F_\lambda^{(b)}(w) \,) \Big]$.  Furthermore, let $R_\lambda^{(b)}$, $0 < \lambda \leq 1$, be probability measures for $F_\lambda^{(b)}$ as given in Definition 2.9.  Then we obtain

$$\int_{C([0,T] \to R)} k(\frac{w(T \cdot)}{T}) \, \nu_{T,b}(dw) = E_{R_\lambda^{(b)}} \Big[ k(w) \Big]. \tag{3.3}$$

From results of the author [4], we get the following.

<u>Lemma 3.1.</u>

$$\lim_{\lambda \to 0} \lambda^2 \log Z_\lambda^{(b)} = \lim_{T \to \infty} \frac{1}{T} \log Z_T(b)$$

$$= \sup\{b \cdot \int_{S'} V(W) \ R(dW) + \int_{S'} U(W) \ R(dW) - H(R); \ R \in P_{st}\}.$$

Here $S'$ denotes the space of tempered distributions on R, $P_{st}$ denotes the
totality of stationary probability measures on $S'$, $H(R)$ denotes the entropy
of R with respect to the Gaussian white noise measure on $S'$, and U and V
are measurable functions defined on $S'$ given by

$$V(W) = \int_0^1 W(s) \ ds \qquad \text{and} \qquad U(W) = -\int_{-\infty}^\infty \delta(\int_0^t W(s) \ ds) \ dt.$$

<u>Proof.</u>  Let $\nu$ denote the Gaussian white noise measure on $S'$, i.e.,

$$\int_{S'} \exp(\sqrt{-1} \ _{S'}\!<W,\phi>_S) \ \nu(dW) = \exp(-\frac{1}{2} \int_R |\phi(x)|^2 dx), \quad \phi \in S, \text{ where } S \text{ is the}$$

space of rapidly decreasing smooth functions on R.  Then we have, for any
$R \in P_{st}$ with $H(R) < \infty$,

$$\frac{1}{T} \log Z_T(b) \qquad\qquad\qquad\qquad\qquad\qquad (3.4)$$

$$= \frac{1}{T} \log E_\nu \left[ \exp(b \cdot \int_0^T W(s)ds - \int_0^T \int_0^T \delta(\int_s^t W(\tau) \ d\tau) \ dt \ ds) \right]$$

$$= \frac{1}{T} \log E_R \left[ \exp(b \cdot \int_0^T W(s)ds - \int_0^T \int_0^T \delta(\int_s^t W(\tau) \ d\tau) \ dt \ ds - \log \frac{dR}{d\nu}\Big|_{G_{(0,T)}}(W)) \right]$$

$$\geq b \cdot \frac{1}{T} E_R \left[ \int_0^T W(s) \ ds \right] - \frac{1}{T} E_R \left[ \int_0^T \int_0^T \delta(\int_s^t W(\tau) \ d\tau) \ dt \ ds \right]$$

$$- \frac{1}{T} E_R \left[ \log \frac{dR}{d\nu}\Big|_{G_{(0,T)}}(W) \right]$$

$$\geq b \cdot \int_{S'} V(W) \ R(dW) + \int_{S'} U(W) \ R(dW) - H(R),$$

where $G_{(0,T)}$ denotes the $\sigma$-algebra over $S'$ generated by $\{_{S'}\!<\cdot,\phi>_S \ ; \ \phi \in S$
and supp $\phi \subset (0,T)\}$.

Now let $U_n(W) = \max\{-\int_{-n}^{n} \delta(\int_0^t W(s)\,ds)dt,\ -n\}$, $n = 1,2,\ldots$, and

$\theta_t: S' \to S'$, $t \in R$, be the shift operator given by $\theta_t u(s) = u(s+t)$, $u \in S'$ and $s \in R$. Then $U_n(W) \downarrow U(W)$ as $n \uparrow \infty$. Thus we have, for $T > 2_n$,

$$Z_T(b) \leqq E_\nu\left[\exp(b \cdot \int_0^T W(s)ds + \int_n^{T-n} U_n(\theta_t W)dt)\right].$$

Then, by results in $[4]$, we see that there exists a $R_n \in P_{st}$ such that

$$\overline{\lim_{T\to\infty}} \frac{1}{T} \log Z_T(b) \leqq b \cdot \int_{S'} V(W)\, R_n(dW) + \int_{S'} U_n(W)\, R_n(dW) - H(R_n).$$

From (3.4) and the fact that $U_n$'s are non-positive we can easily see that $\sup\{H(R_n);\ n = 1,2,\ldots\} < \infty$. Therefore there exist a subsequence $\{n_k\}_{k=1}^{\infty}$ and a $R_o \in P_{st}$ such that $R_{n_k} \to R_o$ in $P_{st}$. Then, by results in $[4]$, we have

$$\overline{\lim_{T\to\infty}} \frac{1}{T} \log Z_T(b) \leqq b \cdot \int_{S'} V(W)\, R_o(dW) + \int_{S'} U_n(W)\, R_o(dW) - H(R_o), \quad n = 1,2,\ldots .$$

$$(3.5)$$

(3.4) and (3.5) lead us to our assertion. This completes the proof.

We will denote $\lim_{\lambda\to 0} \lambda^2 \log Z_\lambda^{(b)}$ by $f(b)$.

Now let $I_b: B \to [-\infty,\infty)$, $b \in R$, be functions for $F_\lambda^{(b)}$ as given in Definition 2.1. Then it is easy to see the following.

<u>Proposition 3.2.</u>    $I_b(u) = b \cdot u(1) + I_o(u)$ for any $u \in B$ and $b \in R$.

The following lemma will play an important role.

<u>Lemma 3.3.</u>    Let $h \in H$ and $0 < \tau < 1$. Let $h_1$, $h_2 \in H$ be given by

$$h_1(t) = \frac{1}{\tau} h(\tau t)$$

and

$$h_2(t) = \frac{1}{1-\tau} \{h(\tau + (1-\tau)t) - h(\tau)\}, \quad 0 \leqq t \leqq 1.$$

Then $I_b(h) \leqq \tau \cdot I_b(h_1) + (1-\tau)I_b(h_2)$. Therefore, if $I_b(h) = f(b)$, then $I_b(h_1) = I_b(h_2) = f(b)$.

__Proof.__  It is sufficient to prove our assertion in the case where $b = 0$ by virtue of Proposition 3.2.  Let $w_1(t;w) = \tau^{-1/2}w(\tau t)$ and $w_2(t;w) = (1-\tau)^{-1/2}\{w(\tau + (1-\tau)t) - w(\tau)\}$, $0 \leq t \leq 1$.  Then $w_1(\cdot;w)$ and $w_2(\cdot;w)$ are independent under $\mu(dw)$ and the probability measures induced by them are equal to $\mu$.  It is easy to see that if $\|\lambda w - h\|_B < \varepsilon$, then $|\tau^{-1/2}\lambda \cdot w_1(\cdot;w) - h_1|_B < 2\varepsilon/\tau$ and $\|(1-\tau)^{-1/2}w_2(\cdot;w) - h_2\|_B < 2\varepsilon/(1-\tau)$.

On the other hand, we obtain

$$\lambda^{-2}F_\lambda^{(0)}(w)$$

$$\leq -\lambda^{-2}\int_0^\tau\int_0^\tau \delta(\lambda(w(t)-w(s)))\,dt\,ds - \lambda^{-2}\int_\tau^1\int_\tau^1 \delta(\lambda(w(t)-w(s)))\,dt\,ds$$

$$= -\lambda^{-2}\cdot\tau^2\int_0^1\int_0^1 \delta(\tau^{1/2}\lambda(w_1(t)-w_1(s)))\,dt\,ds$$

$$-\lambda^{-2}(1-\tau)^2\int_0^1\int_0^1 \delta((1-\tau)^{1/2}\lambda(w_2(t)-w_2(s)))\,dt\,ds$$

$$= (\tau^{-1/2}\lambda)^{-2}F_{\tau^{-1/2}\lambda}^{(0)}(w_1(\cdot;w)) + ((1-\tau)^{-1/2}\lambda)^{-2}F_{(1-\tau)^{-1/2}\lambda}^{(0)}(w_2(\cdot;w)).$$

Therefore we get

$$\lambda^2\log E_\mu\left[\exp(\lambda^{-2}F_\lambda^{(0)}(w)),\ \|\lambda w - h\|_B < \varepsilon\right]$$

$$\leq \tau\cdot(\tau^{-1/2}\lambda)^2\log E_\mu\left[\exp((\tau^{-1/2}\lambda)^{-2}F_{\tau^{-1/2}\lambda}^{(0)}(w)),\ \|\tau^{-1/2}\lambda\cdot w - h_1\|_B < \frac{2\varepsilon}{\tau}\right]$$

$$+ (1-\tau)((1-\tau)^{-1/2}\lambda)^2\log E_\mu\left[\exp(\ ((1-\tau)^{-1/2}\lambda)^{-2}\right.$$

$$\left.\times F_{(1-\tau)^{-1/2}\lambda}^{(0)}(w)),\ \|(1-\tau)^{-1/2}\lambda\cdot w - h_2\|_3 < \frac{2\varepsilon}{1-\tau}\right].$$

Taking $\lim\limits_{\varepsilon\to 0}\overline{\lim}\limits_{\lambda\to 0}$ , we have our assertion.

__Proposition 3.4.__  Suppose that $h \in H$ and $I_b(h) = f(b)$.  If $h(t)$ is differentiable at $t_o$, $0 < t_o < 1$, and $(d/dt)h(t_o) = c$, then $I_b(g_c) = f(b)$.  Here $g_c \in H$ is given by $g_c(t) = ct$, $0 \leq t \leq 1$.

__Proof.__  Let $h_\varepsilon \in H$, $0 < \varepsilon \leq 1$, be given by

$$h_\varepsilon(t) = \frac{1}{\varepsilon(1-t_o)} \{h(t_o + \varepsilon(1-t_o)t) - h(t_o)\}, \quad 0 \leq t \leq 1.$$

Then we can easily see that $I_b(h_\varepsilon) = f(b)$ by Lemma 3.3. Since $h_\varepsilon \to g_c$ in B and $I_b : B \to [-\infty, \infty)$ is upper semi-continuous by Proposition 2.3, we obtain $I_b(g_c) \geq f(b)$. This proves our assertion.

<u>Proposition 3.5.</u>  $f(b) = \max\{b \cdot c + I_o(g_c); \ c \in R\}$, $b \in R$.

<u>Proof.</u>  From Propositions 2.4 and 2.7, we see that there exists an $h \in H$ such that $I_b(h) = f(b)$. Since $h(t)$ is absolutely continuous in $t \in [0,1]$, $h(t)$ is differentiable for almost every $t \in (0,1)$. Thus Proposition 3.4 leads us to our assertion.

<u>Remark 3.6.</u>  Proposition 3.5 is strongly related to Lemma 3.1. This is to some extent the contraction principle as introduced in Donsker-Varadhan $[1]$.
From the proof of Proposition 3.5, we see the following.

<u>Proposition 3.7.</u>  If $I_b(h) = f(b)$, $h \in H$, then $I_b(g_{h'(t)}) = f(b)$ for almost every $t \in (0,1)$.

<u>Proposition 3.8.</u>  (1) $f : R \to R$ is convex.
(2) If $I_{b_o}(g_{c_o}) = f(b_o)$ for some $b_o$, $c_o \in R$, then

$$\frac{d^-}{db} f(b_o) \leq c_o \leq \frac{d^+}{db} f(b_o).$$

<u>Proof.</u>  From Proposition 3.5, assertion (1) is obvious and we see that $f(b) - f(b_o) \geq (b-b_o) \cdot c_o$, $b \in R$. This implies our assertion (2).

From Theorem 2.11, Propositions 3.7 and 3.8, we obtain the following.

<u>Theorem 3.9.</u>  If $f(b)$ is differentiable at $b_o$, then

$$\int_{C([0,T] \to R)} k\left(\frac{w(T \cdot)}{T}\right) \nu_{T,b}(dw) \to k(g_{f'(b_o)})$$

for any continuous functional $k : C([0,1] \to R) \to R$ such that

$$\frac{1}{N^2} \sup\{\log(|k(w)| + 1); \ \max\{|w(t)|; \ 0 \leq t \leq 1\} < N\} \to 0$$

78

as $N \to \infty$.

$f(b)$ is not differentiable at $b = 0$. Therefore we have to study $I_o$ in detail in order to know the asymptotic behaviour of $\nu_{T,o}$. In the rest of this report, we give some observations on the function $I_o : B \to [-\infty, \infty)$.

<u>Proposition 3.10.</u>   For any $h \in H$,

$$I_o(h) \leqq - \{ \max_{0 \leqq t \leqq 1} h(t) - \min_{0 \leqq t \leqq 1} h(t) \}^{-1} - \frac{1}{2} \| h \|_H^2 .$$

<u>Proof.</u>   Let $h \in H$, $m_o = \min_{0 \leqq t \leqq 1} h(t)$, and $m_1 = \max_{0 \leqq t \leqq 1} h(t)$.   Suppose that $\| \lambda w \|_B < \varepsilon$.   Then we have $m_o - \varepsilon < \lambda w(t) + h(t) < m_1 + \varepsilon$.   It is easy to see from (3.2) and Schwarz's inequality that

$$F_\lambda^{(0)}(w + \lambda^{-1}h)$$

$$= -\int_R dx \, \{\int_0^1 \delta(\lambda w(t) + h(t) - x)dt\}^2$$

$$= -(c_1 - c_o + 2\varepsilon)^{-1} \int_{c_o - \varepsilon}^{c_1 + \varepsilon} dx \, \{\int_0^1 \delta(\lambda w(t) + h(t) - x)dt\}^2 \int_{c_o - \varepsilon}^{c_1 + \varepsilon} dx$$

$$\leqq -(c_1 - c_o + 2\varepsilon)^{-1} \{\int_{c_o - \varepsilon}^{c_1 + \varepsilon} dx \int_0^1 \delta(\lambda w(t) + h(t) - x)dt\}^2$$

$$= -(c_1 - c_o + 2\varepsilon)^{-1} .$$

This and Proposition 2.8.(1) lead us to our assertion.

<u>Proposition 3.11.</u>   For any $c \in R$,

$$- \frac{2}{|c|} - \frac{1}{2} c^2 \leqq I_o(g_c) \leqq - \frac{1}{|c|} - \frac{1}{2} c^2 .$$

<u>Proof.</u>   Noting that $F_\lambda^{(0)} \leqq 0$, we have

$$E_\mu \left[ F_\lambda^{(0)}(w + \lambda^{-1}g_c) \mid |\lambda w|_B < \varepsilon \right]$$

$$\geqq E_\mu \left[ F_\lambda^{(0)}(w + \lambda^{-1}g_c) \right] \cdot \mu(|\lambda w|_B < \varepsilon)^{-1}$$

79

It is easy to see that

$$E_\mu\left[F_\lambda^{(0)}(w + \lambda^{-1}g_c)\right]$$

$$= - \int_0^1 \int_0^1 dt\ ds\ (2\pi\lambda^2|t-s|)^{-1/2} \exp(- \frac{c^2(t-s)^2}{2\lambda^2|t-s|})$$

$$\to - \frac{2}{|c|} \quad as \quad \lambda \downarrow 0.$$

This and Proposition 2.8.(2) prove the first inequality.  The second inequality follows immediately from Proposition 3.10.  This completes the proof.

From Proposition 3.11 we see that

$$f(0) \geq \sup_{c \in R} \{- \frac{2}{|c|} - \frac{1}{2} c^2\} = - 3 \cdot 2^{-1/3}.$$

Let $c_o$ and $c_1$ be constants satisfying $0 < c_o < c_1 < \infty$ and

$$- \frac{1}{c_o} - \frac{1}{2} c_o^2 = - \frac{1}{c_1} - \frac{1}{2} c_1^2 = - 3 \cdot 2^{-1/3}.$$

Then Proposition 3.7 implies that, if $I_o(h) = f(0)$, $h \in H$, then

$$c_o \leq |\frac{d}{dt} h(t)| \leq c_1, \quad a.e.\ t \in (0,1).$$

Proposition 3.12.  Suppose that it is true that

$$f(0) > \sup_{c \in R} \{- \frac{2}{|c|} - \frac{1}{2} c^2\} . \qquad (3.6)$$

Then any $h \in H$ with $I_o(h) = f(0)$ must be an increasing or decreasing function.

Proof.  Assume that there exists an $h_o \in H$ with $I_o(h_o) = f(0)$ which is not increasing or decreasing.  Since $I_o(-h_o) = I_o(h_o)$, we may assume that there exist $t_o$ and $t_1$ such that $0 \leq t_o < t_1 \leq 1$ and $h_o(t_o) = h_o(t_1) \leq h_o(t)$, $t_o < t < t_1$.  Let

$$h_1(t) = \frac{1}{t_1 - t_o} (h_o(t_o + (t_1 - t_o)t) - h_o(t_o)), \quad 0 \leq t \leq 1.$$

Then by Lemma 3.3 we see that $I_o(h_1) = f(0)$.

Now let $x = \max\limits_{0 \le t \le 1} h_1(t) > 0$.  It is easy to see that

$$2x = \inf\{|h|_H \; ; \; h \in H, \; h(0) = h(1) = 0, \; x = \max\limits_{0 \le t \le 1} h(t)\}.$$

Thus we have, by Proposition 3.10,

$$I_o(h_1) \le -\frac{1}{x} - 2x^2$$

$$\le \sup_{c \in R} \{-\frac{2}{|c|} - \frac{1}{2} c^2\}$$

$$< f(0).$$

This is a contradiction, therefore we obtain our assertion.

We conjecture that (3.6) is true.  The reason is the following.

Let $R_c$, $c \in R$ and $c \ne 0$, be the stationary probability measures on $S'$ such that

$$\int_{S'} \exp(\sqrt{-1} \;_{S'}\!<W, \phi>_S) \; R_c(dW)$$

$$= \exp(\sqrt{-1} \; c \cdot \int_R \phi(x) \; dx - \frac{1}{2} \int_R \phi(x)^2 \; dx) \quad \text{for any } \phi \in S.$$

Then we see that

$$\int_{S'} U(W) \; R_c(dW) - H(R_c) = -\frac{2}{|c|} - \frac{1}{2} c^2.$$

Thus if $f(0) = \sup\limits_{c \in R} \{-\frac{2}{|c|} - \frac{1}{2} c^2\}$, there exists a $c_o > 0$ such that

$$f(0) = \int_{S'} U(W) \; R_{c_o}(dW) - H(R_{c_o}).$$

Usually equilibrium states are Gibbs states.  Therefore it seems likely that $R_{c_o}$ is a Gibbs state for the potential $U_o$.  However, any $R_c$ is not a Gibbs state.  This is the reason why we conjecture that (3.6) is true.

From Proposition 3.12, if (3.6) is true, then we have

$$c_o^\alpha < \varliminf_{T \to \infty} E_{\nu_{T,0}} \left[\left(\frac{|w(T)|}{T}\right)^\alpha\right] \le \varlimsup_{T \to \infty} E_{\nu_{T,0}} \left[\left(\frac{|w(T)|}{T}\right)^\alpha\right] < c_1^\alpha,$$

for $a > 0$.

REFERENCES

[1] Donsker, M.D. and Varadhan, S.R.S.   Asymptotic evaluation of certain Markov process expectations for large time.  IV, *Comm. Pure Appl. Math. 36*, 183-212 (1983).

[2] Kallianpur, G. and OOdaira, H.   Freidlin-Wentzell type estimates for abstract Wiener spaces. *Sankya, the Indian J. Stat. 40*, 116-137 (1978).

[3] Kuo, H.-H.   *Gaussian Measures in Banach Spaces*, Springer Lecture Notes in Math. vol.463 (1970).

[4] Kusuoka, S.   The variational principle for stationary Gaussian Markov fields, *Theory and Application of Random Fields, Proceedings of the IFIP-WG 7/1 Working Conference, Bangalore 1982*, ed. G. Kallianpur, Springer Lecture Notes in Control and Information vol.49 (1983).

[5] Schilder, M.   Some asymptotic formulas for Wiener integral, *Trans. Amer. Math. Soc. 134*, 63-85 (1968).

[6] Westwater, J.   in *Proceedings of the Bielefeld Encounters in Physics and Mathematics IV - Trends and Developments in the Eighties*. Singapore:  World Scientific Pub. Co. (to be published).


Shigeo Kusuoka
Department of Mathematics
Faculty of Science
University of Tokyo
Tokyo, Japan.

Y LE JAN

# Equilibrium state for a turbulent flow of diffusion

1. Let M be a compact manifold without boundary, of dimension d. Suppose that the space M is filled with a medium of mass distribution $\mu_o$. Assume that the total mass is 1. Suppose that this medium is submitted to a continuous random motion. More precisely, assume the existence of a random continuous family of homeomorphisms $\phi_t(\omega)$, $t \in \mathbb{R}^+$, such that if the initial distribution is $\mu_o$, the distribution at time t is $\phi_t(\omega)(\mu_o) = \mu_t(\omega)$.

Let $F_s$ be the $\sigma$-algebra of random events occurring before time s. If, $\forall t > s$, $\phi_t \circ \phi_s^{-1}(\omega)$ is independent of $F_s$, we say that $\phi_t$ is a free diffusion flow. Actually, it can be viewed as a limiting case for turbulent flows (cf. [3]). If it is only independent of $F_s$ conditionally to $\mu_s$, we may say it is an interacting diffusion flow. The process $\mu_t(\omega)$, which describes the evolution of the mass distribution, is clearly Markovian.

2. The structure of free time homogeneous diffusion flows is known, if one assumes $\phi_{t,\omega}(\omega)$ to be a.s. a diffeomorphism (cf [1] [2] [3]). There is a smooth tensor field $a^{i,j}(x,y) \in T_x(M) \otimes T_y(M)$ and a smooth vector field $b^i(x)$ on M such that

$$\lim_{t \to 0} \frac{1}{t} E((f(\phi_t(x))-f(x))(f(\phi_t(y))-f(y))) = \sum_{i,j} \frac{\partial f}{\partial x^i}(x) \frac{\partial f}{\partial y^i}(y) \, a^{i,j}(x,y),$$

for any $C^2$ function on M, denoted f. And for any g in $C^2(M)$,

$$\lim_{t \to 0} \frac{1}{t} E(g(\phi_t(x))-g(x)) = \frac{1}{2} L_g^a(x) + \sum_{i=1}^{d} b^i(x) \frac{\partial g}{\partial x^i}(x)$$

with

$$L_g^a(x) = \sum_{i,j} a^{i,j}(x,x) \frac{\partial^2 g}{\partial x^i \partial x^j} + \Gamma_j^{i,j}(x) \frac{\partial g}{\partial x^i},$$

and

$$\Gamma_k^{i,j} = \frac{\partial}{\partial y^k} a^{i,j}(x,y) \Big|_{y=x},$$

$a^{i,j}(x,y)$ and $b^i(x)$ are called the local characteristics of the flow.

$a^{i,j}(x,y)$ satisfies the following positivity condition:

$$\forall n, \ \forall x_1,\ldots,x_n \in M, \ \forall \lambda_i^1 \in T^*(x_1)\ldots\lambda_i^n \in T^*(x_n),$$

$$\sum_{1 \leq k,\ell \leq n} \ \sum_{1 \leq i,j \leq d} \lambda_i^k \lambda_j^\ell \, a^{i,j}(x_k,x_\ell) \geq 0.$$

Moreover, if $X(M)$ is the space of smooth vector fields, there exists a $X(M)$-valued Brownian motion $W_t^i(x,\omega)$ such that $a^{i,j}(x_k,x_\ell) = \frac{1}{t}\, E(W_t^i(x_k)W_t^j(x_\ell))$ and, for any $g \in C^2(M)$,

$$g(\phi_t(x)) - g(x) = \sum_{i=1}^{d} \int_0^t \frac{\partial g}{\partial x_i}\,(\phi_s(x))dW^i(\phi_s(x))$$

$$+ \int_0^t \frac{1}{2}\, L_g^a(\phi_s(x)) + \sum_i \int_0^t b^i \frac{\partial g}{\partial x_i}\,(\phi_s(x))ds \qquad (1)$$

where $\displaystyle\sum_i \int_0^t \frac{\partial g}{\partial x^i}\,(\phi_s(x))dW_s^i(\phi_s(x)) = M_t^{g,x}$ is a square integrable martingale characterized by the identity

$$\langle M^{g,x}, W^i(y)\rangle_t = \sum_j \int_0^t \frac{\partial g}{\partial x^j}\,(\phi_s(x)) \, a^{j,i}(\phi_s(x),y)ds.$$

This satisfies

$$\langle M^{g,x}, M^{f,y}\rangle_t = \sum_{i,j} \int_0^t \frac{\partial g}{\partial x^j}\,(\phi_s(x)) \frac{\partial f}{\partial x^i}\,(\phi_s(y)) \, a^{j,i}(\phi_s(x), \phi_s(y))ds.$$

Conversely, given any couple of smooth local characteristics $a^{j,i}(x,y)$ (satisfying the positivity condition) and $b^i$, one can construct the associated $X(M)$ valued Brownian motion $W_t^i$ and then $\phi_t$ by solving the stochastic differential equation (1).

3. Examples of interacting flows of diffusion can be easily constructed by making use of the Girsanov transformation. If $U_i(y,dz)$ $i = 1,\ldots,d$ are bounded kernels on M,

$$M_t = \sum_i \int_0^t \iint \mu_s(dy) \, U_i(y,dz)dW_s(z)$$

is a square integrable martingale with

$$<M>_t = \sum_{i,j} \int_0^t \mu_s(dy)\,\mu_s(dy')\,U_i(y,dz)\,U_j(y',dz')\,a^{i,j}(z,z')$$

$$\exists C > 0, \quad <M>_t \leq Ct.$$

If we set $k^i(x,y) = \int U_j(y,dz)\,a^{i,j}(x,z)$, then, by the change of

probability $\dfrac{d\tilde{P}}{dP}\Big|_{F_t} = e^{M_t - \frac{1}{2}<M>_t}$, we get a diffusion flow satisfying

$$dg(\phi_t(x)) = \sum_i \frac{\partial g}{\partial x_i}\ (\phi_t(x)\ [dW^i(\phi_t(x)) + b^i(\phi_t(x))dt$$

$$+ \int k^i(\phi_t(x),y)\ \mu_t(dy)] + \frac{1}{2} L_g^a(\phi_t(x))dt. \tag{2}$$

4. Assume that, at each point x, the tensor $a^{i,j}(x,x)$ is strictly positive. Let $\lambda$ be the measure associated with the Riemannian structure defined by $a^{i,j}(x,x)$, $x \in M$. Let $\gamma^i_{j,k}$ be the Christoffel symbols. If $\mu_o$ has a smooth density $\rho_o(x)$ with respect to $\lambda$, so has $\mu_t(\omega)$ and its density $\rho_t(\omega)$ satisfies the stochastic partial differential equation

$$\rho_{t,\omega}(x) = -\ \text{div}\ (\int_0^t \rho_s(x)\ d\vec{W}_s(x) + \int_0^t \rho_s(x)\ \vec{b}(x)ds$$

$$-\frac{1}{2} \int_0^t \overrightarrow{\text{grad}}\ \rho_s(x)ds + \frac{1}{2} \int_0^t \rho_s(x)\ \vec{D}(x)ds$$

$$+ \int_0^t \rho_s(x)\ \vec{k}(x,y)\ \rho_s(y)dy) \tag{3}$$

where $D^i(x) = \Gamma^{i,j}_j(x) - a^{j,k}(x,x)\,\gamma^i_{j,k}(x)$. This can easily be deduced from (2) by performing an integration by parts(using the quadratic variation for the martingale term).

<u>Remark.</u> If one makes the covariance $a^{i,j}(x,y)$ converge toward the covariance of a white noise $a^{i,j}(x)\ \delta_{x-y}$, the solution of (3) should converge weakly, in law, toward a solution of the deterministic nonlinear PDE obtained by erasing the martingale in (3).

5.  Perhaps the first question that rises about the process $\mu_t(\omega)$ is its asymptotic behaviour, and the existence of a stationary regime . By analogy with statistical mechanics, let us call a "state" a probability measure on the set K of normalized mass distributions on M(K is compact and metrizable for the vague topology).

The Markov process $\mu_{t,\omega}$ induces a semigroup $\Pi_t$ on K. An equilibrium state should be a probability measure Q on K, such that $Q\Pi_t = Q$. One can try to prove that, given any $\mu_o \in K$,

(i)   $\delta_{\mu_o} \Pi_t \longrightarrow Q$   vaguely a.s.,   as $t \rightarrow \infty$.

(ii)  $\dfrac{1}{t} \displaystyle\int_0^t \delta_{\mu_s}\, ds \longrightarrow Q$   vaguely, a.s.,   as   $t \longrightarrow \infty$.

A state Q can be represented by a projective family of probability measures $Q_n$, defined on $M^n$ as follows. If $F_n(x_1,x_2,\ldots,x_n)$ is a bounded function on $M^n$, set, for any $\mu \in K$,

$$\hat{F}_n(\mu) = \int F_n(x_1,\ldots,x_n)\, d\mu(x_1),\ldots,d\mu(x_n).$$

Then define $Q_n : Q_n(F_n) = Q(\hat{F}_n)$, for all $F_n$. The family $Q_n(dx_1,\ldots,dx_n)$ satisfies the three following properties:

(a)   $Q_n(dx_1,\ldots,dx_n) = Q_n(dx_{\sigma(1)},\ldots,dx_{\sigma(n)})$   (symmetry)

(b)   $\displaystyle\int 1_M(x_n)\, Q_n(dx_1,\ldots,dx_n) = Q_{n-1}(dx_1,\ldots,dx_{n-1})$   (projectivity)

(c)   For all $F_n$   $\left| \displaystyle\int F_n(x_1,\ldots,x_n)\, Q_n(dx_1,\ldots,dx_n) \right| \leq \| \hat{F}_n \|_\infty$   (continuity).

Conversely, it is easy to check that such a family defines a state Q.

We say that Q is *diffuse* iff it is supported by the set of diffuse probability measures on M. Clearly, Q is diffuse iff $Q_2$ does not bear on the diagonal of $M^2$. Then all $Q_n$ are supported by $\widetilde{M}^n = M^n - $ {the union of all diagonals $\{x_i = x_j\}$   $1 \leq i < j \leq n\}$.


6.  From now on, we assume that the flow $\phi_t(\omega)$ is free. It is easy to check that $\forall x_1,x_2,\ldots,x_n \in \widetilde{M}^n$, $\phi_t(x_1),\ldots,\phi_t(x_n)$ is a diffusion process on $\widetilde{M}^n$ with generator

$$\forall F_n \in C_o^2(\tilde{M}^n), \quad A^n F_n = \frac{1}{2} \sum_{k,\ell=1}^{n} \sum_{i,j=1}^{d} a^{i,j}(x_k,x_\ell) \frac{\partial^2 F_n}{\partial x_k^i \, \partial x_k^j}$$

$$+ \sum_{k=1}^{n} \sum_{i,j=1}^{d} (b^i(x_k) + \Gamma_j^{i,j}(x_k)) \frac{\partial F_n}{\partial x_k^i} \, .$$

Let $\Pi_t^n$ be its transition kernel. Clearly, $\Pi_t^n$ induces $\Pi_t^{n-1}$ on $\tilde{M}^n$, is symmetric, and $|\Pi_t^n(F_n)| \leq \| \hat{F}_n \|_\infty$ .

For all $F_n$ we have

$$\widehat{\Pi_t^n F_n}(\mu) = \Pi_t(\hat{F}_n)(\mu)$$

$$= E\left( \int F_n(\phi_t(x_1),\ldots,\phi_t(x_n)) \, \mu(dx_1),\ldots,\mu(dx_n)\right).$$

<u>Definition.</u>  We say that the flow $\phi_t$ is nonsingular iff, for all $n$ and $t$, $\Pi_t^n(x,dy)$ has a strictly positive density $\Pi_t^n(x,y)$ with respect to any Lebesguean measure.  In particular, this condition is satisfied as soon as, for all $n$, the generator $A^n$ is elliptic, or, more generally, satisfies the Hörmander condition

$$A^n = \sum_{i=1}^{N} L_i^2 \, ,$$

where $L_i$ are vector fields on $M^n$ such that, if $L$ is the Lie algebra generated by the $L_i$, $L(x) = T_{\tilde{M}^n}(x)$ for all $x \in \tilde{M}^n$.

<u>Example.</u>(the simplest non-trivial one).  Take $M = (\mathbb{R}/_{2\pi\mathbb{Z}})^d$.  For any $x = (x^1,x^2,\ldots,x^d) \in M$, $T_x(M)$ has a natural basis $\partial/\partial x^1,\ldots,\partial/\partial x^d$ as soon as we choose an orientation on each circle $\mathbb{R}/_{2\pi\mathbb{Z}}$ of the product.

Let $B^{i,j,k}$, $i,j \leq d$, $k \in \mathbb{Z}$, be a countable family of independent Brownian motions.  Let $\alpha_k$, $k \in \mathbb{N}$, be a sequence of real numbers such that

$$\sum_{k=0}^{\infty} \alpha_k^2 = 1 \quad \text{and} \quad \sum_{k=0}^{\infty} k^m \alpha_k^2 < \infty \quad \text{for all } m \in \mathbb{N}.$$

Set $\rho(z) = \sum_{k=0}^{\infty} \alpha_k^2 \cos(kz)$, and

$$W_t^i(x) = \frac{1}{\sqrt{d}} \sum_{k,j} (\alpha_k \cos(kx^j)\, B_t^{i,j,k} + \alpha_k \sin(kx^j)\, B_t^{i,j,-k}.$$

The covariance of $W_t^i$ can be written $a^{i,j}(x,y) = \delta_{i,j}\, p(x-y)$ with

$$p(x-y) = \frac{1}{d} \sum_{j=1}^{d} \rho(x^j - y^j).$$

Taking $b^i = 0$, we can construct an associated flow of diffeomorphisms $\phi_t$ satisfying (1).

$- A^n$ is elliptic as soon as $\alpha_K \neq 0$ for all $k$. Indeed,

$$\sum_{\ell,k=1}^{n} \sum_{j=1}^{n} \lambda_j^\ell \lambda_j^k\, p(x_k - x_\ell) = 0$$

iff

$$\forall k \in N,\ h,\ j = 1,2,\ldots,d \quad \alpha_k \sum_{\ell=1}^{n} \lambda_j^\ell \cos(kx_\ell^h) = 0$$

and

$$\alpha_k \sum_{\ell=1}^{n} \lambda_j^\ell \sin(kx_\ell^h) = 0.$$

If $\alpha_k \neq 0$ for all $k$, we have $\sum_{\ell=1}^{n} \lambda_j^\ell\, e^{ikx_\ell^h} = 0 \quad \forall h,j,k$, and we have to prove that $\lambda_j^\ell = 0$ for all $\ell,j$. We can fix $j$ and omit it in the following:

(a)    $d = 1$. Then $\ell \neq m \Rightarrow x_\ell \neq x_m$ and the matrix

$$A = (e^{ik\, x_\ell}) \qquad \begin{array}{l} k = 0,\ldots,n-1 \\ \ell = 1,\ldots,n \end{array}$$

is invertible.

(b)    $d \neq 1$. Then $\ell \neq m \Rightarrow x_\ell^h \neq x_m^h$ for at least one index $h$. Define the matrix

$$A^h = (e^{ikx_\ell^h}) \qquad \begin{array}{l} k = 0,\ldots,n-1 \\ \ell = 1,\ldots,n \end{array}$$

and the $nd \times n$ matrix, $A = (A^1 \ldots A^d)$. We have $(\lambda^1 \ldots \lambda^n)A = 0 \Rightarrow \lambda = 0$ iff $A$ has rank $n$. So we have to show that the horizontal lines $H_1,\ldots,H_n$ of $A$ are independent.

$\forall \ell_1, \ldots, \ell_p$, if there is $h \in \{1, \ldots, d\}$ such that $x_{\ell_1}^h \neq x_{\ell_2}^h \ldots \neq x_{\ell_p}^h$, the vectors $H_{\ell_1} \ldots H_{\ell_p}$ are independent. Looking at the first coordinate $x_\ell^1$, one can divide the system $\{H_1 \ldots H_n\}$ into independent subsystems, each associated with one of its values. Applying the same process to each subsystem with the second coordinate, etc, we finally obtain the independence of the $H_\ell$.

If we set, for all $j \in \{1, \ldots, d\}$, $k \in \mathbb{Z}$, $\ell \in \{1, \ldots n\}$,

$$L^{k,\ell,j,h} = \cos (k\, x_\ell^j)\, \frac{\partial}{\partial x^h} \quad \text{if} \quad k \geq 0$$

$$= \sin (k\, x_\ell^j)\, \frac{\partial}{\partial x^h} \quad \text{if} \quad k < 0,$$

then $A^n = \Sigma (L^{k,\ell,j,h})^2$.

The ellipticity condition means that, for all $(x_1, \ldots, x_n) = x$, the vectors $\alpha_{|k|}\, L^{k,\ell,j,h}(x)$ span $T_n(\widetilde{M}^n)$. We have shown that the vectors $L^{k,\ell,j,h}$ span $T_x(\widetilde{M}^n)$ and that, in consequence, the ellipticity condition is satisfied as soon as $\alpha_k \neq 0$ for all $k$.

Let us now give a weaker condition for nonsingularity involving the Hörmander condition. Let $L_{k_1, k_2}$ be the Lie algebra generated by the vectors $L^{k_1, \ell, j, h}$ and $L^{k_2, \ell, j, h}$. Then $L_{k_1, k_2}(x) = T_x(M^n)$ as soon as $k_1$ and $k_2$ are relatively prime. In consequence, the Hörmander condition is checked as soon as $\alpha_{|k_1|}$ and $\alpha_{|k_2|}$ are non-zero.

<u>Proof.</u> From the preceding result, we know it is enough to show that $L_{k_1, k_2} \to L^{k,\ell,j,h} \quad \forall k, \ell, j, h$.

Set $B^{k,\ell,j,h} = \cos (k\, x_\ell^j)\, \frac{\partial}{\partial x^h} + i\, \sin(k x_\ell^j)\, \frac{\partial}{\partial x_\ell^h}$, and note that

$$[B^{k,\ell,j,h}, B^{k',\ell,j,j}] = i(k - k')\, B^{k+k',\ell,j,j}$$

$$[B^{k,\ell,j,h}, B^{k',\ell,j,j}] = i\, k\, B^{k+k',\ell,j,h}.$$

Then $B^{n_1 k_1 + n_2 k_2, \ell, j, h} \in L_{k_1, k_2}$, $\forall n_1, n_2$, and we can conclude.

7. <u>Equilibrium for free flows.</u>  Let $Q^1$ be the degenerate equilibrium $\int_M dx \, \varepsilon_{\delta_x}$  ($\varepsilon_\mu = \delta_\mu$ the Dirac measure at $\mu$), where all the mass is concentrated at a random point.  Let $\Delta$ be the set of all Dirac measure $\delta_x$, $x \in M$.   $\Delta$ is a subset of $K$.

<u>Theorem:</u>   Suppose that $\phi_t$ is nonsingular.  Convergence I and II hold toward

    (a) a diffuse equilibrium if $\Pi_t^2$ is positive recurrent and if $\mu_o$ is diffuse;

    (b) the degenerate equilibrium $Q^1$ iff $\Pi_t^2$ is transient or null recurrent.

    (c) Moreover, if $\Pi_t^2$ is transient, $\delta_{\mu_t}(\omega)$ converges weakly to $0$ on $K-\Delta$, almost certainly.

<u>Proof.</u>   From the nonsingularity hypothesis, we know that all n-point diffusions are $m^{\otimes n}$-irreducible and aperiodic (m being any Lebesguean measure on M).  Therefore (cf. [5]), they are positive recurrent, transient or null recurrent.

<u>Lemma.</u>   If $\Pi_t^2$ is positive recurrent,  all  $\Pi_t^n$ are positive recurrent.

Since all "n-point processes" are irreducible and aperiodic, they are positive recurrent as soon as there is a compact $C_n \subseteq \tilde{M}^n$, such that

$$\limsup_{t \to \infty} \frac{1}{t} \int_0^t \chi_{C_n} (\phi_s(x_1), \ldots, \phi_s(x_n)) ds > 0 \text{ for some } (x_1, \ldots, x_n) \in \tilde{M}^n.$$

However, for any $1 \le k < \ell \le n$, the process $(\phi_t(x_k), \phi_t(x_\ell))$ is positive recurrent in $\tilde{M}^2$, by hypothesis.  For any $\varepsilon > 0$, let $C_2$ be compact in $\tilde{M}^2$ such that $Q_2(C_2) \ge 1 - (\varepsilon/2)$.  Then, $\exists \Gamma > 0$, such that

$$P\left( \frac{1}{t} \int_0^t \chi_{C_2} (\phi_s(x_k), \phi_s(x_\ell)) ds \ge 1 - \varepsilon \right) \ge 1-\varepsilon, \quad \text{for all } t > T.$$

90

Choose $\varepsilon$ such that $\eta = \frac{n(n-1)}{2} \varepsilon < 1$; then if $P_{k,\ell}$ is the projection of $\tilde{M}^n$ on $\tilde{M}^2$ defined by the $k^{th}$ and $\ell^{th}$ coordinates of $M^n$ and if $C_n = \underset{k,\ell}{\cap} \, p_{k,\ell}^{-1}(C_2)$, we have

$$P\left( \frac{1}{t} \int_0^t \chi_{C_n}(\phi_t(x_1),\ldots,\phi_t(x_n)) \geq 1-\eta \right) \geq 1-\eta$$

and we can conclude, since $C_n$ is compact in $\tilde{M}^n$ (being contained in the complement of a neighbourhood of the diagonals $\{x_k = x_\ell\}$). Therefore, if $\Pi_t^2$ is positive recurrent, each n-point diffusion has an invariant measure $Q_n << m^{\otimes n}$. The associated state $Q$ is a diffuse equilibrium for $\Pi_t$.

By Orey's theorem, one has $\lim_{t\to\infty} \mu_o^n \otimes \Pi_t^n(F_n) = Q_n(F_n)$ for any diffuse initial distribution $\mu_o \in K$ and bounded function $F_n$ defined on $\tilde{M}^n$. Since the algebra of all $\hat{F}_n$, $F_n \in C^b(\tilde{M}^n)$ is dense in $C(K)$, this implies that convergence (I) holds. Let us prove convergence (II).

Since the n-points process in recurrent Harris,

$$\lim_{t\to\infty} \frac{1}{t} \int_0^t F_n(\phi_s(x_1),\ldots,\phi_s(x_n))ds = Q_n(F_n) \quad a.s.$$

for all $(x_1,\ldots,x_n) \in \tilde{M}^n$ and $F_n$ bounded on $\tilde{M}^n$, then

$$\lim_{t\to\infty} \frac{1}{t} \int_0^t \hat{F}_n(\mu_s)ds = \int \mu_o^{\otimes n}(dx_1,\ldots,dx_n) \, F_n(\phi_s(x_1),\ldots,\phi_s(x_n))$$

converges toward $Q_n(F_n) = Q(\hat{F}_n) \quad a.s.$

Let d be any metric on M. Set

$$U_\varepsilon^n = \{(x_1,\ldots,x_n) \in M^n, \quad \underset{i,j \leq n}{\inf} \, d(x_i,x_j) > \varepsilon\}.$$

If $\Pi_t^2$ is transient or null recurrent we have, for n = 2,

$$\lim_{t\to\infty} \frac{1}{t} \int \chi_{U_\varepsilon^n}(\phi_t(x_1), \phi_t(x_2)) = 0 \quad \text{and} \quad \lim_{t\to\infty} \mu_o^{\otimes n} \Pi_t^n \chi_{U_\varepsilon^n} = 0 \qquad (*)$$

(cf. [5]).

Moreover, since $\chi_{U_\varepsilon^n}(x_1,\ldots,x_n) \leq \underset{i,j}{\Sigma} \, \chi_{U_\varepsilon^2}(x_i,x_j)$, then (*) holds for general n. Statement (b) follows easily.

Statement (c) follows from the fact that all n-point diffusions are transient, and that continuous functions on K which are zero on $\Delta$ are generated by functions $\hat{F}_n$, $F_n \in C(M^n)$, $F_n = 0$ on $M^n - \overset{\sim}{M}{}^n$.

Example.  Let us return to the previous example.  In this case,

$$\forall F \in C_o^2(\overset{\sim}{M}{}^2), \quad A^2 F(x,y) = \frac{1}{2} \sum_{i=1}^{d} \left( \frac{\partial^2 F}{\partial x^i} + \frac{\partial^2 F}{\partial y^i} + 2p(x-y) \frac{\partial^2 F}{\partial x^i \partial y^i} \right).$$

One checks that $\Pi_t^2$ has $\dfrac{1}{1-p(x-y)} \, dx^1 \ldots dx^d \, dy^1 \ldots dy^d$ as invariant measure, and that it is symmetric with respect to this measure.  Indeed:

$$\forall G \in C_o^2(\overset{\sim}{M}{}^2), \quad \int_{\overset{\sim}{M}{}^2} - A^2 F . G \, \frac{1}{1-p} \, dx \, dy$$

$$= \frac{1}{2} \sum_i \int_{\overset{\sim}{M}{}^2} \left( \frac{\partial F}{\partial x^i} \frac{\partial G}{\partial x^i} + \frac{\partial F}{\partial y^i} \frac{\partial G}{\partial y^i} + 2p(x-y) \frac{\partial F}{\partial x^i} \frac{\partial G}{\partial y^i} \right) dx \, dy$$

$$+ \sum_i \int \frac{\partial F}{\partial x^i} \, G \left( \frac{\partial}{\partial y^i} \left( \frac{p}{1-p} \right) + \frac{\partial}{\partial x^i} \left( \frac{1}{1-p} \right) \right) dx \, dy$$

$$+ \sum_i \int \frac{\partial F}{\partial y_i} \, G \left( \frac{\partial}{\partial x^i} \left( \frac{p}{1-p} \right) + \frac{\partial}{\partial y^i} \frac{1}{1-p} \right) dx \, dy \tag{4}$$

However, the two last terms are zero since, for example,

$$\frac{\partial}{\partial y^i} \left( \frac{p}{1-p} \right) = \frac{\partial}{\partial y^i} \left( \frac{1}{1-p} - 1 \right) = \frac{\partial}{\partial y^i} \left( \frac{1}{1-p} \right) = - \frac{\partial}{\partial x^i} \left( \frac{1}{1-p} \right) .$$

Therefore, $\Pi_t^2$ is the semigroup associated with the Dirichlet form (4).

Now $\int_{\overset{\sim}{M}{}^2} \dfrac{1}{1-p(x-y)} \, dx^1 \ldots dx^d \, dy^1 \ldots dy^d$ is finite iff $d \geq 3$.  Therefore *a diffuse equilibrium exists for* $d \geq 3$.  For $d = 1$ or $2$, $\Pi_t^2$ is null recurrent or transient and $\mu_{t,\omega}$ converges towards $Q^1$.

Remark $(d \geq 3)$.  In general, three-point diffusion is not symmetric with respect to Q (the equilibrium is irreversible).

92

REFERENCES

[1] Baxendale, P., *Brownian Motions in the Diffeomorphism Group I.*
    Univ. Aberdeen (1982).

[2] Harris, T.E., Brownian motion on the homeomorphisms of the plane.
    *Ann. Prob. 9,* 232, 254 (1981).

[3] Kesten, H. and Papanicolaou, G.C., A limit theorem for turbulent
    diffusion. *Comm. Math. Phys.* 65, 97–128 (1979).

[4] Le Jan, Y. and Watanabe, S., Stochastic flows of diffeomorphisms.
    *Proceedings of the Katata Conference 1983.* (to appear).

[5] Revuz, D., *Markov Chains.* North Holland (1975); and Lois du tout ou
    rien... *Ann. Inst. Henri Poincare 19,* 9–24 (1983).

Y. Le Jan,
Laboratoire de Probabilites,
Universite Pierre et Marie Curie,
75005 PARIS.

F MARTINELLI & E SCOPPOLA

# Absence of absolutely continuous spectrum in the Anderson model for large disorder or low energy

1. Let us consider the following discrete Hamiltonian on $1^2(Z^d)$:

$$H(v) = -\Delta + v \tag{1}$$

where $\Delta$ is the finite difference Laplacian on $Z^d$,

$$(\Delta\psi)(n) = \sum_{\substack{m\in Z^d \\ |n-m| = 1}} (\psi(m) - \psi(n)), \; n \in Z^d \tag{2}$$

and $v = \{v(n)\}_{n\in Z^d}$ are i.i.d. random variables with common distribution

$d\lambda(v) = (d\lambda/dv)dv, \quad |(d\lambda/dv)|_\infty = \delta^{-1} < +\infty.$

Thus the random potential $v$ belongs to the probability space
$\Omega = \Pi(R, d\lambda(v(n))$ with product measure $dP(V) = \prod_n d\lambda(v_n).$

Such a Hamiltonian was introduced by Anderson [1] to describe the behaviour of an electron in a crystal with randomly distributed impurities.

In a recent paper Fröhlich and Spencer [2] proved that for large $\delta$ or low energies the Green's function of the Hamiltonian (1) decays exponentially at large distances with probability one. This in turn implies that the system does not diffuse.

In this note we show how to derive from their result the absence of an absolutely continuous part of the spectrum of $H$ for large disorder $\delta$ or low energy E.

More precisely we will prove the following:

<u>Theorem 1.</u>   There exist positive constants $\delta_o$, $E_o$ such that, if $\delta > \delta_o$, then $\sigma_{a.c.}(H(v)) = \emptyset$ P.a.s. and, if $d\lambda(v)$ is Gaussian,

$$(\{E; \; |E| \geq E_o\} \cap \sigma_{a.c.}(H(v)) = \emptyset \;\; P.a.s.$$

where $\sigma_{a.c.}(H(v))$ denotes the absolutely continuous part of the spectrum of $H(v)$.

In order to prove Theorem 1 we give first a result on the exponential decay of the Green's function, proved by Fröhlich and Spencer in a form suitable for our purposes.

Theorem 2.    (i) For any $p > 0$ and any $m > 0$, there exists a constant $C_{p,m}$ such that, if $|E| + \delta \geq C_{p,m}$, then the following event holds with probability at least $1 - \ell^{-p}$:  there exists a set $A_\ell$ containing the origin such that:

$$\frac{1}{2} \ell \leq \min_{b \in A_\ell} \text{dist}(0,b) \leq \max_{b \in A_\ell} \text{dist}(0,b) \leq \ell$$

and, for all $\epsilon \neq 0$,

$$|(H_{A_\ell}(v) - E - i\epsilon)^{-1}(x,y)| \leq e^{-m|x-y|}$$

for all $x,y$ satisfying $|x-y| \geq \ell^{3/4}$.

(ii) $E \notin \sigma(H_{A_\ell}(v))$ P.a.s.

Here $H_{A_\ell}(v)$ denotes the restriction of the Hamiltonian $H(v)$ to the set $A_\ell$ with Dirichlet boundary conditions at the bondary of $A_\ell$, $\partial A_\ell = \{(n,n');$ $|n-n'| = 1\ n \in A, n' \notin A_\ell\}$.

We are now in a position to prove Theorem 1.  Let us fix the energy $E$ in such a way that $|E| + \delta \geq C_{p,m}$ and let us fix a sequence of length scales $\ell_n = 2^n$. We then denote by $\Omega_{E,\delta}$ the set of realizations of the potential $v$ for which there exists an integer $n_o(E,\delta,v)$ such that, for any $n \geq n_o(E,\delta,v)$, there exists a set $A_{\ell_n}$ satisfying conditions (i), (ii) of Theorem 2.  Using Theorem 2 and the Borel-Cantelli lemma, we obtain:

$$P(\Omega_{E,\delta}) = 1. \tag{3}$$

The following result is now an easy consequence of (3) and Theorem 2.

Lemma 1.    For any $v \in \Omega_{E,\delta}$, the equation

$$(H(v) - E)\psi = 0 \tag{4}$$

has no polynomially bounded solution.

Proof of Lemma 1.    Fix $v \in \Omega_{E,\delta}$ and assume that (4) has a polynomially bounded solution $\psi_E$.  Then, for any $n \geq n_o(E,\delta,v)$, $\psi_E$ satisfies

$$\psi_E(x) = \sum_{(z,z') \in \partial A_{\ell_n}} (H_{A_{\ell_n}} - E)^{-1} (x,z) \, \psi_E(z') \qquad (5)$$

for any $x \in A_{\ell_n}$. Using now the polynomial boundedness of $\psi_E$ and the exponential decay of $(H_{A_{\ell_n}} - E)^{-1} (x,z)$ for $|x-z| \geq \ell_n^{3/4}$, we get that

$$|\psi E^{(x)}| \leq C \, e^{-m\ell_n^{3/4}} \, \ell_n^{\alpha} \qquad (6)$$

for some $C > 0$, $\alpha > 0$ and any $x$ such that dist $(x, \partial A) \leq \ell_n^{3/4}$. The arbitrariness of n together with (6) imply now that $\psi_E = 0$.

The proof of Theorem 1 now follows the argument used by Pastur [3] for the one-dimensional case. We first need the following result (see, e.g. Berezanskii [4]) concerning the solutions of the equation

$$(H(v) - E) \, \psi_E = 0. \qquad (7)$$

<u>Lemma 2.</u>   Let $\rho_v$ be the spectral measure of $H(v)$. Then for $\rho_v$-almost all E the solutions of (7) satisfy

$$|\psi_E(x)| \leq C(1 + x)^{\frac{d}{2} + \eta} \qquad \forall \eta > 0 \text{ and some } C > 0.$$

For definiteness let us now consider the case $\delta \geq \delta_o = C_{p,m}$, the case $d\lambda(v)$ Gaussian and $|E| \geq E_o = C_{p,m}$ being similar.

It is well known (see, e.g., [5]) that $\sigma_{a.c.}(H(v))$ is P-almost surely independent of v, so let us assume that, with probability one, $\sigma_{a.c.}(H(v)) \neq \emptyset$. This, together with Lemma 2, implies that there is a set $\bar{\Omega}$ with $P(\bar{\Omega}) = 1$ such that, for any $v \in \bar{\Omega}$, the set $B_v = \{E; (H(v) - E)\psi = 0$ has a polynomially bounded solution$\}$ has positive Lebesgue measure. Let us now introduce the space $M = \Omega \times R$ and define on M the measure $\tilde{P} = P \times \mu$ where $\mu$ denotes the Lebesgue measure. In M we now consider the set

$$M_o = \{(v,E) \in M; (H(v) - E)\psi = 0 \quad \text{has a polynomially}$$
$$\text{bounded solution}\}$$

Clearly, $M_o \supset \{(v,E) \in M; v \in \bar{\Omega} \text{ and } E \in B_v\}$. Thus $\tilde{P}(M_o) > 0$. On the other hand, by Fubini's theorem and Lemma 1 we have

$$\tilde{P}(M_o) = \int d\ P(v)\ \int d\mu(E)\ \chi((v,E) \in M_o)$$

$$= \int d\mu(E)\ \int dP(v)\ \chi((v,E)\ M_o) = 0.$$

Thus $\sigma_{a.c.}(H(v)) = $    P-almost surely.

Remark.    A result similar to Theorem 2 has been proved in $[6]$ for a continuous version of the model considered here.  Since Lemma 2 extends also to the continuum (see, e.g., Simon $[7]$), our proof applies without modification also to the situation considered in $[6]$.


## REFERENCES

[1]  Anderson, P.    *Phys. Rev. 109*, 1492 (1958).

[2]  Fröhlich, J. and Spencer, T.    Absence of diffusion in the Anderson
     Tight binding model for large disorder or low energy, *Comm. Math.
     Phys. 88*, 151-184 (1983).

[3]  Pastur, L.A.    Spectral properties of disordered systems in the one body
     approximation. *Comm. Math. Phys. 75*, 179 (1980).

[4]  Berezanskii, Ju.M.    *Expansion in Eigenfunctions of Self Adjoint
     Operators*. Translations of Mathematical Monographs, Vol. 17,
     A.M.S. (1968).

[5]  Kirsch, W. and Martinelli, F.    On the ergodic properties of the
     spectrum of general random operators; *J. reine angew. Math. 334*
     (1982).

[6]  Holden, H. and Martinelli, F.    On the absence of diffusion near the
     bottom of the spectrum for a random Schrödinger operator on
     $L^2(R^\nu)$. *Comm. Math. Phys. 93*, 197 (1984).

[7]  Simon, B.    Schrödinger semigroups. *Bull. Amer. Math. Soc. 7*,
     447-526 (1983).



F. Martinelli,                          E. Scoppola,
Dipartimento di Matematica,             Dipartimento di Fisica,
Università di Matematica,               Università "La Sapienza",
Povo,                                   Roma.
Trento.

CH-E PFISTER

# Gibbs random fields and symmetry breakdown

<u>INTRODUCTION</u>

The notion of a Gibbs random field was introduced in statistical mechanics by Dobrushin [1], in order to characterize those probability measures which correspond to equilibrium states of lattice models, such as the Ising model. A slightly different but equivalent characterization was formulated by Lanford and Ruelle [2]. If a random field describes an equilibrium state, then all its conditional probability distributions inside any finite region, under the condition that its values are known outside the region, are determined by the Hamiltonian through the well-known Gibbs formula. Therefore the Hamiltonian specifies completely a consistent family of conditional probability distributions and the basic problem of statistical mechanics has two parts:  the first one is to find all random fields, whose conditional probability distributions are equal to those fixed by the Hamiltonian (these random fields are called Gibbs random fields); the second part is to investigate the properties of these random fields.  This is the most interesting part of the problem.  However, only in few cases is it possible to answer completely the first part of the problem.  The main topic of this paper is to study this question for a class of lattice models by investigating the relation between symmetry breakdown and the existence of several Gibbs random fields.

In Section 1 the lattice models which are studied later on are described in detail and basic notions such as symmetry group and Hamiltonian are introduced.  Finally, several well-known examples of lattice models are given.  In Section 2, Gibbs random fields are defined precisely and the notion of symmetry breakdown is introduced.  Some properties of Gibbs random fields are recalled.  Section 3 contains the main part of the paper and in Section 4 aspects of the problem of symmetry breakdown other than those treated in Section 3 are briefly discussed.

There is an extensive literature on the subject.  We mention only the very interesting monograph by Sinai [3].

# 1. LATTICE MODELS

Perhaps the best way to start is to introduce the *sample space* of the random fields. All random fields are indexed by the elements of a lattice $\mathbb{L}$, which will be always the d-dimensional integer lattice $\mathbb{Z}^d$. The sample space $\Omega$ consists of all functions defined on $\mathbb{Z}^d$ with values in a fixed measurable space X, called the *state space*. The sample space $\Omega$ is a measurable space with the product $\sigma$-algebra denoted by $\underline{F}$. The state space X is the group $T = \mathbb{R}/2\pi\mathbb{Z}$, $\mathbb{Z}_n$ or a direct sum of these groups. In particular, X may be any finite Abelian group. In all these cases X is an Abelian compact topological group with the $\sigma$-algebra of Borel subsets. X may also be the product of T or $\mathbb{Z}_n$ with the unit interval $[0,1]$. For the sake of simplicity we consider the case $X = T$. The sample space $\Omega$ is called in statistical mechanics the *configuration space*. An element of $\Omega$, $\vartheta = \{\vartheta(x)\in T, x \in \mathbb{Z}^d \}$, is called a *configuration*. Let $\Lambda$ be any subset of $\mathbb{Z}^d$ and let $\overline{\Lambda} = \mathbb{Z}^d \backslash \Lambda$. The restriction of a configuration $\vartheta$ to the subset $\Lambda$ is denoted by $\vartheta(\Lambda)$ and the set of all possible $\vartheta(\Lambda)$ by $\Omega(\Lambda)$. For any $\Lambda$ we identify $\Omega$ and $\Omega(\Lambda) \times \Omega(\overline{\Lambda})$, in particular, $\vartheta$ and $(\vartheta(\Lambda), \vartheta(\overline{\Lambda}))$. $\underline{F}(\Lambda)$ is the $\sigma$-algebra generated by all cylinder sets with bases in $\Lambda$.

The notion of symmetry is very important in physics. In the present setting there are two groups which act on the sample space $\Omega$. There is the group $\mathbb{Z}^d$ whose action on $\Omega$ is

$$(T_y\vartheta)(x) = \vartheta(x-y), \quad y \in \mathbb{Z}^d . \tag{1.1}$$

On the other hand, the sample space itself is an Abelian compact topological group G with the group operation defined pointwise. The action of G on $\Omega$ is

$$(T_\phi\vartheta)(x) = \vartheta(x) + \phi(x), \quad \phi \in G. \tag{1.2}$$

The dual group of G is $G^{\wedge}$. It is isomorphic to the group $\Gamma$ consisting of all $\mathbb{Z}$-valued functions m defined on $\mathbb{Z}^d$ with finite supports. The group operation in $\Gamma$ is defined pointwise. Indeed, all elements of $G^{\wedge}$ are of the form

$$\chi_m(\vartheta) = \exp(im\vartheta) \tag{1.3}$$

with

$$m\vartheta = \sum_x m(x)\vartheta(x) . \tag{1.4}$$

The group $\mathbb{Z}^d$ acts also on $\Gamma$ by

$$(T_y m)(x) = m(x+y), \quad y \in \mathbb{Z}^d . \tag{1.5}$$

The value $\vartheta(x)$ of the random field at x represents the state of some physical entity located at x, which is called here spin in regard to the physical meaning of the models discussed at the end of this section. These spins interact and the interaction is described by the Hamiltonian. The joint interaction energy of the spins in some finite subset V of $\mathbb{Z}^d$ is given by a real-valued function $\Phi(V)$ on $\Omega(V)$. The Hamiltonian is the formal expression

$$H = \sum_V \Phi(V) \tag{1.6}$$

where the sum is extended to all finite subsets of $\mathbb{Z}^d$. Since we can consider the Fourier decomposition of $\Phi(V)$, it is here more convenient to express the Hamiltonian as

$$H = - \sum_{m \in \Gamma} J(m) \chi_m . \tag{1.7}$$

We always suppose that

$$J(T_x m) = J(m), \quad \forall x \in \mathbb{Z}^d \tag{1.8}$$

and the summability condition

$$\sum_{\substack{m: \\ \mathrm{supp}\, m \ni 0}} |J(m)| < \infty . \tag{1.9}$$

The expression (1.7) is purely formal, but for any finite set $\Lambda$

$$H_\Lambda(\vartheta) = - \sum_{\substack{m: \\ \mathrm{supp}\, m \subset \Lambda}} J(m) \chi_m(\vartheta) \tag{1.10}$$

and

$$W_\Lambda(\vartheta) = - \sum_{\substack{m: \\ \mathrm{supp}\, m \cap \Lambda \neq \emptyset \\ \mathrm{supp}\, m \cap \bar{\Lambda} \neq \emptyset}} J(m) \chi_m(\vartheta) \tag{1.11}$$

are well-defined functions. $H_\Lambda$ represents the interaction energy of the spins inside $\Lambda$ and $W_\Lambda$ the interaction energy between the spins in $\Lambda$ and those in $\overline{\Lambda}$. By our hypothesis (1.9) the Hamiltonian is translation invariant or $\mathbb{Z}^d$-invariant in the sense that

$$J(T_y m)\chi_{T_y m} = J(m)\chi_m \circ T_y. \tag{1.12}$$

$\mathbb{Z}^d$ is called the *spatial symmetry group*.

H is also invariant under the subgroup S of G defined by

$$S = \{\phi \in G : \chi_m(\phi) = 1 \quad \forall\, m \text{ with } J(m) \neq 0\} \tag{1.13}$$

since $\chi_m(\phi+\vartheta) = \chi_m(\vartheta)\chi_m(\phi)$. S is called the *internal symmetry group*.

Before giving the examples, let us consider the case with $X = T\times[0,1]$. The value of the random field at x is $\eta(x) = (\vartheta(x), r(x))$ with $\vartheta(x) \in T$ and $0 \leq r(x) \leq 1$. The action on $\Omega$ of the group G defined above is now

$$(T_\phi \eta)(x) = (\vartheta(x) + \phi(x), r(x)). \tag{1.14}$$

Let A be any function defined on $\mathbb{Z}^d$ with values in the positive integers and with finite support. Let $\rho$ be a bounded positive monotone function on $[0,1]$ and $\rho_A$ be the function on $\Omega$ defined by

$$\rho_A(\eta) = \prod_{x \in \text{supp } A} \rho(r(x))^{A(x)}. \tag{1.15}$$

The Hamiltonian is formally

$$H = - \sum_{A,m} J(A,m)\rho_A \chi_m \tag{1.16}$$

with

$$J(T_x A, T_x m) = J(A,m) \tag{1.17}$$

and

$$\sum_{\substack{A: \\ \text{supp } A \ni 0}} \sum_{\substack{m: \\ \text{supp } m \ni 0}} |J(A,m)| < \infty . \tag{1.18}$$

The spatial symmetry group is again $\mathbb{Z}^d$ and the internal symmetry group is the same as before.

Example 1. *Planar rotor model or* XY *model.* The state space is T. Let <xy> denote a two-point subset of $\mathbb{Z}^d$ such that

$$\| x-y \|^2 = \sum_{i=1}^{d} (x_i - y_i)^2 = 1. \tag{1.19}$$

The Hamiltonian is

$$H = - \sum_{<xy>} J \cos(\vartheta(x) - \vartheta(y)). \tag{1.20}$$

The internal symmetry group S is

$$\{\phi \in G : \phi(x) = \psi, \forall x\} \overset{\sim}{=} T. \tag{1.21}$$

Example 2. $\mathbb{Z}_n$-_model_. The state space is the discrete subgroup $T_n$ of T of order n, which is isomorphic to $\mathbb{Z}_n$. The Hamiltonian is again given by (1.20). Another possibility, known in physics as the Potts model, is

$$H = - \sum_{<xy>} \sum_{q=0}^{n-1} \frac{J}{n} \cos q(\vartheta(x) - \vartheta(y)). \tag{1.22}$$

Since the second sum

$$\frac{1}{n} \sum_{q=0}^{n-1} \cos q(\vartheta(x) - \vartheta(y)) = \delta(\vartheta(x), \vartheta(y)) \tag{1.23}$$

gives the Kronecker $\delta$-function, H can be written as

$$H = - \sum_{<xy>} J\delta(\vartheta(x), \vartheta(y)). \tag{1.24}$$

The internal symmetry group S is according to our definition isomorphic to $\mathbb{Z}_n$. However, it is clear from (1.24) that the Hamiltonian is invariant under a larger symmetry group, which is isomorphic to the group of the permutations of n elements. If $J > 0$ then only the symmetry group S is important, as is shown by the results of Section 4.

Example 3. _Ising model_. This is a special case of Example 2 with $n = 2$. Since

$$\cos \vartheta(x) \equiv \sigma(x) = \pm 1, \tag{1.25}$$

the Hamiltonian is usually written in terms of the $\sigma$-variables (1.25)

$$H = - \sum_{<xy>} J\sigma(x)\sigma(y). \tag{1.26}$$

Example 4. *Ashkin-Teller model.* The state space is $T_2 \times T_2 \cong \mathbb{Z}_2 \times \mathbb{Z}_2$.

The value of the random field at x is $\eta(x) = (\vartheta(x), \vartheta'(x))$. Using the $\sigma$ variables $\cos \vartheta(x) = \sigma(x)$ and $\tau$-variables $\cos \vartheta'(x) = \tau(x)$, the Hamiltonian is

$$H = - \sum_{\langle xy \rangle} \lambda_1 \sigma(x)\sigma(y) + \lambda_2 \tau(x)\tau(y) + \mu\sigma(x)\sigma(y)\tau(x)\tau(y). \qquad (1.27)$$

The internal symmetry group has four elements: $\{g_o, g_1, g_2, g_3\}$ with $g_o = \mathrm{id}$, $g_1^2 = g_2^2 = g_3^2 = g_o$ and $g_1 g_2 = g_3$.

Example 5. *Triangular Ising model.* We have chosen as lattice $\mathbb{Z}^d$. But this is not a necessary condition. In this example the lattice $\mathbb{L}$ is the two-dimensional triangular lattice and the state space is again $X = T_2 \cong \mathbb{Z}_2$ as in the Ising model.

Let $\Delta = \{x,y,z\}$ be an elementary triangular cell of $\mathbb{L}$ and let $\sigma(\Delta) = \sigma(x)\sigma(y)\sigma(z)$. The Hamiltonian is

$$H = - \sum_{\Delta} J\sigma(\Delta). \qquad (1.28)$$

The sum in (1.28) is extended to all elementary cells of $\mathbb{L}$. The spatial symmetry group is $\mathbb{L}$ and the internal symmetry group S has four elements $\{g_o, g_1, g_2, g_3\}$ with $g_o = \mathrm{id}$, $g_1^2 = g_2^2 = g_3^2 = g_o$ and $g_1 g_2 = g_3$. This group is isomorphic to the internal symmetry group of Example 4. There is, however, an important difference. The elements of the symmetry group of Example 4 are all fixed points of the action of the spatial group $\mathbb{Z}^d$. This is not the case in Example 5 and the smallest non-trivial subgroup of S which is stable under the action of $\mathbb{L}$ is S itself. This fact has physical consequence explained in Section 4.

## 2. GIBBS RANDOM FIELDS AND SYMMETRY BREAKDOWN

Let H be a fixed Hamiltonian. The normalized Haar measure on the state space X is $\mu$ and the product measure on $\Omega(\Lambda)$ is

$$\mu(d\vartheta(\Lambda)) = \prod_{x \in \Lambda} \mu(d\vartheta(x)). \qquad (2.1)$$

For any finite $\Lambda \subset \mathbb{Z}^d$,

$$H_\Lambda(\vartheta(\Lambda) \,|\, \vartheta(\overline{\Lambda})) = H_\Lambda(\vartheta) + W_\Lambda(\vartheta) \qquad (2.2)$$

is a continuous function, and using the Gibbs formula we can define a probability measure on $\Omega(\Lambda)$

$$P_\Lambda(d\vartheta(\Lambda)|\vartheta(\overline{\Lambda})) = Z(\Lambda|\vartheta)^{-1}\exp\{-\beta H_\Lambda(\vartheta(\Lambda)|\vartheta(\overline{\Lambda}))\ \mu(d\vartheta(\Lambda)). \tag{2.3}$$

Here $\beta$ is a nonnegative real parameter, the inverse temperature, and $Z(\Lambda|\vartheta)$ is a normalizing constant, the partition function

$$Z(\Lambda|\vartheta) = \int_{\Omega(\Lambda)} \exp(-\beta H_\Lambda(\vartheta(\Lambda)|\vartheta(\overline{\Lambda}))\mu(d\vartheta(\Lambda)). \tag{2.4}$$

The measure (2.3) defines uniquely a probability measure on $(\Omega,\ \underline{\underline{F}})$, which is concentrated on the subset

$$\{\phi \in \Omega:\ \phi(\overline{\Lambda}) = \vartheta(\overline{\Lambda})\}. \tag{2.5}$$

The expectation value of an integrable function f with respect to this measure is denoted by

$$<f>_\Lambda(\vartheta). \tag{2.6}$$

Let $I_B$ be the indicator function of a set $B \in F$. For any fixed $(H,\beta)$ we can define a family of Markov kernels $K_\Lambda$ on $\Omega\times\underline{\underline{F}}$, indexed by the finite subsets $\Lambda$ of $\mathbb{Z}^d$, by setting

$$K_\Lambda(\vartheta,B) = <I_B>_\Lambda(\vartheta). \tag{2.7}$$

These kernels have the following properties:

$$K_\Lambda(\cdot,B) \text{ is } F(\overline{\Lambda})\text{-measurable} \tag{2.8}$$

and, for any pair $\Lambda_1 \subset \Lambda_2$, we have $\underline{\underline{F}}(\overline{\Lambda}_1) \supset \underline{\underline{F}}(\overline{\Lambda}_2)$ and

$$K_{\Lambda_2}(\vartheta,B) = \int K_{\Lambda_2}(\vartheta,d\phi)K_{\Lambda_1}(\phi,B). \tag{2.9}$$

Therefore this family of Markov kernels forms a consistent family of conditional probability distributions.

Definition.   A Gibbs random field (with respect to $(H,\beta)$) is any probability measure $\nu$ on $(\Omega,F)$ such that, for any finite $\Lambda \subset \mathbb{Z}^d$ and any $B \in \underline{\underline{F}}$,

$$E_\nu\left[I_B|\underline{\underline{F}}(\overline{\Lambda})\right] = <I_B>_\Lambda(\cdot) \qquad \nu\text{-a.s.} \tag{2.10}$$

Let $\Delta(H,\beta)$ be the set of all Gibbs random fields, which is never empty in

the present setting.  Clearly $\Delta(H,\beta)$ is a convex set.  Moreover, if
$\nu \in \Delta(H,\beta)$, then $T_x \nu$ and $T_\phi \nu$ also belong to $\Delta(H,\beta)$ for any $x \in \mathbb{Z}^d$ and any
$\phi \in S$.  The action of $\mathbb{Z}^d$ or S on the space of measures is by definition

$$T_x \nu(f) = \nu(f \circ T_x). \tag{2.11}$$

In particular, if $\Delta(H,\beta)$ consists of one measure only, then this measure is
necessarily $\mathbb{Z}^d$-invariant and S-invariant.  In other words, the Gibbs random
field has the same symmetry as the Hamiltonian.  On the other hand, if there
exists a measure $\nu \in \Delta(H,\beta)$ which is not $\mathbb{Z}^d$-invariant or S-invariant, then
we say that the *symmetry* of the Hamiltonian *is broken*.  This phenomenon
occurs in all examples given in the last section.  For $\beta$ small enough there
is always a unique Gibbs random field.  On the other hand, if $\beta$ is large
enough, $J > 0$, $d \geq 3$ in the first example and $d \geq 2$ in the other examples,
then there exist Gibbs random fields which break the symmetry of the
Hamiltonian.

One method for constructing a Gibbs random field is to consider sequences
of measures $<\cdot>_{\Lambda_n} (\vartheta_n)$ (see (2.6)) with $\Lambda_n \uparrow \mathbb{Z}^d$ in the sense that
eventually $\Lambda_n \supset D$ for any finite subset D of $\mathbb{Z}^d$.  Indeed, if such a
sequence converges to a measure $\nu$ in the weak topology of measures, then $\nu$ is
a Gibbs random field.  The following theorems ensure that this method gives
essentially all Gibbs random fields, which is a very satisfactory result from
the physical viewpoint, and that any Gibbs random field has a unique extremal
decomposition.

<u>Theorem 1.</u>    The closed convex hull (in the weak topology of measures) of the
set of all elements of $\Delta(H,\beta)$ obtained by the above method is $\Delta(H,\beta)$.

<u>Theorem 2.</u>    The set $\Delta(H,\beta)$ is a Choquet simplex.  The set $\Delta_i(H,\beta)$ of all
$\mathbb{Z}^d$-invariant elements of $\Delta(H,\beta)$ is also a Choquet simplex.

We finally make a remark on the case $X = T \times [0,1]$.  The measure $\mu$ in (2.1)
is now the product measure of the Haar measure on T and of any probability
measure on $[0,1]$ not concentrated at 0.

Beside the monograph of Sinai [3] we recommend a paper of Föllmer [4]
concerning the general structure of $\Delta(H,\beta)$.

## 3. $\mathbb{Z}^d$- INVARIANT GIBBS RANDOM FIELDS AND SYMMETRY BREAKDOWN OF THE INTERNAL SYMMETRY GROUP

### 3.1. General discussion

The general structure of $\Delta(H,\beta)$ is given by Theorems 1 and 2. If we want to investigate further $\Delta(H,\beta)$ it is necessary to restrict the class of Hamiltonians. From now on we make the following hypothesis (F) (for ferromagnetism) :

Hypothesis (F):    $J(m) = J(-m) \geq 0$.

(In the case $X = Tx[0,1]$ we have of course $J(A,m) = J(A,-m) \geq 0$.) It is therefore convenient to write the Hamiltonian as

$$H(\vartheta) = - \sum_m J(m) \cos m\vartheta. \tag{3.1}$$

If we have $J > 0$ in Examples 1 to 5 then the hypothesis (F) is verified. Let H be given. It is natural to study $\Delta(H,\beta)$ by varying parameter $\beta$, since $\beta$ is the inverse temperature. For any value of $\beta$ it is possible to construct a particular Gibbs random field $\nu_o$ by the method indicated in Theorem 1, which has the following properties: $\nu_o$ is an extremal element of the convex set $\Delta(H,\beta)$ and it is $\mathbb{Z}^d$-invariant. Let $S(\beta)$ be the subgroup of the internal symmetry group S defined by

$$S(\beta) = \{\phi \in S : T_\phi \nu_o = \nu_o\}. \tag{3.2}$$

Since $\nu_o$ is $\mathbb{Z}^d$-invariant, $S(\beta)$ is a subgroup of S which is stable under the action of $\mathbb{Z}^d$. (The group S is stable under the action of $\mathbb{Z}^d$ because the Hamiltonian is $\mathbb{Z}^d$-invariant.) At small enough $\beta$, we have of course $S(\beta) = S$ since there is a unique Gibbs random field. On the other hand, for $\beta$ large enough, $S(\beta)$ is in general a proper subgroup of S. We know that it is the case in the examples of Section 2. The most important property of $\nu_o$ is that *all* Gibbs random fields are $S(\beta)$-invariant. They may be invariant under a larger symmetry group but not a smaller one. In other words, the question of the symmetry breakdown of S can be investigated by considering only the particular Gibbs random field $\nu_o$. Let us suppose that $S(\beta)$ is a proper subgroup of S. There is a natural action of the quotient group $S/S(\beta)$ on $\Delta(H,\beta)$ because all elements of $\Delta(H,\beta)$ are $S(\beta)$-invariant. Let g denote an element of $S/S(\beta)$ and let $\nu_g$ be the measure obtained by the action of g on

$\nu_o$. Clearly, $\nu_g$ is again an extremal element of $\Delta(H,\beta)$. (But $\nu_g$ is not necessarily $\mathbb{Z}^d$-invariant since g need not be a fixed point of the natural action of $\mathbb{Z}^d$ on $S/S(\beta)$.) Moreover, if $\lambda$ is any $\mathbb{Z}^d$-invariant probability measure on $S/S(\beta)$, then

$$\nu = \int_{S/S(\beta)} \lambda(dg)\nu_g \tag{3.3}$$

is a $\mathbb{Z}^d$-invariant Gibbs random field. In particular, if dg is the normalized Haar measure on $S/S(\beta)$,

$$\tilde{\nu} = \int_{S/S(\beta)} dg\nu_g \tag{3.4}$$

is a $\mathbb{Z}^d$-invariant and S-invariant Gibbs random field by construction.

<u>Definition.</u>  The inverse temperature $\beta$ is *regular* if $\tilde{\nu}$ is the unique Gibbs random field which is $\mathbb{Z}^d$-invariant and S-invariant.

The main results of this section are summarized in Theorem 3.

<u>Theorem 3.</u>  Let H be a Hamiltonian verifying hypothesis (F) and let the state space X be as in Section 2.

(a) For all $\beta$, $S(\beta)$ is a subgroup of the internal symmetry group S, which is stable under the action of $\mathbb{Z}^d$. All Gibbs random fields are $S(\beta)$-invariant.

(b) If $\beta_1 \leq \beta_2$ then $S(\beta_2)$ is a subgroup of $S(\beta_1)$.

(c) The set of *non*-regular $\beta$ is at most countable.

(d) If $\beta$ is regular, then all $\mathbb{Z}^d$-invariant, resp. periodic, Gibbs random fields of $\Delta(H,\beta)$ are given by (3.3) with $\lambda$ any $\mathbb{Z}^d$-invariant, resp. periodic, probability measure on $S/S(\beta)$.

(e) If $\beta$ is regular, there is a one-to-one correspondence between the $\mathbb{Z}^d$-ergodic probability measures on $S/S(\beta)$ and the extremal points of the Choquet simplex $\Delta_i(H,\beta)$ (i.e. the ergodic Gibbs random fields).

Before looking at the main parts of the proof it is perhaps useful to come back to the examples of Section 2. We suppose of course that $J > 0$. We also suppose that the dimension of the lattice $d \geq 2$. Indeed, if $d = 1$ there is a unique Gibbs random field for all values of $\beta \in [0,\infty)$.

Planar rotor model:   In this model we know that either $S(\beta) = S \cong T$ or $S(\beta)$ is trivial ($S(\beta)$ contains only one element). Moreover, if $S(\beta) = S$, then $\beta$ is regular and therefore there is a unique $\mathbb{Z}^d$- invariant Gibbs random field [5], [6]. It is well-known in physics that in general a continuous symmetry group cannot be broken if $d = 2$. There are several various proofs of this fact in statistical mechanics [7]. Therefore if $d = 2$ there is no symmetry breakdown of S in the planar rotor model and hence a unique $\mathbb{Z}^d$- invariant Gibbs random field. On the other hand, if $d \geq 3$ there is symmetry breakdown of S if $\beta$ is large enough [8]. By Theorem 3 there is a unique $\beta_c$ such that $S(\beta) = S$ if $\beta < \beta_c$ and $S(\beta)$ is trivial if $\beta > \beta_c$. In the last case, all extremal points of $\Delta_i(H,\beta)$, i.e. all ergodic Gibbs random fields, are known if $\beta$ is regular. They are parametrized by $\alpha \in [0,2\pi)$. Moreover, there are no periodic Gibbs random fields. These results were first proven in [9]. Very likely all $\beta$ are regular (this is proven if $d = 2$) and if $S = S(\beta)$ there is a unique Gibbs random field.

Potts model with n components:   Let $d \geq 2$ and n be large enough. Then there is a $\beta_c(n)$ such that $\beta_c(n)$ is *non*-regular and, if $\beta > \beta_c$, $S(\beta)$ is trivial. In particular, at $\beta_c$ there are at least $(n + 1)$ ergodic Gibbs random fields [10]. For all regular $\beta > \beta_c$ there are exactly n ergodic Gibbs random fields and no periodic Gibbs random fields. More recently Baxter proved that this situation occurs for $n > 4$ [11]. Presumably, $\beta_c$ is the only non-regular point, for $\beta < \beta_c$ there is a unique Gibbs random field and at $\beta_c$ there are exactly $n + 1$ ergodic Gibbs random fields.

Ising model.   Clearly in this model $S(\beta)$ is either S or trivial. This implies the existence of a unique $\beta_c$ (for $d \geq 2$) such that, for all $\beta > \beta_c$, there are at least two ergodic Gibbs random fields. If $d = 2$, all $\beta$ are regular and above $\beta_c$ there are exactly two ergodic Gibbs random fields. This was first proven in [12].

Ashkin-Teller model.   If the coupling constants $\lambda_1 > 0$, $\lambda_2 > 0$ and $\mu > 0$ are suitably chosen, one can prove the existence of two different points $\beta_c$ and $\beta_c' > \beta_c$ such that, for all regular $\beta$, the following situation holds. If $\beta < \beta_c$, then $S(\beta) = S$ and there is a unique ergodic Gibbs random field. If $\beta_c < \beta < \beta_c'$, then $S(\beta)$ is one of the non-trivial sub-groups $S_i = \{g_o,g_i\}$ of S. The group S is only partially broken since all Gibbs random fields are

$S(\beta)$-invariant (this last result being true for all $\beta$ satisfying $\beta_c < \beta < \beta_c'$). There are two ergodic Gibbs random fields which are naturally labelled by the elements of the quotient group $S/S(\beta)$. For $\beta > \beta_c'$, $S(\beta)$ is trivial and there are four ergodic Gibbs random fields. For details see [13].

Triangular Ising Model. The internal symmetry group $S$ is isomorphic to the internal symmetry group of the Ashkin-Teller model. But here $S(\beta)$ is either $S$ or is trivial as we mentioned already in Section 2. Therefore there exists a unique $\beta_c$ with $S(\beta) = S$ for $\beta < \beta_c$ and $S(\beta)$ trivial for $\beta > \beta_c$. Since the group elements $g_1$, $g_2$ and $g_3$ are not fixed points of the action of $\mathbb{L}$, we get in this case extremal periodic Gibbs random fields $\nu_1$, $\nu_2$ and $\nu_3$ for $\beta > \beta_c$. Moreover, if $\beta$ is regular there are exactly two ergodic Gibbs random fields $\nu_o$ and $\nu' = \frac{1}{3}(\nu_1 + \nu_2 + \nu_3)$. We have therefore a simple example of an ergodic Gibbs random field $\nu'$, which is not an extremal Gibbs random field.

## 3.2. Main steps of the proof of Theorem 3

The main tool in the proof of Theorem 3 is the use of correlation inequalities. This technique has been very successful in statistical mechanics. In our present situation it consists of exploiting the positivity property $J(m) = J(-m) > 0$. Hence the basic paper on which our proof relies is the very important and beautiful paper by Ginibre [14]. All correlation inequalities used here can be found in [14] or can be proven using the method of [14]. The proof is divided into three steps. In the first step we use correlation inequalities to establish the main Lemma 4. In the second step we use the Pontryagin duality theorem. Finally, in the third step we use the Choquet simplex structure of $\Delta(H,\beta)$ (Theorem 2).

First step. Let us construct the Gibbs random field $\nu_o$. Let $\Lambda$ be any finite subset of $\mathbb{Z}^d$ and let the configuration $\vartheta(x) = 0$ for all $x \in \mathbb{Z}^d$ be denoted by $0$. Let $<\cdot>_\Lambda(0)$ be the measure (2.6) which is concentrated on $\{\vartheta \in \Omega : \vartheta(x) = 0, \ \forall x \in \overline{\Lambda}\}$.

Lemma 1. (a) Let $\Lambda_1 \subset \Lambda_2$ and $m \in \Gamma$. Then

$$<\cos m\phi>_{\Lambda_1}(0) \geq <\cos m\phi>_{\Lambda_2}(0). \qquad (3.5)$$

(b) For any $\vartheta \in \Omega$, $m \in \Gamma$ and parameter $\alpha$

$$<\cos(m\phi-\alpha)>_\Lambda (\vartheta) \leq <\cos m\phi>_\Lambda (0). \tag{3.6}$$

The first part of Lemma 1 follows directly from the paper $\boxed{14}$. The second part is proven in $\boxed{6}$. Let us apply the correlation inequalities of Lemma 1. Clearly, from (3.5) we get the existence of the limit

$$\lim_{\Lambda \uparrow \mathbb{Z}^d} <\cos m\phi>_\Lambda (0) \equiv \nu_o(\cos m\phi). \tag{3.7}$$

Since by the symmetry $\phi \to -\phi$ we have, for any $\Lambda$,

$$<\sin m\phi>_\Lambda (0) = 0 \tag{3.8}$$

we have in fact

$$\lim_{\Lambda \uparrow \mathbb{Z}^d} <\chi_m>_\Lambda (0) = \nu_o(\chi_m). \tag{3.9}$$

Therefore this implies the existence of the measure $\nu_o$ on $(\Omega, \underline{F})$. Let $\Lambda + x = \{y+x, y \in \Lambda\}$. By definition,

$$<\chi_m>_\Lambda (0) = <\chi_m \circ T_x>_{\Lambda+x} (0). \tag{3.7}$$

But from the monotonicity property (3.5)

$$\lim_{\Lambda \uparrow \mathbb{Z}^d} <\chi_m>_\Lambda (0) = \lim_{\Lambda \uparrow \mathbb{Z}^d} <\chi_m>_{\Lambda+x} (0). \tag{3.8}$$

Therefore $\nu_o$ is $\mathbb{Z}^d$-invariant. Finally, let us prove the extremality of $\nu_o$. Using Theorem 1 and (3.6), we have

$$\nu_o(\cos m\phi) \geq \nu(\cos m\phi) \tag{3.9}$$

for any Gibbs random field $\nu$. If we can prove the implication

$$\nu_o(\cos m\phi) = \nu(\cos m\phi) \Rightarrow \nu(\sin m\phi) = 0 \tag{3.10}$$

then clearly from (3.9) we must have that $\nu_o$ is an extremal point of $\Delta(H,\beta)$. It turns out that (3.10) is an easy consequence of (3.6). Indeed, by Theorem 1,

$$\nu_o(\cos m\phi) \geq \nu(\cos(m\phi-\alpha)). \tag{3.11}$$

Therefore if $\nu_o(\cos m\phi) = \nu(\cos m\phi)$ we have

$$\nu_o(\cos m\phi) \geq \cos \alpha \; \nu(\cos m\phi) + \sin \alpha \; \nu(\sin m\phi)$$

$$= \cos \alpha \; \nu_o(\cos m\phi) + \sin \alpha \; \nu(\sin m\phi). \qquad (3.12)$$

Hence

$$\pm \nu(\sin m\phi) \leq \frac{1-\cos \alpha}{\sin \alpha} \; \nu_o(\cos m\phi) \qquad (3.13)$$

with $0 < \alpha < \pi$. Letting $\alpha$ going to zero we get (3.10). We have therefore constructed for all $\beta$ an extremal Gibbs random field $\nu_o$ which is $\mathbb{Z}^d$- invariant. We need also the following property of $\nu_o$ which is a direct consequence of [14].

<u>Lemma 2.</u> For any $\beta$ and any m such that $J(m) > 0$, $\nu_o(\chi_m) > 0$.

The next result is the key step in the proof of Lemma 4. It is proven in [9], [15].

<u>Lemma 3.</u> Let $\nu$ be any Gibbs random field. Let $\Omega'$ be a second copy of the sample space $\Omega$, whose elements are denoted by $\phi'$. Let $\rho = \nu_o \times \nu$ be the product measure on the product space $\Omega \times \Omega'$. Then

$$\rho((\cos m\phi \pm \cos m\phi') \exp(\pm\lambda \cos n\phi \cos n\phi')) \geq 0 \qquad (3.14)$$

for any m, n $\in \Gamma$ and positive $\lambda$.
From Lemma 3 we get

<u>Lemma 4.</u> (a) If $\nu_o(\chi_m) > 0$ and $\nu_o(\chi_n) > 0$ then $\nu_o(\chi_{m\pm n}) > 0$.

(b) Let $\nu$ be any Gibbs random field. If $\nu_o(\chi_m) = \nu(\chi_m)$ and

$$\nu_o(\chi_n) = \nu(\chi_n) > 0, \text{ then } \nu_o(\chi_{m\pm n}) = \nu(\chi_{m\pm n}).$$

<u>Proof.</u> Let us prove (b). With the notation of Lemma 3 we have

$$\rho((\cos m\phi - \cos m\phi') \exp(\pm\lambda \cos n\phi \cos n\phi')) \geq 0. \qquad (3.15)$$

For small $\lambda$,

$$\exp(\pm\lambda \cos n\phi \cos n\phi') = 1 \pm \lambda \cos n\phi \cos n\phi' + O(\lambda^2). \qquad (3.16)$$

Therefore we can write (3.15) as

$$0 \leq \pm \lambda\nu_o(\cos n\phi)\left[\nu_o(\cos m\phi \cos n\phi) - \nu(\cos m\phi \cos n\phi)\right] + O(\lambda^2).$$

$$(3.17)$$

Dividing by $\lambda$ and letting $\lambda \to 0$, we get

$$\nu_o(\cos n\phi \cos m\phi) = \nu(\cos n\phi \cos m\phi). \qquad (3.18)$$

But this can be written as

$$\nu_o(\cos(m+n)\phi) + \nu_o(\cos(m-n)\phi) = \nu(\cos(m+n)\phi) + \nu(\cos(m-n)\phi). \qquad (3.19)$$

Using (3.11) with $\alpha = 0$ and (3.10), we get the desired result.

We have now completed the first part of the proof.

<u>Second step.</u>   Let $I$ be the subgroup of $\Gamma$, which is generated by $m$ with $J(m) > 0$.   Let $I(\beta)$ be the set of all elements $m \in \Gamma$ such that $\nu_o(\chi_m) > 0$.

<u>Lemma 5.</u>   (a) $I(\beta)$ is a subgroup of $\Gamma$ and $I(\beta) \supset I$.

(b) $S$ is the annihilator of $I$, i.e. the subgroup

$$\{\phi \in G : \chi_m(\phi) = 1, \ \forall m \in I\}.$$

(c) $S(\beta)$ is the annihilator of $I(\beta)$.

(d) $(S/S(\beta))^{\wedge} \cong I(\beta)/I$.

<u>Proof.</u>   (a) follows from Lemmas 2 and 4.   Parts (b) and (c) are evident. The last part follows from the Pontryagin duality theorem.

<u>Third step.</u>   Let $\Delta_o(H,\beta)$ be the set of all Gibbs random fields such that, for all $m$ with $J(m) > 0$,

$$\nu_o(\chi_m) = \nu(\chi_m). \qquad (3.20)$$

By construction $\nu_\phi = T_\phi \nu_o$ (see (2.11)) is an element of $\Delta_o(H,\beta)$ and is an extremal point of $\Delta(H,\beta)$.   For all $m \in \Gamma$,

$$\nu_\phi(\chi_m) = \chi_m(\phi)\, \nu_o(\chi_m). \qquad (3.21)$$

Therefore, if $\nu_\phi = \nu_\psi$, we must have

$$\chi_m(\phi-\psi) = 1, \quad \forall m \in I(\beta). \qquad (3.22)$$

This implies that $\phi-\psi \in S(\beta)$.   Let $g$ be an element of $S/S(\beta)$ and $\nu_g$ be the measure obtained by the action of $g$ on $\nu_o$.

$\underline{\text{Lemma 6.}}$  Let $\nu$ be a Gibbs random field in $\Delta_o(H,\beta)$.  Then there is a unique probability measure $\lambda$ on $S/S(\beta)$ such that (see (3.3))

$$\nu = \int_{S/S(\beta)} \lambda(dg)\nu_g. \tag{3.23}$$

$\underline{\text{Proof.}}$  Since $\nu_g$ is an extremal point of $\Delta(H,\beta)$, and $\nu_{g_1} \neq \nu_{g_2}$ if $g_1 \neq g_2$, a representation such as (3.23) is necessarily unique by Theorem 2.  By the same theorem, $\nu$ has an extremal decomposition in $\Delta(H,\beta)$.  By (3.9) almost all extremal components of $\nu$ are in $\Delta_o(H,\beta)$.  Let $\rho$ be such a component.  By Lemma 4(b) we have, for all $m \in I$,

$$\rho(\chi_m) = \nu_o(\chi_m). \tag{3.24}$$

Let $\rho_g$ be the measure obtained by the action of $g$ on $\rho$ and let (see (3.4))

$$\tilde{\rho} = \int_{S/S(\beta)} dg\, \rho_g. \tag{3.25}$$

From (3.6) we have $\tilde{\rho}(\chi_m) = 0$ for all $m \notin I(\beta)$.  Let $m \in I(\beta)\setminus I$.  Then, by Lemma 5(d),

$$\int_{S/S(\beta)} dg\, \chi_m(g) = 0 \tag{3.26}$$

and therefore $\tilde{\rho}(\chi_m) = 0$.  But this means that $\tilde{\rho} = \tilde{\nu}$ (see (3.4)) and, by the uniqueness of the extremal decomposition, we must have $\rho = \nu_g$ for some $g \in S/S(\beta)$.  This being true for almost all $\rho$ of the extremal decomposition of $\nu$, we get the result (3.23).

The proof of Theorem 3 is finished by using the next lemma, whose proof is based on another characterization of the $\mathbb{Z}^d$- invariant Gibbs random fields as tangent functionals to the pressure.

$\underline{\text{Lemma 7.}}$  (a) The set of non-regular $\beta$ is at most countable.

(b) If $\beta$ is regular, then all $\mathbb{Z}^d$- invariant, resp. periodic, Gibbs random fields are in $\Delta_o(H,\beta)$.


$\underline{\text{Remarks.}}$  (1) Theorem 3 was first proven in the cases $X = \mathbb{Z}_2$ and $X = \mathbb{Z}_2 \times [0,1]$ from the works of Slawny [16], Lebowitz [17], [18] and Bricmont, Lebowitz, Pfister [19].  The general case is proven in Pfister [15].

(2) The characterization of the Gibbs random fields as tangent functionals
to the pressure is fully developed in the monograph of Israel [20].

(3) The reader interested in the correlation inequalities may consult [21]
and [22].

## 4. GIBBS RANDOM FIELDS WHICH ARE NOT $\mathbb{Z}^d$- INVARIANT OR PERIODIC

Let us suppose that $\beta$ is *regular*, since it is the generic case. This is
true for the models considered in this section, if $\beta$ is large enough. In
the proof of Theorem 3 we have considered a convex subset $\Delta_o(H,\beta)$ (see (3.20))
of $\Delta(H,\beta)$. The set of all extremal elements of $\Delta_o(H,\beta)$ is the $S/S(\beta)$ - orbit
of $\nu_o$. Any element of $\Delta_o(H,\beta)$ has a unique extremal decomposition in
$\Delta_o(H,\beta)$ which coincides with its unique extremal decomposition in $\Delta(H,\beta)$
(see Lemma 6). Since we have supposed that $\beta$ is regular, all $\mathbb{Z}^d$- invariant
Gibbs random fields are in $\Delta_o(H,\beta)$. In general, there are Gibbs random
fields which are not $\mathbb{Z}^d$- invariant in $\Delta_o(H,\beta)$; for example, all periodic
Gibbs random fields are in $\Delta_o(H,\beta)$ (see, e.g., Example 5). A natural
question is whether there are other Gibbs random fields. In other words, is
$\Delta_o(H,\beta)$ equal or not equal to $\Delta(H,\beta)$? Of course, if such Gibbs random
fields exist, they are neither $\mathbb{Z}^d$- invariant nor periodic.

We can answer this question for the Ising model. For the case where the
dimension of the lattice $d \geq 3$, Dobrushin has constructed a countable family
of extremal Gibbs random fields which are not in $\Delta_o(H,\beta)$, provided $\beta$ is
large enough [23]. Later, van Beijeren has found a clever and simple proof
of the existence of these random fields using correlation inequalities [24].
On the other hand, Gallavotti has proven that the method of Dobrushin does
not give new Gibbs random fields if $d = 2$ [25]. The obvious conjecture
that $\Delta_o(H,\beta) = \Delta(H,\beta)$ if $d = 2$, was proven only recently by Higuchi [26] and
Aizenman [27].

In the Ising model there are exactly two extremal Gibbs random fields in
$\Delta_o(H,\beta)$, which are $\mathbb{Z}^d$- invariant, provided $\beta$ is large enough. The physical
interpretation of those fields is that they represent the pure phases of the
system, one being positively magnetized, i.e. $\nu_o(\cos \phi(x)) = m^* > 0$, and the
other negatively magnetized, $\nu_\pi(\cos \phi(x)) = -m^*$. Usually the model is
described in terms of the $\sigma(x)$-variables (see (1.25)), and instead of $\nu_o$ and
$\nu_\pi$ one uses the notation $\nu_+$ and $\nu_-$. The new Gibbs random fields constructed
by Dobrushin have the properties

114

$$\lim_{x_1 \to +\infty} \nu(\sigma(x_1,0,0)) = m^* \tag{4.1}$$

and

$$\lim_{x_1 \to -\infty} \nu(\sigma(x_1,0,0)) = -m^*. \tag{4.2}$$

They are asymptotically equal to $\nu_+$ or $\nu_-$ if $|x_1| \to \infty$. The region of $\mathbb{Z}^d$ where

$$|\nu(\sigma(x)| \ll m^* \tag{4.3}$$

is roughly a plane with finite thickness. For $d = 2$, Gallavotti has proven that such a region cannot exist, and that $\nu = \frac{1}{2} (\nu_+ + \nu_-)$. These questions are discussed, e.g., in [28].

It is, of course, possible to extend the results of Dobrushin, Gallavotti and van Beijeren to other models with a *discrete* internal symmetry group. On the other hand, the proofs of Higuchi and Aizenman do not extend even to the Ising model with next nearest interactions.

In the case of the planar rotor model, which is the simplest model with a continuous internal symmetry group, the situation is quite different. It can be shown that random fields $\nu$, like those considered by Dobrushin and Gallavotti in the Ising model, are always $\mathbb{Z}^d$- invariant for any dimension $d$ [9]. In [9] arguments are given in favour of the existence of Gibbs random fields which are not in $\Delta_o(H,\beta)$ provided that $d \geq 4$. If $d = 3$, it is likely that $\Delta(H,\beta) = \Delta_o(H,\beta)$. However, it seems very difficult to get a proof of this conjecture. Roughly speaking, the Gibbs random fields, whose existence is conjectured in [9] for $d \geq 4$, are of the following type. These random fields are invariant under the subgroup of $\mathbb{Z}^d$

$$\{a \in \mathbb{Z}^d : a_1 = a_2 = 0\}. \tag{4.4}$$

If we consider the restriction of the random fields on the region

$$D = \{x \in \mathbb{Z}^d : x_3 = \ldots = x_d = 0\} \tag{4.5}$$

then, for $y \in D$, with $y_2 = \mathrm{tg}\alpha\, y_1$,

$$\lim_{y_1 \to \infty} \nu(\cos \phi(y)) = \cos(\alpha+c) \; \nu_o(\cos \phi(y)) \tag{4.6}$$

$$= \cos(\alpha+c) \; \nu_o(\cos \phi(0))$$

where $c$ is a constant independent of $\alpha$.

REFERENCES


[1]  Dobrushin, R.L.   Gibbsian random fields for lattice systems with pairwise interactions. *Funct. Anal. Applic.* 2, 31-43 (1968) - The problem of the uniqueness of a Gibbsian random field and the problem of phase transitions. *Funct. Anal. Applic.* 2, 44-57 (1968) - Gibbsian random fields. The general case. *Funct. Anal. Applic.* 3, 27-35 (1969).

[2]  Lanford, O.E. and Ruelle, D.  Observables at infinity and states with short range correlations in statistical mechanics. *Commun. math. Phys.* 13, 194-215 (1969).

[3]  Sinai, Ya G.  Theory of phase transitions : rigorous results. Oxford-New York-Toronto-Sydney-Paris-Frankfurt: Pergamon (1982).

[4]  Föllmer, H.  Phase transition and Martin boundary, in *Séminaire de Probabilités* IX, 305-317, Universite de Strasbourg, Lecture Notes in Mathematics 465. Berlin-Heidelberg-New York: Springer (1975).

[5]  Bricmont, J., Fontaine, J.R. and Landau L.J.  On the uniqueness of the equilibrium state for plane rotators. *Commun. math. Phys. 56,* 281-296 (1977).

[6]  Messager, A., Miracle-Sole S. and Pfister C.E.  Correlation inequalities and uniqueness of the equilibrium state for the plane rotator ferromagnetic model. *Commun. math. phys. 58,* 19-29 (1978).

[7]  Mermin, N.D.  Absence of ordering in certain classical systems. *J. Math. Phys. 8,* 1061-1064 (1967).

Dobrushin, R.L. and Shlosman, S.B.  Absence of breakdown of continuous symmetry in two-dimensional models of statistical physics. *Commun. math. Phys. 42,* 31-40 (1975).

Pfister, C.E.  On the symmetry of the Gibbs states in two-dimensional
     lattice systems. *Commun. math. Phys. 79*, 181-188 (1981).

Simon, B. and Sokal, A.D.  Rigorous energy-entropy arguments *J. Stat.
     Phys. 25*, 679-194 (1981).

Fröhlich, J. and Pfister, C.E.  On the absence of spontaneous symmetry
     breaking and of crystalline ordering in two-dimensional systems.
     *Commun. math. Phys. 81*, 277-298 (1981).

[8]  Fröhlich, J., Simon, B. and Spencer, T.  Infrared bounds, phase
     transitions and continuous symmetry breaking. *Commun. math.
     Phys. 50*, 79-85 (1976).

[9]  Fröhlich, J. and Pfister, C.E.  Spin waves, vortices and the structure
     of equilibrium states in the classical XY model. *Commun. math.
     Phys.* (1983).

[10]  Kotecky', R. and Shlosman, S.B.  First-order phase transitions in
     large entropy models. *Commun. math. Phys. 83*, 493-515 (1982).

[11]  Baxter, R.J.  Magnetization discontinuity of the two-dimensional Potts
     model. *J. Physics A 15*, 3329-3340 (1982).

[12]  Messager, A. and Miracle-Sole, S.  Equilibrium states of the two-
     dimensional Ising model in the two-phase region. *Commun. math.
     Phys. 40*,  187-196 (1975).

[13]  Pfister, C.E.  Phase transitions in the Ashkin-Teller model. *J. Stat.
     Phys. 29*, 113-116 (1982).

[14]  Ginibre, J.  General formulation of Griffith's inequalities. *Commun.
     math. Phys. 16*, 310-328 (1970).

[15]  Pfister, C.E.  Translation invariant equilibrium states of ferromagnetic
     abelian lattice systems. *Commun. math. Phys. 86*, 375-390 (1982).

[16]  Slawny, J.  Low temperature behaviour of ferromagnets. *Commun. math.
     Phys. 35*, 297-305 (1974).

[17]  Lebowitz, J.L.  Coexistence of phases in Ising ferromagnets. *J. Stat.
     Phys. 16*, 463-476 (1977).

[18] Lebowitz, J.L.  Number of phases in one component ferromagnets.  In *Mathematical Problems in Theoretical Physics Proceedings, Rome 1977*, 68-80.  Lecture Notes in Physics, 80.  Berlin, Heidelberg, New York: Springer (1978).

[19] Bricmont, J., Lebowitz, J.L. and Pfister, C.E.  Periodic Gibbs states of ferromagnetic spin systems. *J. Stat. Phys. 24*, 269-277 (1981).

[20] Israel, R.B.  Convexity in the theory of lattice gases.  Princeton, New Jersey: Princeton University Press (1979).

[21] Cartier, P.  Inégalités de corrélation en mécanique statistique. *Séminaire Bourbaki no 431*, 242-264 (1973).  Lecture Notes in Mathematics, 383. Berlin, Heidelberg, New York: Springer (1974).

[22] Shlosman, S.B.  Correlation inequalities and their applications. *Soviet Mathematics 15*, 79-101 (1981).

[23] Dobrushin, R.L.  Gibbs state describing coexistence of phases for a three-dimensional Ising model. *Theory Probab. Applic. 17*, 582-600 (1972) - Investigation of Gibbsian states for three-dimensional lattice systems. *Theory Probab. Applic. 18*, 253-271 (1973).

[24] van Beijeren, H.  Interface sharpness in the Ising model. *Commun. math. Phys. 40*, 1-6 (1975).

[25] Gallavotti, G.  The phase separation line in the two-dimensional Ising model. *Commun. math. Phys. 27*, 103-136 (1972).

[26] Higuchi, Y.  On the absence of non-translation invariant Gibbs states for the two-dimensional Ising model. *Colloquia Mathematica Societatis Janos Bolyai 27*, Random Fields, Esztergom (1979).

[27] Aizenman, M.  Translation invariance and instability of phase coexistence in two-dimensional Ising model. *Phys. Rev. Lett. 43*, 407 (1979); *Commun. math. Phys. 73*, 83-94 (1980).

[28] Pfister, C.E.  Interface and surface tension in Ising model.  In *Scaling and Self-similarity in Physics*. Ed. J. Frohlich, Boston-Basel-Stuttgart: Birkhaüser (1983) 139-161.


Charles-Ed. Pfister,
Département de Mathématiques,
Ecole Polytechnique Fédérale,
CH-1015 Lausanne, Switzerland.

M RÖCKNER & N WIELENS

# Dirichlet forms – closability and change of speed measure

## 0. INTRODUCTION

There is a well-known correspondence between Hunt processes on the Euclidean space $\mathbb{R}^d$ and Dirichlet spaces on $L^2(\mathbb{R}^d ; m)$, where m, the speed measure, is a positive Radon measure with $\text{supp}[m] = \mathbb{R}^d$ (cf. [5], [13]).

If $(F,E)$ is a Dirichlet space on $\mathbb{R}^d$ (for the precise definition cf. Section 1) related to a Hunt process with continuous sample paths, such that $C_o^\infty(\mathbb{R}^d)$ is dense in $F$, then by a famous result of Beurling and Deny the form $E$ can be represented on $C_o^\infty(\mathbb{R}^d)$ by

$$E(u,v) = \sum_{i,j=1}^{d} \int D_i u \; D_j v \; d\mu_{ij} + \int u \, v \, dk, \tag{0.1}$$

where $(\mu_{ij})_{i,j}$ is a non-negative definite matrix of signed Radon measures on $\mathbb{R}^d$ and k is a positive Radon measure on $\mathbb{R}^d$ (cf. [5], [14]). ($D_i$ denotes differentiation with respect to $x_i$.)

Given, conversely, a symmetric form $E$ by (0.1) with domain $D(E) = C_o^\infty(\mathbb{R}^d)$, a natural question arises: is it possible to construct an associated Hunt process? In other words, using the correspondence mentioned above: what conditions on the measures $\mu_{ij}$ and k imply that there is an appropriate regular Dirichlet space $(F, E')$, such that the restriction of $E'$ to $D(E)$ is equal to $E$? The crucial property leading to the answer to this question is the closability of the form $E$ on $L^2(\mathbb{R}^d ; m)$ for some speed measure m.

The aim of this paper is to survey recent work on the closability of symmetric forms and to present some new ideas on this subject.

In Section 1 we introduce the basic notions and summarize the one-dimensional case which is completely solved (cf. 1.1, [7] and [12] resp.) The following sections deal with the multidimensional case.

First we treat a special case of the form (0.1), namely

$$E(u,v) = \sum_{i=1}^{d} \int D_i u \; D_i v \; d\mu_i$$
$$D(E) = C_o^\infty(D), \tag{0.2}$$

119

where D is an open set in $\mathbb{R}^d$ and $\mu_i$, $1 \leq i \leq d$, are absolutely continuous with respect to the Lebesgue measure $\lambda^d$.

In Section 2 we summarize a result of Silverstein [15] concerning the closability and the representation of a certain class of forms in the case $d = 2$. Furthermore, we briefly describe the method of Albeverio et al. [1], by which they gain closability criteria for the form (0.2).

A new concept to show the closability of the form (0.2) is developed in Section 3: instead of studying the existence of the closure of this form, we consider a form $E'$, which essentially is the maximal extension of $E$, and give some conditions for $E'$ to be closed. As a consequence, we can partially generalize Hamza's theorem (cf. Section 1, [7] resp.) to the multidimensional case (see 3.4). Concerning the uniqueness of the closed extension of (0.2) we refer to [16].

Section 4 deals with the closability of the form

$$E(u,v) = \sum_{i,j=1}^{d} \int D_i u \; D_j v \; a_{ij} \; d\lambda^d$$

$$\mathcal{D}(E) = C_o^\infty(D),$$

(0.3)

where $a_{ij}$; $i,j \in \{1,\ldots,d\}$, are locally $\lambda^d$-integrable functions on D such that the matrix $(a_{ij}(x))_{i,j}$ is symmetric and non-negative definite for every $x \in D$. Using the results of Section 3, we can replace the usual ellipticity condition (cf. [5] §2.1; 1.b) by a weaker one (cf. 4.2, see also 4.3).

In the last section we study the question: for what kind of speed measures $\mu$ is the form (0.3) closable on $L^2(D;\mu)$, if it is closable with respect to some speed measure m? In the transient case we give a characterization of these measures $\mu$ (cf. 5.1, 5.5 and 5.9).

## 1. BASIC DEFINITIONS AND THE ONE-DIMENSIONAL CASE

The notation of this paper is essentially based on the representation given in [5]. We recall the main notions: let X be a locally compact, second countable Hausdorff space, and let $M(X)$ denote the set of all Radon measures on X and

$$M^+(X) := \{\mu \in M(X) : \mu \geq 0\}.$$

Given $\mu \in M^+(X)$, let $L^2(X;\mu)$ be the space of all (classes of) $\mu$-square integrable, real functions on X with inner product

$$(u,v)_\mu = \int u(x)\ v(x)\ \mu(dx) \qquad (u,v \in L^2(X;\mu)).$$

Set $\|\cdot\|_\mu := (\ ,\ )_\mu^{1/2}$. Let $m \in M^+(X)$ with $X = \text{supp}[m]$ (i.e. the support of m). Let $E$ be a symmetric form on $L^2(X;m)$ (in the sense of [5]), i.e. $E$ is a non-negative definite, symmetric, bilinear form on $L^2(X;m)$ with domain $D(E)$, where $D(E)$ is a dense linear subspace of $L^2(X;m)$.

$E$ is said to be *closed* if the space $D(E)$ is complete with respect to the inner product

$$E_1(u,v) := E(u,v) + (u,v)_m, \qquad u,v \in D(E).$$

We say that $E$ is *closable* if the following condition is satisfied: given a sequence $(u_n)_{n\in\mathbb{N}}$ in $D(E)$ such that $E(u_n-u_k,\ u_n-u_k) \xrightarrow[k,n\to\infty]{} 0$ (i.e. $(u_n)$ is an $E$-Cauchy sequence and $(u_n,u_n)_m \xrightarrow[n\to\infty]{} 0$, then $E(u_n,u_n) \xrightarrow[n\to\infty]{} 0$.

Given two symmetric forms $E^{(1)}$ and $E^{(2)}$ on $L^2(X;m)$, $E^{(2)}$ is called an *extension of* $E^{(1)}$ if

$$D(E^{(1)}) \subset D(E^{(2)}) \quad\text{and}\quad E^{(2)} = E^{(1)} \quad\text{on}\quad D(E^{(1)}) \times D(E^{(1)}).$$

Clearly, a symmetric form on $L^2(X;m)$ is closable iff there exists a closed extension.

The following condition is sufficient for the closability of a symmetric form on $L^2(X;m)$ (cf. [5] (1.1.3)) :

$$
\begin{aligned}
&\text{Given a sequence } (u_n)_{n\in\mathbb{N}} \text{ in } D(E) \text{ such that} \\
&(u_n,u_n)_m \xrightarrow[n\to\infty]{} 0, \text{ then } E(u_n,v) \xrightarrow[n\to\infty]{} 0 \text{ for every } v \in D(E).
\end{aligned}
\tag{1.1}
$$

Hence, in particular, the form $E$ defined by

$$E(u,v) = (u,Sv)_m$$

$$D(E) = D(S),$$

where S is a non-negative definite symmetric operator with domain $D(S)$ in $L^2(X;m)$, is a closable form on $L^2(X;m)$.

Given a symmetric form $E$ on $L^2(X;m)$, we say that *the unit contraction operates on* $E$ if, for any $u \in D(E)$, the function $v := (0 \vee u) \wedge 1 \in D(E)$ and

$E(v,v) \leq E(u,u)$.

We now define the *Markov property* of $E$ which is more useful and more general than the above notion. However, the two notions are equivalent if $E$ is a closed form (cf. [5] Theorem 1.4.1).

A symmetric form $E$ on $L^2(X;m)$ is said to be *Markovian* if the following condition is satisfied: for each $\varepsilon > 0$, there exists a real function $\phi_\varepsilon(t)$, $t \in \mathbb{R}$, such that:

(i) $\phi_\varepsilon(t) = t$ for every $t \in [0,1]$;

(ii) $-\varepsilon \leq \phi_\varepsilon(t) \leq 1 + \varepsilon$ for every $t \in \mathbb{R}$;

(iii) $0 \leq \phi_\varepsilon(t') - \phi_\varepsilon(t) \leq t' - t$ whenever $t < t'$; and

(iv) if $u \in \mathcal{D}(E)$, then $\phi_\varepsilon \circ u \in \mathcal{D}(E)$ and $E(\phi_\varepsilon \circ u, \phi_\varepsilon \circ u) \leq E(u,u)$.

A *Dirichlet form* is defined as a closed Markovian form. The pair $(\mathcal{D}(E),E)$ is then called a *Dirichlet space*.

A symmetric form $E$ on $L^2(X;m)$ is called *regular* if the space $\mathcal{D}(E) \cap C_o(X)$ is dense in $(\mathcal{D}(E),E_1)$ and dense in $(C_o(X), \|\ \|_\infty)$. $((C_o(X), \|\ \|_\infty)$ denotes the space of all continuous functions on X with compact support endowed with the norm $\|u\|_\infty := \sup\{|u(x)|: x \in X\}$, $u \in C_o(X)$.)

$E$ has the *local property*, if the following condition is satisfied: if $u,v \in \mathcal{D}(E)$ such that supp$[u\,m]$ and supp$[v\,m]$ are disjoint and compact, then $E(u,v) = 0$.

Suppose now that X is a domain D in $\mathbb{R}^d$, $d \geq 1$. Let $C_o^\infty(D)$ denote the set of all infinitely differentiable functions on D with compact support. By the Beurling-Deny formula (cf. e.g. [5] Theorem 2.2.2) any local closable Markovian form $E$ on $L^2(D;m)$ with $\mathcal{D}(E) = C_o^\infty(D)$ can be expressed uniquely as

$$E(u,v) = \sum_{i,j=1}^{d} \int D_i u \ \ D_j v \ \ d\mu_{ij} + \int u\,v\,dk, \tag{1.2}$$

where $\mu_{ij}$ $(1 \leq i, j \leq d)$ are Radon measures on D such that, for any $\xi = (\xi_1,\ldots,\xi_d) \in \mathbb{R}^d$ and any compact set $K \subset D$,

$$\sum_{i,j=1}^{d} \xi_i \xi_j \ \mu_{ij}(K) \geq 0, \qquad \mu_{ij}(K) = \mu_{ji}(K), \qquad 1 \leq i, j \leq d.$$

$k$ is a positive Radon measure on D, called the *killing measure*.

As indicated in the introduction, we want to treat the reversal: what forms on $C_o^\infty(D)$ represented by (1.2) are local, closable and Markovian?

122

Since locality and Markov property are always satisfied (cf. [5] Example 1.2.1), the only problem is the closability.

In Sections 1-4 we restrict ourselves to the case k = 0 i.e.,

$$E(u,v) = \sum_{i,j=1}^{d} \int D_i u \; D_j v \; d\mu_{ij}, \quad \mathcal{D}(E) = C_o^\infty(D). \tag{1.3}$$

(The case $k \neq 0$ is treated only in 4.5 and in Section 5.)

In the one-dimensional case one has a complete characterization of the closability of the form (1.3). For a detailed representation see Spönemann, Rullkötter [12], 13.2:

__Theorem 1.1.__ Let I be an open interval in $\mathbb{R}$ and let $\mu$ and m be positive Radon measures on I such that $\mathrm{supp}\,[m] = I$; then the form $E$ defined by

$$E(u,v) = \int_I u'v'd\mu, \quad \mathcal{D}(E) = C_o^\infty(I), \text{ is closable on } L^2(I;m) \text{ if and only if the}$$

following conditions are satisfied:

(i) $\mu$ is absolutely continuous with respect to $\lambda^1$;

(ii) the density function $\rho$ is locally $\lambda^1$-integrable and vanishes $\lambda^1$-a.e. on the set $I \backslash R(\rho)$, where $R(\rho)$ is the *regular set* of $\rho$; i.e.,

$$R(\rho) := \{x \in I \,|\, \text{there exists a neighbourhood V of x in I,}$$

$$\text{such that } \int_V \rho^{-1} \, d\lambda^1 < \infty\} .$$

This theorem is an extension of a well-known result of Hamza [7], 3.1-3.3, who treated the closability of the form (1.2) on $L^2(I, \lambda^1)$.

The problem of the closability of the form (1.3) in the multi-dimensional case, which is treated in the following sections, is more complicated, since the measures $\mu_{ij}$ $(1 \leq i, j \leq d)$ are not necessarily absolutely continuous (cf. [5], §2.1, $(2^o)$).


## 2. CLOSABILITY OF $\sum_{i=1}^{d} \int D_i u \, D_i v \, d\mu_i$ — SUMMARY OF KNOWN RESULTS

First we refer to a result by Silverstein [15] in the case d = 2, where he studies the closability of the general form

$$E(u,v) := \sum_{i,j=1}^{d} \int D_i u \ D_j v \ d\nu_{ij}$$

$$\mathcal{D}(E) := C_o^1(\mathbb{R}^2),$$

which is assumed to satisfy

$$E(u,v) = \int_{-\infty}^{\infty} \int_{-\infty}^{\infty} \nabla u(x,y) \cdot \nabla v(x,y) \ |y|^{1-\alpha} d\lambda^1(x) \ d\lambda^1(y)$$

for some $\alpha \in \, ]0,2[$ if $u,v \in C_o^1(\mathbb{R}^2 \setminus \{(x,y) \in \mathbb{R}^2 : y = 0\})$.

Silverstein proves that $(\mathcal{D}(E),E)$ is closable on $L^2(\mathbb{R}^2; m)$ for any $m \in M^+(\mathbb{R}^2)$ with supp $m = \mathbb{R}^2$ iff $(\mathcal{D}(E),E)$ is closable on $L^2(\mathbb{R}^2; m_o)$ with $m_o := |y|^{1-\alpha} \lambda^2$. Furthermore, he proves that, if this is true, then $E$ is necessarily of the form

$$E(u,v) = \int_{-\infty}^{\infty} \int_{-\infty}^{\infty} \nabla u(x,y) \cdot \nabla v(x,y) \ |y|^{1-\alpha} \ d\lambda^1(x) \ d\lambda^1(y)$$

$$+ \int_{-\infty}^{\infty} \frac{\partial}{\partial x} u(x,0) \ \frac{\partial}{\partial x} v(x,0) \ \rho(x) \ d\lambda^1(x)$$

for general $u,v \in C_o^1(\mathbb{R}^2)$, where $\rho$ is a certain non-negative locally integrable function on $\mathbb{R}$ satisfying certain properties listed in [15]. For another example in $\mathbb{R}^2$ see also [4] §2 Ex.2, [9] resp. .

Now we come to another basic concept to show the closability of forms. This concept is mainly based on partial integration. The ideas are essentially taken from [1].

In the sequel, let D be an open subset in $\mathbb{R}^d$, $d \geq 1$, and $m \in M^+(D)$ with supp$[m]$ = D.

Remark 2.1.   We recall that a symmetric form $E$ on $L^2(D;m)$ is closable iff there exists a closable linear operator T on $L^2(D;m)$ such that

$$E(u,v) = \int Tu \ Tv \ dm$$

$$\mathcal{D}(E) = \mathcal{D}(T)$$

(cf. Theorem 1.3.1 in [5]).   Furthermore, we recall that the operator T is closable iff its adjoint is densely defined or, equivalently, iff the map

$$u \longmapsto \int Tu\, v\, dm \text{ is continuous from } (\mathcal{D}(T), \|\ \|_m) \text{ to } \mathbb{R} \text{ for all } v \text{ in a}$$

dense subset of $(L^2(D;m), \|\ \|_m)$.

Fix $i \in \{1,\ldots,d\}$. We want to study the closability of the following form on $L^2(D;m)$:

$$E(u,v) := \int D_i u\, D_i v\, d\mu$$

$$\mathcal{D}(E) := C_o^\infty(D), \tag{2.1}$$

where $\mu \in M^+(D)$.

The following lemma is an easy consequence of Remark 2.1 and Riesz's representation theorem.

<u>Lemma 2.2.</u>  Let $\mu = m$ and assume that the map $u \longmapsto \int D_i u\, dm$ is continuous from $(C_o^\infty(D), \|\ \|_m)$ to $\mathbb{R}$. Then the form (2.1) is closable on $L^2(D;m)$.

The following theorem can be found in $\boxed{1}$. The proof is included for completeness.

<u>Theorem 2.3.</u>  Let $E$ be of the form (2.1) for some $\mu \in M^+(D)$. Assume that there exist non-negative locally $\lambda^d$-square-integrable functions $\phi$ and $\Psi$ on $D$, such that $\mu = \phi^2 \lambda^d$ and $m = \Psi^2 \lambda^d$, $\Psi > 0$ $\lambda^c$-a.e. on $D$ and such that the following conditions hold for their distribution derivatives $D_i\phi$ and $D_i\Psi$ on some open subset $U$ of $D$ with $m(D\backslash U) = 0$:

(i) $D_i\Psi$ is locally $\lambda^d$-integrable on $U$;

(ii) $D_i\phi$ and $\phi/\Psi\, D_i\Psi$ are locally $\lambda^d$-square integrable on $U$.

Then $E$ is closable on $L^2(D;m)$.

<u>Proof.</u>  Since, for $u,v \in C_o^\infty(D)$,

$$E(u,v) = \int (\tfrac{\phi}{\Psi} D_i u)\, (\tfrac{\phi}{\Psi} D_i v)\, dm,$$

and since $C_o^\infty(U)$ is dense in $(L^2(D;m), \|\ \|_m)$, it suffices to show that the map

$$L : u \longmapsto \int (\tfrac{\phi}{\Psi} D_i u)\, v\, dm = \int (D_i u)\, v\, \phi\, \Psi\, d\lambda^d$$

is continuous on $(C_o^\infty(D), \| \ \|_m)$ for every $v \in C_o^\infty(U)$ (cf. Remark 2.1). But if $u \in C_o^\infty(D)$ and $v \in C_o^\infty(U)$, then by assumption we may integrate by parts and obtain (using $[6]$, (7.18))

$$|L(u)| = |\int u\,v\,\psi\,D_i\phi\,d\lambda^d + \int u\,v\,\phi\,D_i\psi\,d\lambda^d + \int u\,\phi\,\psi\,D_i v\,d\lambda^d|$$

$$\leq \sup_U \,(|D_i v|+|v|) \int_{\mathrm{supp}[v]} |u| \left| \frac{D_i\phi}{\psi} + \frac{\phi D_i\psi}{\psi^2} + \frac{\phi}{\psi} \right| \, dm.$$

But by assumption the function $((D_i\phi/\psi) + (\phi D_i\psi/\psi^2) + (\phi/\psi))$ is locally m-square integrable on U, hence the assertion follows.

<u>Corollary 2.4.</u>   Let $\psi$ and $\phi_i$, $1 \leq i \leq d$, be non-negative, locally $\lambda^d$-square integrable functions, such that $\psi > 0$ $\lambda^d$-a.e. on D and $m = \psi^2 \lambda^d$. Assume that, for every $i \in \{1,\ldots,d\}$, the following conditions hold for their distribution derivatives $D_i\phi_i$ and $D_i\psi$ on some open subset U of D with $m(D\backslash U) = 0$:

(i) $D_i\psi$ is locally $\lambda^d$-integrable on U;

(ii) $D_i\phi_i$ and $\phi_i/\psi\,D_i\psi$ are locally $\lambda^d$-square integrable on U.   Then the form

$$E(u,v) := \sum_{i=1}^d \int D_i u\,D_i v\ \phi_i^2\,d\lambda^d$$

$$D(E) \quad := C_o^\infty(D)$$

is closable on $L^2(D;m)$.

<u>Proof.</u>   The assertion is an immediate consequence of Theorem 2.3, since the sum of closable forms is a closable form.

<u>Remark 2.5.</u>   In the case $d = 1$ the condition that $\phi'$ is locally $\lambda^d$-square integrable in Corollary 2.4 implies that $\phi^2$ is absolutely continuous. Hence clearly $\phi^2 = 0$ on $D\backslash R(\phi^2)$ in accordance with Theorem 1.1

## 3. CLOSABILITY OF $\sum_{i=1}^{d} \int D_i u \, D_i v \, d\mu_i$ – NEW RESULTS

Let U and D be non-empty open subsets of $\mathbb{R}^d$, $d \geq 1$, $U \subset D$. For $\mu \in M^+(U)$ consider the following condition:

(A) There exists a non-negative, locally $\mu$-square integrable function g on U such that

$$g \cdot \mu = \lambda^d .$$

Given $\mu \in M^+(U)$ satisfying (A), then every version of an element in $L^2(U;\mu)$ is $\lambda^d$-measurable on U. In the sequel we will not distinguish between classes of functions with respect to $\mu$ and certain versions, as far as no confusion is possible.

<u>Lemma 3.1.</u>  Let $\mu \in M^+(U)$, satisfying (A) with density function g. Let $\phi \in C_o^\infty(U)$. Then there exists a constant $c > 0$ such that

$$\int |u| \, |\phi| \, d\lambda^d \leq c \left( \int u^2 \, d\mu \right)^{1/2} \quad \text{for every} \quad u \in L^2(U;\mu) .$$

<u>Proof.</u>  Applying Hölder's inequality to the measure $|\phi| \cdot \mu$ we obtain

$$\int |u| \, |\phi| \, d\lambda^d \leq \left( \sup_{x \in U} |\phi(x)| \right)^{1/2} \left( \int g^2 |\phi| \, d\mu \right)^{1/2} \left( \int u^2 \, d\mu \right)^{1/2} .$$

Given $\mu \in M^+(U)$ satisfying (A), then Lemma 3.1 in particular implies that each $u \in L^2(U;\mu)$ is locally $\lambda^d$-integrable on U.

Let $m \in M^+(D)$ such that $m_{|U}$, the restriction of m to U, satisfies (A), and let $\rho$ be a non-negative locally $\lambda^d$-integrable function on U. We define, for $1 \leq i \leq d$,

$$H_i^{1,2}(D,U,m,\rho) := \{ u \in L^2(D;m) : D_i u \text{ is locally } \lambda^d\text{-integrable on}$$
$$U \text{ such that } D_i u \in L^2(U;\rho\lambda^d) \} ,$$

where the derivative $D_i u$ is taken in the sense of distributions.

Fix $i \in \{1,\ldots,d\}$. For $u,v \in H_i^{1,2}(D,U,m,\rho)$ we define

$$\langle u,v \rangle := \int u\ v\ dm + \int_U D_i u\ D_i v\ \rho\ d\lambda^d,$$

$$\| u \| := \langle u,u \rangle^{1/2}.$$

$\langle , \rangle$ is well defined, since $m_{|U}$ satisfies (A). Hence $(H_i^{1,2}(D,U,m,\rho),\ \langle , \rangle)$ is a pre-Hilbert space. Now we want to give a sufficient condition for this space to be complete; i.e., to be a Hilbert space.

<u>Theorem 3.2.</u>   Let $m \in M^+(D)$ such that $m_{|U}$ satisfies (A) and let $\rho$ be a non-negative, locally $\lambda^d$-integrable function on U. If $\rho^{-1}$ is locally $\lambda^d$-integrable on U, then $(H_i^{1,2}(D,U,m,\rho),\langle , \rangle)$ is complete.

<u>Proof.</u>   Let $(u_n)_{n \in \mathbb{N}}$ be a Cauchy sequence in $(H_i^{1,2}(D,U,m,\rho),\ \| \ \|\ )$. Then there exists $u \in L^2(D;m)$ and $h \in L^2(U;\rho\lambda^d)$ such that

$$\lim_{n \to \infty} u_n = u \text{ in } L^2(D;m)$$

and

$$\lim_{n \to \infty} D_i u_n = h \quad \text{in} \quad L^2(U;\rho\lambda^d).$$

Let $\phi \in C_o^\infty(U)$. Then by Lemma 3.1

$$\int_U h\ \phi\ d\lambda^d = \lim_{n \to \infty} \int_U D_i u_n\ \phi\ d\lambda^d = -\lim_{n \to \infty} \int_U u_n\ D_i \phi\ d\lambda^d$$

$$= -\int_U u\ D_i \phi\ d\lambda^d.$$

It follows that $D_i u = h$ as distributions on U. Consequently, since $h$ is locally $\lambda^d$-integrable on U by Lemma 3.1,

$$u \in H_i^{1,2}(D,U,m,\rho) \quad \text{and} \quad \lim_{n \to \infty} \| u-u_n \| = 0.$$

<u>Corollary 3.3.</u>   Let $m \in M^+(D)$ such that $m_{|U}$ satisfies (A). Assume that $\text{supp}[m] = D$. Let $\rho$ be a non-negative function on U such that $\rho$ and $\rho^{-1}$ are locally $\lambda^d$-integrable on U.

 (i) If $m(D \setminus U) = 0$, then the symmetric form

$$E(u,v) := \int_U D_i u \, D_i v \, \rho \, d\lambda^d$$

$$\mathcal{D}(E) := C_o^\infty(U)$$

is closable on $L^2(D;m)$.

(ii) If there exists a measure $\mu \in M^+(D)$ such that $\mu_{|U} = \rho\lambda^d$, then the symmetric form

$$E'(u,v) := \int_U D_i u \, D_i v \, \rho \, d\lambda^d$$

$$\mathcal{D}(E') := C_o^\infty(D)$$

is closable on $L^2(D;m)$.

Proof. Consider the symmetric form

$$\tilde{E}(u,v) := \int_U D_i u \, D_i v \, \rho \, d\lambda^d$$

$$\mathcal{D}(\tilde{E}) := H_i^{1,2}(D,U,m,\rho).$$

Then $(\mathcal{D}(\tilde{E}),\tilde{E})$ is an extension of $(\mathcal{D}(E),E)$, $(\mathcal{D}(E'),E')$ respectively, which is closed on $L^2(D;m)$ by Theorem 3.2. Hence $(\mathcal{D}(E),E)$ and $(\mathcal{D}(E'),E')$ are closable on $L^2(D;m)$.

To finish this section we want to show that Corollary 3.3 (ii) generalizes one part of Hamza's theorem (cf. 1.1 with $m = \lambda^d$) to the multidimensional case.

If $\rho$ is a non-negative, locally integrable function on D, we define, as in the one-dimensional case,

$$R(\rho) := \{x \in D : \text{there exists an open neighbourhood } V \text{ of } x$$
$$\text{in } D \text{ such that } \int_V \rho^{-1} \, d\lambda^d < \infty\},$$

where $\rho^{-1}(x) = +\infty$, if $x \in \{\rho=0\}$.

$R(\rho)$ is called the *regular set of* $\rho$ and $S(\rho) := D\backslash R(\rho)$ the *singular set of* $\rho$.

Theorem 3.4. Let $\rho$ be a non-negative, locally $\lambda^d$-integrable function on D such that $\rho = 0$ $\lambda^d$-a.e. on $S(\rho)$. Let $m \in M^+(D)$. Assume that there exists an open subset U of D with $\lambda^d(R(\rho)\backslash U) = 0$ such that $g \cdot m\big|_{U \cap R(\rho)} = \lambda^d$ for some locally $\lambda^d$-integrable function g on $U \cap R(\rho)$. Then the form

$$E(u,v) := \int D_i u \ D_i v \ \rho \ d\lambda^d$$

$$\mathcal{D}(E) := C_o^\infty(D).$$

is closable on $L^2(D;m)$.

<u>Proof.</u>   By assumption we have

$$\int D_i u \ D_i v \ \rho \ d\lambda^d = \int_{U \cap R(\rho)} D_i u \ D_i v \ \rho \ d\lambda^d.$$

Furthermore, $U \cap R(\rho)$ is open and $\rho^{-1}$ is locally $\lambda^d$-integrable on $R(\rho)$. Hence the assertion follows by Corollary 3.3 (ii).

<u>Examples.</u>   Let $\rho$ be a non-negative, locally $\lambda^d$-integrable function on D.

  (i) If $\rho^{-1}$ is locally $\lambda^d$-integrable on D, then $S(\rho) = 0$
      (cf. also [16], 4.2.3).

(ii) If $\rho$ is lower semi-continuous on D, then $\rho = 0$ on $S(\rho)$.


Since the sum of closable forms is a closable form, we have, by Theorem 3.4:

<u>Corollary 3.5.</u>   Let $\rho_i$, $1 \leq i \leq d$, satisfy the conditions on $\rho$ in Theorem 3.4. Let $V := \bigcup_{i=1}^{d} R(\rho_i)$ and $m \in M^+(D)$.  Assume that there exists an open subset U of D with $\lambda^d(V\backslash U) = 0$ such that $g \cdot m\big|_{U \cap V} = \lambda^d$ for some locally $\lambda^d$-integrable function g on $U \cap V$.  Then the form

$$E(u,v) := \sum_{i=1}^{d} \int D_i u \ D_i v \ \rho_i \ d\lambda^d,$$

$$\mathcal{D}(E) := C_o^\infty(D)$$

is closable on $L^2(D;m)$.

This generalizes a result of Brasche obtained in [2], Section 4, Satz 4.

130

4. **CLOSABILITY OF** $\sum\limits_{i,j=1}^{d} \int D_i u \, D_j v \, a_{ij} \, d\lambda^d$

In this section let D be an open subset of $\mathbb{R}^d$, $d \geq 1$, and $m \in M^+(D)$ with supp$[m]$ = D.

Now we want to study the closability of the form

$$E(u,v) = \sum_{i,j=1}^{d} \int D_i u \, D_j v \, a_{ij} \, d\lambda^d.$$

$$D(E) \quad := C_o^\infty(D)$$

(4.1)

on $L^2(D;m)$, where $a_{ij}$; $i,j \in \{1,\ldots,d\}$ are locally $\lambda^d$-integrable functions on D, such that the matrix $(a_{ij}(x))_{i,j}$ is symmetric and non-negative definite for every $x \in D$. For $m = \lambda^d$ the following result can be found in $[5]$ §2.1 $(1^o_.a)$.

<u>Proposition 4.1.</u> Let $E$ be of the form (4.1). Assume that there exists a non-negative locally m-integrable function g on D such that $\lambda^d = g \cdot m$ and that the distribution derivatives $D_i \, a_{ij}$ of $a_{ij}$, $1 \leq i$, $j \leq d$, are locally $\lambda^d$-integrable on D. If, in addition, $g^{1/2} \, a_{ij}$ and $g^{1/2} \, D_i \, a_{ij}$ are locally $\lambda^d$-square integrable on D for all $i,j \in \{1,\ldots,d\}$,* then $E$ is closable on $L^2(D;m)$.

<u>Proof.</u> Define the linear operator

$$Su \quad := g \sum_{i,j=1}^{d} (a_{ij} \, D_i D_j u + D_i \, a_{ij} \, D_j u)$$

$$D(S) := C_o^\infty(D).$$

Then, by assumption, S is a symmetric linear operator on $L^2(D;m)$ such that, for $u,v \in C_o^\infty(D)$, $E(u,v) = -(u,Sv)_m$. Hence the assertion follows (cf. (1.1)).

The following closability criterion, which is due to Sato, is a generalization of Proposition 4.1, if $D = \mathbb{R}^d$ and $m = \lambda^d$. For the proof, see $[5]$ §2.1 $(1^o_.c)$.

---

* There is a misprint in $[5]$ §2.1 $(1^o_.a)$: $L^1_{loc}$ should be replaced by $L^2_{loc}$.

$\underline{\text{Proposition 4.2.}}$   Let $E$ be of the form (4.1) with $D = \mathbb{R}^d$ and $m = \lambda^d$.
Assume that, for $i,j \in \{1,\ldots,d\}$, the function $a_{ij}$ and, if $i \neq d$, also its
distribution derivative $D_i a_{ij}$ are locally $\lambda^d$-square integrable on $\mathbb{R}^d$.  If,
in addition, the distribution derivative $D_d a_{dj}$ is locally $\lambda^d$-square
integrable on $\mathbb{R}^d \setminus \{(x_1,\ldots,x_d) \in \mathbb{R}^d : x_d = 0\}$ for every $j \in \{1,\ldots,d\}$, then
$E$ is closable on $L^2(\mathbb{R}^d ; \lambda^d)$.

Now we want to generalize $(1.b)$ in $\boxed{5}$ §2.1.  The proof of the following
lemma is essentially the same as that of $(1.b)$ in $\boxed{5}$ §2.1.

$\underline{\text{Lemma 4.3.}}$   Let $\rho_i$, $1 \leq i \leq d$, be locally $\lambda^d$-integrable functions on D,
strictly positive $\lambda^d$-a.e. on D, such that the form

$$E'(u,v) = \sum_{i=1}^{d} \int D_i u \, D_i v \, \rho_i \, d\lambda^d$$

$$\mathcal{D}(E') = C_o^\infty(D)$$

is closable on $L^2(D;m)$.  (For sufficient conditions, cf. Sections 2 and 3.)
Let $E$ be of the form (4.1) such that, for $\lambda^d$-almost every $x \in D$,

$$\sum_{i,j=1}^{d} a_{ij}(x)\xi_i\xi_j \geq \sum_{i=1}^{d} \rho_i(x) \, \xi_i^2 \quad \text{for every } \xi = (\xi_1,\ldots,\xi_d) \in \mathbb{R}^d.$$

Then $E$ is closable on $L^2(D;m)$.

$\underline{\text{Proof.}}$   Let $(u_n)_{n \in \mathbb{N}}$ be an $E$-Cauchy sequence in $C_o^\infty(D)$ such that
$\lim\limits_{n \to \infty} \|u_n\|_m = 0$.  Then, by assumption, $(u_n)_{n \in \mathbb{N}}$ is an $E'$-Cauchy sequence and
consequently $\lim\limits_{n \to \infty} E'(u_n,u_n) = 0$.  Therefore, there exists a subsequence
$(u_{n_k})_{k \in \mathbb{N}}$ of $(u_n)_{n \in \mathbb{N}}$ such that

$$\lim_{k \to \infty} D_i u_{n_k} = 0 \qquad \lambda^d\text{-a.e. on D}, \qquad 1 \leq i \leq d.$$

Since, by Fatou's lemma,

$$E(u_n,u_n) = \int \lim_{k \to \infty} \sum_{i,j=1}^{d} \left( D_i(u_n - u_{n_k}) \, D_j(u_n - u_{n_k}) \, a_{ij} \right) d\lambda^d$$

$$\leq \liminf_{k \to \infty} E(u_n - u_{n_k}, u_n - u_{n_k}),$$

132

the assertion follows.

For $A \subset D$ we denote by $\partial A$ and $\overset{o}{A}$ the topological boundary, interior respectively.

Now we are prepared to prove the generalization of $(1\overset{o}{.}b)$ in $[5]$ §2.1.

<u>Proposition 4.4.</u>   Let $E$ be of the form (4.1).  Assume that there exists an open subset U of D with $\lambda^d(D\setminus U) = 0$ such that $g \cdot m = \lambda^d$ for some locally $\lambda^d$-integrable function g on U and such that, for every compact subset K of U, there exists a constant $\delta_K > 0$ such that, for $\lambda^d$-almost every $x \in K$,

$$\sum_{i,j=1}^{d} a_{ij}(x)\, \xi_i \xi_j \geq \delta_K |\xi|^2 \quad \text{for every} \quad \xi = (\xi_1, \ldots, \xi_d) \in \mathbb{R}^d .$$

Then $E$ is closable on $L^2(D;m)$.

<u>Proof.</u>   Let $(K_n)_{n \in \mathbb{N}}$ be a sequence of compact subsets of U increasing to U such that $K_n \subset \overset{o}{K}_{n+1}$ and $\lambda^d(\partial K_n) = 0$ for every $n \in \mathbb{N}$.  Set $K_o := \emptyset$.  Given $n \in \mathbb{N}$, then, by assumption, there exists $\delta_n > 0$ such that, for $\lambda^d$-almost every $x \in K_n \setminus \overset{o}{K}_{n-1}$,

$$\sum_{i,j=1}^{d} a_{ij}(x)\, \xi_i \xi_j \geq \delta_n |\xi|^2 \quad \text{for every} \quad \xi = (\xi_1, \ldots, \xi_d) \in \mathbb{R}^d .$$

Let, for $n \in \mathbb{N}$, $B_n := \overset{o}{K}_n \setminus K_{n-1}$ and define $\rho : D \to \mathbb{R}$ by $\rho := \sum_{n=1}^{\infty} \delta_n 1_{B_n}$, where $1_{B_n}$ is the characteristic function of $B_n$ on D.  Then $\rho$ is lower semi-continuous and $\lambda^d(R(\rho)\setminus U) \leq \lambda^d(D\setminus U) = 0$.  Hence, by Corollary 3.5 the form

$$E(u,v) := \sum_{i=1}^{d} \int D_i u\, D_i v\, \rho\, d\lambda^d ,$$

$$D(E) \quad := C_o^{\infty}(D)$$

is closable on $L^2(D;m)$.  Let $x \in U \setminus \bigcup_{n=1}^{\infty} \partial K_n$.  Then there exists exactly one $n \in \mathbb{N}$ such that $x \in B_n \subset K_n \setminus \overset{o}{K}_{n-1}$.  Hence, for $\lambda^d$-almost every $x \in D$,

$$\sum_{i,j=1}^{d} a_{ij}(x)\xi_i \xi_j \geq \rho(x)\, |\xi|^2 \quad \text{for every} \quad \xi = (\xi_1, \ldots, \xi_d) \in \mathbb{R}^d .$$

133

Thus, since $\rho > 0$ $\lambda^d$-a.e. on D, the assertion follows by Lemma 4.3.

At the end of this section we want to treat briefly  a  simple case where
the killing measure, k, introduced in Section 1, is non-zero.

<u>Proposition 4.5.</u>    Let $m \in M^+(D)$ with supp[m]= D.  Let $E$ be a symmetric form
on $L^2(D;m)$ with domain $C_o^\infty(D)$.  Let c be a non-negative locally m-square
integrable function on D.  If $E$ is closable on $L^2(D;m)$, then the form

$$E_c(u,v) := E(u,v) + \int u\, v\, c\, dm$$

$$\mathcal{D}(E_c) := C_o^\infty(D)$$

is closable on $L^2(D;m)$.

<u>Proof.</u>    Let $(u_n)_{n\in\mathbb{N}}$ be an $E_c$-Cauchy sequence in $C_o^\infty(D)$ such that
$\lim_{n\to\infty} \| u_n \|_m = 0$.  Then there exists a subsequence $(u_{n_k})_{k\in\mathbb{N}}$ of $(u_n)_{n\in\mathbb{N}}$ such
that $\lim_{k\to\infty} u_{n_k} = 0$ m-a.e. .  Hence, by Fatou's lemma,

$$\int |u_n|^2 c\, dm \leq \liminf_{k \to \infty} \int |u_n - u_{n_k}|^2 c\, dm.$$

This implies $\lim_{n\to\infty} \int |u_n|^2 c\, dm = 0$.  Since $(u_n)_{n\in\mathbb{N}}$ is $E$-Cauchy and $E$ is
closable on $L^2(D;m)$, we have $\lim_{n\to\infty} E(u_n,u_n) = 0$.  It follows that
$\lim_{n\to\infty} E_c(u_n,u_n) = 0$.

## 5.  <u>PROPER SPEED MEASURES OF TRANSIENT DIRICHLET SPACES</u>

This section is related to §5.5 "Random Time Changes" in $\begin{bmatrix}5\end{bmatrix}$.  In a sense it
can be considered as the analytic analogue.

Let D be an open set in $\mathbb{R}^d$, $d \geq 1$, and $m \in M^+(D)$ with $\text{supp}\begin{bmatrix}m\end{bmatrix} = D$.  We
are interested in the question:

> Given a symmetric form $E$ on $L^2(D;m)$ with domain $C_o^\infty(D)$, which is
> closable.  What conditions on an arbitrary measure $\mu \in M^+(D)$        (5.1)
> with $\text{supp}\begin{bmatrix}\mu\end{bmatrix} = D$ imply, that $E$ is also closable on $L^2(D;\mu)$?

We can answer this question in the transient case.  First we need some
notation and preparations.

Let X be a locally compact, second countable space and $m \in M^+(X)$ such
that $\mathrm{supp}\,[m] = X$.  Let $(F,E)$ be a regular Dirichlet space on $L^2(X;m)$.  We
assume that $(F,E)$ is *transient*; i.e., (cf. [5] §1.5) there exists a
bounded m-integrable function g, strictly positive m-a.e. on X such that

$$\int |u|\,g\; dm \leq \sqrt{E(u,u)} \quad \text{for every } u \in F.$$

Denote by $(F_e,E)$ the associated *extended Dirichlet space with reference
measure* m  (cf. [5] Theorem 1.5.2).

In the sequel we shall make strict use of results of [5] Chapter III,
though they are stated there for $(F, E_1)$ instead of $(F_e,E)$.  But all
assertions we need here have their so-called 0-order versions (cf. the
remarks made at the end of each section in chapter III of [5]).

Given $A \subset X$, let $\mathrm{Cap}_o(A)$ denote the capacity of A as defined in [5] §3.1
(in particular p.67).  We recall that, by [5] Theorem 3.1.3, each element
$u \in F_e$ admits a quasi-continuous modification in the restricted sense
denoted by $\tilde{u}$ (see also [5] Lemma 3.1.4).  Denote by $\overset{\sim}{F}_e$ the set of all quasi-
continuous functions in the restricted sense belonging to $F_e$.  We define the
vector space of measures of bounded energy by

$$M_E := \{\mu \in M: \text{ there exists a constant } c > 0 \text{ such that}$$

$$\int |u|\,d|\mu| \leq c\,\sqrt{E(u,u)} \quad \text{for every } u \in F_e \cap C_o(X)\}$$

(where $|\mu| = \max(\mu,-\mu)$ in the lattice vector space M).  Let

$$M_E^+ := \{\mu \in M_E : \mu \geq 0\}.$$

Given $\mu \in M_E^+$, we denote by $U\mu$ the associated potential in $F_e$ given by Riesz's
representation theorem (cf. [5] §3.2) and we set

$$\|\mu\|_E := E(U\mu,U\mu)^{1/2}.$$

Let

$$M_p := \{\mu \in M^+(X) : \mu(N) = 0 \text{ for every Borel subset N of X such}$$

$$\text{that } \mathrm{Cap}_o(N) = 0\}.$$

Definition 5.1.   Let $\mu \in M^+(X)$.  Then $\mu$ is called a *proper speed measure of* $(F,E)$ if, for every Borel measurable $\tilde{v} \in \tilde{F}_e$,

$$\tilde{v} = 0 \ \mu\text{-a.e., iff } \tilde{v} = 0 \text{ quasi-everywhere,}$$

where "quasi-everywhere", abbreviated "q.e.", means: except for a set of capacity zero.  Let $M_o((F,E))$ denote the set of all proper speed measures of $(F,E)$.

By [5] Lemma 3.1.4, we know that given any locally m-integrable function $g$ strictly positive m-a.e. on X, then $g \cdot m$ is a proper speed measure of $(F,E)$.  Obviously, $M_o((F,E)) \subset M_p$.


Remark 5.2.   Let $\mu \in M_o((F,E))$.  Then

(i)  $\mathrm{supp}[\mu] = X$.

(ii) Let V be an open subset of X.  Then, for every Borel measurable $\tilde{v} \in \tilde{F}_e$,

$$\tilde{v} = 0 \ \mu\text{-a.e. on } V, \quad \text{iff } \tilde{v} = 0 \text{ q.e. on } V.$$

Proof.   (i) The assertion follows immediately by the regularity of $E$.

(ii) Let $\tilde{v} \in \tilde{F}_e$, $\tilde{v}$ Borel measurable.  Assume that $\mu(\{\tilde{v} \neq 0\} \cap V) = 0$.  We may assume that $\tilde{v}$ is bounded.  Let $K \subset \{\tilde{v} \neq 0\} \cap V$, K compact.  By [5] Lemma 1.4.2 (ii) there exists $u \in F \cap C_o(X)$ such that

$$u = 1 \ \text{ on } \ K,$$
$$\mathrm{supp} \ u \subset V.$$

Let $\tilde{w} := (u \wedge \tilde{v}) \cdot u$.  Then, by [5] Theorem 1.4.2 (ii), we have $w \in F$ and

$$\mu(\{\tilde{w} \neq 0\}) \leq \mu(\{\tilde{v} \neq 0\} \cap V\}) = 0.$$

Therefore, by assumption,

$$\mathrm{Cap}_o(K) \leq \mathrm{Cap}_o(\{\tilde{w} \neq 0\}) = 0.$$

Consequently, since $\mathrm{Cap}_o$ is a Choquet capacity,

$$\mathrm{Cap}_o(\{\tilde{v} \neq 0\} \cap V) = 0.$$

The converse assertion is trivial, since $\mu \in M_p$.

Now we want to characterize $M_o((F,E))$.  First a characterization of $M_p$:

$\underline{\text{Proposition 5.3.}}$ Let $\mu \in M^+(X)$. Then the following assertions are equivalent:

(i) $\mu \in M_p$.

(ii) There exists a bounded, $\mu$-integrable function g on X, strictly positive, such that $g \cdot \mu \in M_E^+$.

$\underline{\text{Proof.}}$ (ii) $\Rightarrow$ (i) : $[5]$ Lemma 3.2.2.

(i) $\Rightarrow$ (ii) : Assume (i). By $[5]$ Theorem 3.2.3 there exists an increasing sequence $(K_n)_{n \in \mathbb{N}}$ of compact sets such that, for $A := X \setminus \bigcup_{n=1}^{\infty} K_n$, we have $\mu(A) = 0$ and

$$1_{K_n} \cdot \mu \in M_E^+ \quad \text{for every} \quad n \in \mathbb{N}.$$

Define $g : X \to \mathbb{R}$ by

$$g := \sum_{n=1}^{\infty} \alpha_n \, 1_{K_n} + 1_A$$

with

$$\alpha_n := 2^{-n}(\mu(K_n) + \| 1_{K_n} \cdot \mu \|_E + 1)^{-1} \quad \text{for} \quad n \in \mathbb{N}.$$

Then $0 < g \leq 1$ on X and $\int g \, d\mu \leq 1$.

Set, for $N \in \mathbb{N}$,

$$\mu_N := \sum_{n=1}^{N} \alpha_n \, (1_{K_n} \cdot \mu).$$

Then $\mu_N \in M_E^+$ and $(\mu_N)_{N \in \mathbb{N}}$ vaguely converges to $g \cdot \mu$. Since

$$\| \mu_N \|_E \leq \sum_{n=1}^{N} \alpha_n \, \| 1_{K_n} \cdot \mu \|_E \leq 1 \quad \text{for every } N \in \mathbb{N},$$

we conclude by $[13]$ Lemma 3.4 that $g \cdot \mu \in M_E^+$.

We introduce the notion of a "strict potential" (which is well known in axiomatic potential theory). For the existence, see [11] 2.5.

<u>Definition 5.4.</u>  Let $\mu \in M_E^+$. The potential $p := U\mu \in F_e$ is called *strict* iff, for any $u,v \in F_e$ such that $u,v \geq 0$ m-a.e. and

$$E(p,u) = E(p,v),$$
$$E(U\nu,u) \leq E(U\nu,v) \quad \text{for every} \quad \nu \in M_E^+,$$

it follows that $u = v$ m-a.e.


Now we are prepared to characterize $M_o((F,E))$.


<u>Theorem 5.5.</u>  Let $\mu \in M^+(X)$. Then the following assertions are equivalent:

(i) $\mu \in M_o((F,E))$.

(ii) There exists a bounded, $\mu$-integrable function $g$ on $X$, strictly positive, such that $g \cdot \mu \in M_E^+$ (i.e., $\mu \in M_p$) and the potential $U(g \cdot \mu)$ is strict.


<u>Proof.</u>  (i) $\Rightarrow$ (ii) : Let $\mu \in M_o((F,E))$. Since $\mu \in M_p$, we can choose $g$ as in Proposition 5.3. It remains to be shown that $p := U(g \cdot \mu)$ is strict. Let $u,v \in F_e$, $u,v \geq 0$ m-a.e. such that

$$E(p,u) = E(p,v),$$
$$E(U\nu,u) \leq E(U\nu,v) \quad \text{for} \quad \text{every} \quad \nu \in M_E^+.$$

We conclude, by $\boxed{5}$ Theorem 3.2.2,

$$\int (\tilde{v}-\tilde{u})\,d\nu \geq 0 \quad \text{for every} \quad \nu \in M_E^+,$$

where $\tilde{v},\tilde{u}$ are quasi-continuous Borel versions of $v,u$ respectively.  Since $1_{\{\tilde{v}<\tilde{u}\}} \cdot \nu \in M_E^+$ for every $\nu \in M_E^+$, it follows that

$$0 \geq \int (\tilde{v}-\tilde{u})\,1_{\{\tilde{v}<\tilde{u}\}}\,d\nu \geq 0,$$

hence $\nu(\{\tilde{v}<\tilde{u}\}) = 0$ for every $\nu \in M_E^+$. Thus, by $\boxed{5}$ Theorem 3.3.2,

$$\tilde{v} \geq \tilde{u} \quad \text{q.e.} \quad \text{on X,}$$

hence

$$\tilde{v} \geq \tilde{u} \quad \mu\text{-a.e.} \quad \text{on X.}$$

138

Therefore $E(u,p) = E(v,p)$ and $[5]$ Theorem 3.2.2 imply that $\overset{\sim}{v} = \overset{\sim}{u}$ $\mu$-a.e. on X. Hence, by assumption,

$$\overset{\sim}{v} = \overset{\sim}{u} \quad \text{q.e.} \quad \text{on} \quad X,$$

consequently $v = u$ m-a.e. on X, and (ii) is shown.

(ii) $\Rightarrow$ (i) : Assume (ii). Let $\overset{\sim}{v} \in \overset{\sim}{F}_e$, $\overset{\sim}{v}$ Borel measurable. Since $\mu \in M_p$ by Proposition 5.3, it remains to be shown: if $\mu(\{\overset{\sim}{v} \neq 0\}) = 0$, then $Cap_0(\{\overset{\sim}{v} \neq 0\}) = 0$.

Let $\mu(\{\overset{\sim}{v} \neq 0\}) = 0$. We may assume, that $\overset{\sim}{v} \geq 0$ on X. Since $E(v,U\nu) \geq 0$ for every $\nu \in M_E^+$ (cf. $[5]$ Theorem 3.2.1) and $E(v,U(g\cdot\mu)) = \int \overset{\sim}{v}\, g\, d\mu = 0$, it follows by assumption, that $v = 0$ m-a.e..

Thus, by $[5]$ Lemma 3.1.4, $\overset{\sim}{v} = 0$ q.e. .


Before coming back to problem (5.1) we state two lemmas, which are useful to show that some Dirichlet space $(F,E)$ is transient. The proofs are obvious, so they are omitted.

<u>Lemma 5.6.</u>   Let $(F,E)$ be a Dirichlet space on $L^2(X;m)$. Let $D$ be a linear subspace of $F$, dense with respect to $E_1$. Assume that there exists a bounded, m-integrable function g on X, strictly positive m-a.e. on X such that

$$\int |u|\, g\, dm \leq E(u,u)^{1/2} \quad \text{for every} \quad u \in D.$$

Then $(F,E)$ is transient.

<u>Lemma 5.7.</u>   Let $(F,E)$ be a Dirichlet space on $L^2(X;m)$. Let $D \subset F$ and U be an open subset of X. Then the following assertions are equivalent:

(i) For every $K \subset U$, K compact, there exists a constant $\gamma_K > 0$ such that

$$\int_K |u|\, dm \leq \gamma_K\, E(u,u)^{1/2} \quad \text{for every} \quad u \in D.$$

(ii) There exists an m-integrable function g on X, strictly positive on U, such that $g|_U \in C_\infty(U)$ (where $C_\infty(U)$ denotes the set of all continuous functions on U vanishing at infinity) and

$$\int |u|\, g\, dm \leq E(u,u)^{1/2} \quad \text{for every} \quad u \in D.$$

139

Suppose now that X is an open set D in $\mathbb{R}^d$, $d \geq 1$. Let $m \in M^+(D)$ with supp$[m]$ = D. Let $E'$ be of the form (4.1). Let c be a non-negative locally m-square integrable function on D. Consider the form

$$E(u,v) := E'(u,v) + \int u \, v \, c \, dm.$$

$$\mathcal{D}(E) \quad := C_o^\infty(D).$$

$$(5.2)$$

<u>Proposition 5.8.</u>  Let $m = \lambda^d$ and let $E$ be of the form (5.2).  Assume that $E$ is closable on $L^2(D;\lambda^d)$ and that one of the following conditions is satisfied:

(i) There exists a constant $\gamma > 0$ such that for $\lambda^d$-almost every $x \in D$

$$\sum_{i,j=1}^{d} a_{ij}(x) \, \xi_i \xi_j \geq \gamma \, |\xi|^2 \quad \text{for every} \quad \xi = (\xi_1,\ldots,\xi_d) \in \mathbb{R}^d$$

and $\mathbb{R}^d \backslash D$ is not polar (in the classical sense), if $d \leq 2$.

(ii) There exists an open subset U of D with $\lambda^d(D\backslash U) = 0$ and $c^{-1}$ is locally $\lambda^d$-integrable on U.

Then $(\mathcal{D}(\bar{E}), \bar{E})$, the closure of $(\mathcal{D}(E),E)$ in $L^2(D;\lambda^d)$, is a regular, transient Dirichlet space.

<u>Proof.</u>  It remains to show the transience.

Assume that (i) is satisfied.  Since the classical Dirichlet space (denoted by $(\frac{1}{2} D, H_o^1(D))$ in $[5]$ Example 1.2.3) is transient, if $d \geq 3$ or, if $\mathbb{R}^d \backslash D$ is not polar (cf. $[10]$ 3.7 and $[8]$, 8.33), there exists a constant $\gamma_K > 0$ such that for every $u \in C_o^\infty(D)$

$$E(u,u)^{1/2} \geq \gamma^{1/2} \, ( \sum_{i=1}^{d} \int (D_i u)^2 \, d\lambda^d)^{1/2} \geq \gamma_K \int |u| \, d\lambda^d.$$

Assume that (ii) is satisfied.  Let $K \subset U$, K compact.  Then, by Hölder's inequality applied to the measure $(1_K \cdot c^{-1}) \cdot \lambda^d$, it follows that, for every $u \in C_o^\infty(D)$,

140

$$E(u,u)^{1/2} \geq \left( \int_K |u|^2 c^2 \frac{1}{c} \, d\lambda^d \right)^{1/2}$$

$$\geq \gamma_K^{-1} \int_K |u| \, d\lambda^d,$$

where $\gamma_K := \left( \int_K \frac{1}{c} \, d\lambda^d \right)^{1/2} + 1$. Hence the assertions follow by Lemmas 5.6 and 5.7.

Now we are prepared to solve problem (5.1) in the transient case.

__Theorem 5.9.__ Let $E$ be of the form (5.2). Assume that $(\mathcal{D}(E),E)$ is closable on $L^2(D;m)$ and that the closure $(F,\overline{E})$ in $L^2(D;m)$ is a transient Dirichlet space. (For sufficient conditions see Section 4 and Proposition 5.8.) Then the following assertions are equivalent:

(i) $\mu \in M_o(F,\overline{E}))$.

(ii) $(\mathcal{D}(E),E)$ is closable on $L^2(D;\mu)$ and its closure $(F^\mu,\overline{E}^\mu)$ in $L^2(D;\mu)$ is a transient Dirichlet space.

__Proof.__ Let $(F_e,\overline{E})$ be the extended Dirichlet space with reference measure m, which is associated to $(F, \overline{E})$.

(i) $\Rightarrow$ (ii): Let $\mu \in M_o((F,\overline{E}))$. By Theorem 5.5 there exists a bounded, $\mu$-integrable function g on X, strictly positive, such that $g \cdot \mu \in M_E^+$. Hence, by [5] Theorem 3.2.2,

$$\tilde{F}_e \subset L^1(X; g \cdot \mu)$$

$$\int \tilde{u} \, g \, d\mu \leq \|g \cdot \mu\|_E \, \overline{E}(u,u)^{1/2} \quad \text{for every } \tilde{u} \in \tilde{F}_e. \tag{5.3}$$

Let $(u_n)_{n \in \mathbb{N}}$ be an $E$-Cauchy sequence in $C_o^\infty(D)$ such that $\lim\limits_{n \to \infty} \|u_n\|_\mu = 0$. Let $u \in F_e$ such that $(u_n)_{n \in \mathbb{N}}$ converges to u in $(F_e,\overline{E})$. By (5.3) we obtain that $(u_n)_{n \in \mathbb{N}}$ converges to a quasi-continuous Borel version $\tilde{u}$ of u in $L^1(X, g \cdot \mu)$. Hence there exists a subsequence $(u_{n_k})_{k \in \mathbb{N}}$ of $(u_n)_{n \in \mathbb{N}}$ such that

$$\lim_{k \to \infty} u_{n_k} = \overset{\sim}{u} \quad g \cdot \mu\text{-a.e.}$$

$$\lim_{k \to \infty} u_{n_k} = 0 \quad \mu\text{-a.e.}$$

Consequently, $\overset{\sim}{u} = 0$, $\mu$-a.e., and thus, by assumption, $\overset{\sim}{u} = 0$, q.e. . This implies $\lim_{n \to \infty} E(u_n, u_n) = 0$, and therefore $(\mathcal{D}(E), E)$ is closable on $L^2(D; \mu)$.

It remains to show the transience. But this is an easy consequence of Lemma 5.6 and (5.3), since $\|g \cdot \mu\|_E > 0$ by Theorem 5.5. Thus (ii) is proved.

(ii) $\Rightarrow$ (i) : Assume (ii). Let $(F_e^\mu, \overline{E}^\mu)$ be the extended Dirichlet space with reference measure $\mu$, which is associated to $(F^\mu, \overline{E}^\mu)$. Problem 3.3.2 (p. 72) in $[5]$ implies that the notion of capacity with respect to $(F_e, \overline{E})$ is identical to the notion of capacity with respect to $(F_e^\mu, \overline{E}^\mu)$. Hence, by $[5]$ Theorem 3.1.4, it easily follows that

$$\overset{\sim}{F}_e = \{f : X \to \overline{\mathbb{R}} : \text{ there exists an } E\text{-Cauchy sequence } (u_n)_{n \in \mathbb{N}} \text{ in}$$
$$C_o^\infty(D) \text{ such that } \lim_{n \to \infty} u_n = f \quad \text{q.e.}\}$$

$$= \overset{\sim}{F}_e^\mu.$$

Consequently, by definition, $M_o((F, \overline{E})) = M_o((F^\mu, \overline{E}^\mu))$. Since clearly $\mu \in M_o((F^\mu, \overline{E}^\mu))$, assertion (i) is shown.

## Acknowledgement

We are grateful to Prof. Dr S. Albeverio, who suggested that we should write this paper. We thank him for his interest and helpful discussions.

REFERENCES

[1] Albeverio, S., Høegh-Krohn, R. and Streit, L. Energy forms, Hamiltonians, and distorted Brownian paths, *J. Math. Phys.*, _18_(5), 907-917 (1977).

[2] Brasche, J.F. *Störungen von Schrödingeroperatoren durch Masse*, Staatsexamensarbeit, Univ. Bielefeld (1983).

[3] Dynkin, E.B. *Markov Processes*, Vols. I and II, Berlin-Heidelberg-New York: Springer (1965).

[4] Fukushima, M. On the generation of Markov processes by symmetric forms; *Proc. 2nd Japan-USSR Symp. on Probability Theory*, Lecture Notes in Math., Vol. 330, 46-97, Berlin-Heidelberg-New York: Springer (1973).

[5] Fukushima, M. *Dirichlet Forms and Markov Processes*, Amsterdam-Oxford-New York: North Holland (1980).

[6] Gilbarg, D. and Trudinger, N.S. *Elliptic Partial Differential Equations of Second Order*, Berlin-Heidelberg-New York: Springer (1977).

[7] Hamza, M.M. *Détermination des Formes de Dirichlet sur* $\mathbb{R}^n$, Thèse 3e cycle, Orsay (1975).

[8] Helms, L.L. *Introduction to Potential Theory*, New York: Wiley-Interscience (1969).

[9] Ikeda, N. and Watanabe, S. The local structure of a class of diffusions and related problems, *Proc. 2nd Japan-USSR Symp. on Probability Theory*, Lecture Notes in Math., Vol.330, 124-169, Berlin-Heidelberg-New York: Springer (1973).

[10] Röckner, M. Self-adjoint harmonic spaces and Dirichlet forms. *Hiroshima Math.* **14**, 55-66 (1984).

[11] Röckner, M. Generalized Markov Fields and Dirichlet Forms, *Acta Appl. Math.* (to appear).

[12] Rullkötter, K. and Spönemann, U. *Dirichletformen und Diffusionsprozesse*, Diplomarbeit, Univ. Bielefeld (1983).

[13] Silverstein, M.L. *Symmetric Markov Processes*, Lecture Notes in Math., Vol. 426, Berlin-Heidelberg-New York: Springer (1974).

[14] Silverstein, M.L. *Boundary Theory for Symmetric Markov Processes*, Lecture Notes in Math., Vol. 516, Berlin-Heidelberg-New York: Springer (1976).

[15] Silverstein, M.L. On the closability of Dirichlet forms, *Z. Wahrsch. verw. Gebiete, 51*, 185-200 (1980).

[16]   Wielens, N. Eindeutigkeit von Dirichletformen und wesentliche
       Selbstadjungiertheit von Schrödingeroperatoren mit stark
       singulären Potentialen, Diplomarbeit, Univ. Bielefeld (1982)
       *J. Funct. Anal.* (to appear).

Michael Röckner,
Fakultät für Mathematik,
Universität Bielefeld,
Universitätsstr. 1,
4800 Bielefeld 1,
West Germany.

Norbert Wielens,
Fakultät für Mathematik,
Universität Bielefeld,
Universitätsstr. 1,
4800 Bielefeld 1,
West Germany.

A S SZNITMAN

# A fluctuation result for nonlinear diffusions

## 1. INTRODUCTION

Much work has been published on the subject of nonlinear diffusion; the authors include Braun-Hepp [1], Dawson [2], Dobrusin [3], McKean [6], Marchioro-Pulvirenti [7], Tanaka [11], Tanaka-Hitsuda [13].

Let us describe the problem. Let $Y^\ell(a,b)$ $\ell \in [0,m]$ be $m+1$ $\mathbb{R}^d$-valued vector fields on $\mathbb{R}^d \times \mathbb{R}^d$ uniformly bounded and Lipschitz, let $(B^\ell_t)$ $\ell \in [1,m]$ be $\ell$ real valued independent Brownian motions, and $X(0)$ be an $\mathbb{R}^d$-valued random variable independent of the $(B^\ell)$: it is possible (see McKean [6], Tanaka [11]) to prove the existence and uniqueness (trajectorial and in law) of the $(\overline{X}_t)_{t \geq 0}$ solution of

$$
\begin{cases}
d\overline{X}_t = \sum_{\ell=0}^{m} \int Y^\ell(\overline{X}_t,y) \, u_t(dy) \, dB^\ell_t \\[2mm]
\overline{X}_o = X(0), \quad u_t(dy) \text{ the law of } \overline{X}_t
\end{cases}
\tag{1.1}
$$

(with the convention $dB^o_t = dt$).

The process defined by (1.1) is interesting in that it describes the asymptotic individual behaviour of each fixed coordinate process on $C(\mathbb{R}_+, \mathbb{R}^d)^N$ (N going to infinity) when one studies the laws of the interacting diffusions

$$
\begin{cases}
dX^i_t = \sum_{\ell=0}^{m} \frac{1}{N} \sum_{j=1}^{N} Y^\ell(X^i_t,X^j_t) \, dB^{\ell,i}_t \quad \text{(for } \ell \neq 0 \; (B^{\ell,i}) \text{ are independent real} \\
\qquad\qquad\qquad\qquad\qquad\qquad\qquad \text{valued Brownian motions and} \\
\qquad\qquad\qquad\qquad\qquad\qquad\qquad dB^{o,i} = dt) \\
(X^i_o) \text{ is } u_N\text{-distributed.}
\end{cases}
\tag{1.2}
$$

where $u_N$ satisfies a condition of chaos that we describe below (for instance, one can take $u_N = u^{\otimes N}$).

Let us first give a definition: let E be a separable metric space, $v$ a distribution on E. A sequence $(v_N)$ of symmetric probabilities on $E^N$ is said to be v-chaotic if

$$\forall f_1 \ldots f_k \in C_b(E) \quad \lim_{N \to \infty} \; <v_N, \; f_1 \otimes \ldots \otimes f_k \otimes 1 \ldots \otimes 1> \; = \; \prod_1^k <v, f_i>. \qquad (1.3)$$

If M(E) denotes the probabilities on E, one can show (see Tanaka [12], Sznitman [10]) that (1.3) is equivalent to

$$Z_N = \frac{1}{N} \sum_i \varepsilon_{X^i} \text{ converges in law towards the constant } v \text{ when N goes} \qquad (1.4)$$

to infinity.

(The $X^i$ are the canonical coordinates on $E^N$, and $Z_N$ is a M(E) (= set of probabilities on E) valued random variable defined on the probability space $(E^N, v_N)$.)

One can describe the link between (1.1) and (1.2) in the following way (see McKean [6], Tanaka [11], Sznitman [10]): if the sequence $u_N$ in (1.2) is u-chaotic, then the laws $P_N$ of the $(X^i_{\cdot}) i \in [1,N]$ are $\overline{P}$-chaotic where $\overline{P}$ is the law of the process $(\overline{X}_{\cdot})$ defined by (1.1). This result, in view of (1.4), is a type of law of large numbers, and one then tries to investigate the behaviour of the fluctuations. One may try to study

$$\eta^N_t = \frac{1}{\sqrt{N}} \sum_i (\varepsilon_{X^i_t} - u_t) \text{ as a } S'\text{-valued process.} \qquad (1.5)$$

See, for instance, Tanaka-Hitsuda [13], and Dawson [2] in the case where the initial law $u_N$ is the invariant probability for the $(X^i) i \in [1,N]$.

Here we are going to study the fluctuation field

$$T^f_N = \frac{1}{\sqrt{N}} \sum_i (f(X^i_{\cdot}) - E_{\overline{P}}[f]) \qquad f \in C^1_b(C([0,1], \mathbb{R}^d))$$

(we work with time bounded by 1).

For the case m=d and $Y^\ell_i(.,.) = \delta_{i\ell}$, see Tanaka [11], Kusuoka-Tamura [5], Sznitman [10].

We consider the case of a complete dependence of the Y (a,b) on a,b. The technique of Sznitman [10] does not work in this case, but in fact the ideas of the proof we present here are very similar to ideas present in Tanaka [12].

Take independent copies $(B^{\ell,i}_{\cdot}), X^i(o), i \geq 1$ and define the independent

146

processes $\overline{X}^i$ $i \geq 1$ satisfying (1.1) (with $(B^{\ell,i}_{\cdot})$ and $X^i(0)$ instead of $(B^{\ell}_{\cdot})$ and $X(0)$). Construct, for each $N \geq 1$, $(X^i)$ $i \in [1,N]$ with $(B^{\ell,i}_{\cdot})$, $X^i$ $i \in [1,N]$. Then one has

$$T_N^f = \frac{1}{\sqrt{N}} \sum_i (f(X^i) - E_{\overline{P}}[f]) = \frac{1}{\sqrt{N}} \sum_i (f(\overline{X}^i) - E_{\overline{P}}[f]) + \frac{1}{\sqrt{N}} \sum_i (f(X^i) - f(\overline{X}^i)).$$

The behaviour of the first term of the right-hand side is described by a classical central limit theorem; in order to study the last term, we introduce U. On $W \times C([0,1],\mathbb{R}^d)$ (W is the space on which we construct $\overline{X}(\cdot)$ satisfying (1.1)), the solution of

$$\begin{cases} dU_t(w,x) = \sum_0^m \left[ \int_W \frac{\partial Y^{\ell}}{\partial a} (\overline{X}_s(w), \overline{X}_s(\overset{\sim}{w})) . U_s(w,x) \right. \\ \qquad\qquad + \frac{\partial Y^{\ell}}{\partial b} (\overline{X}_s(w), \overline{X}_s(\overset{\sim}{w})) . U_s(\overset{\sim}{w},x) \ dP(\overset{\sim}{w}) \\ \qquad\qquad \left. + Y^{\ell}(\overline{X}_s(w),x_s) \right] dB^{\ell}_s \\ U_o = 0 \end{cases} \qquad (1.7)$$

(for a more precise definition of U, see Lemma 2.1) and prove that

$$X^i = \overline{X}^i + \frac{1}{N} \sum_{j \neq i} (U(w^i,\overline{X}_s(w^j))) - \int_C U(w^i,x) d\overline{P}(x)) + o\left(\frac{1}{\sqrt{N}}\right) . \qquad (1.8)$$

This allows us then to show that

$$\{T_N^f, f \in C_{b,o}^1\} \quad (C_{b,o}^1 = \{f \in C_b^1(C(0,1;\ \mathbb{R}^d)), \ E_{\overline{P}}[f] = 0 \ )$$

converge in the finite distribution sense towards a centered gaussian field $\{T^f,\ f \in C_{b,o}^1\}$ with covariance

$$cov(T^f,T^g) = E_{\overline{P}}\left[(I+\overline{\Pi})f \ (I+\overline{\Pi})g\right] \qquad (1.9)$$

where $\overline{\Pi}f(x) = \int f'(\overline{X}_{\cdot}(w_1)).(U_{\cdot}(w_1,x) - \int U_{\cdot}(w_1,x) d\overline{P}(x)) \ dP(w_1)$.


## 2. THE FLUCTUATION RESULT

Consider $(W,F,(F_t)_{0 \leq t \leq 1},\ (B^{\ell}_t)_{0 \leq t \leq 1},\ \ell \in [1,m],P)$ a complete probability space endowed with a filtration $(F_t)$ such that each $F_t$ contains the P-negligible sets of $F$, $F_t = \bigcap_{u>t} F_u$, and with $m$ real-valued $F_t$-Brownian motions

$(B_t^\ell)_{0\le t\le 1}$, $1 \le \ell \le m$.

We suppose we are given $(m+1)$ $C_b^2$ $\mathbb{R}^d$-valued vector fields on $\mathbb{R}^d \times \mathbb{R}^d$: $Y^\ell(a,b)$, $\ell \in [0,m]$, and an $F_o$-measurable $\mathbb{R}^d$-valued $X(0)$ with distribution $u$.

We define $(\bar{X}_t)_{0\le t\le 1}$ on $W$, solution of (1.1). On $(\Omega,Q) = (W^{\mathbb{N}^*}, P^{\otimes\mathbb{N}^*})$ we consider for $N \ge 1$ the processes $(X_t^{i,N})_{0\le t\le 1}$, $1 \le i \le N$, satisfying

$$dX_t^{i,N} = \sum_0^m \frac{1}{N} \sum_{j=1}^N Y^\ell(X_t^{i,N}, X_t^{j,N}) dB_t^{\ell,i} \tag{2.1}$$

$$X_o^{i,N} = X^i(0)$$

(with the obvious notation $B^{\ell,i}(\omega) = B^\ell(w_i)$, $X^i(0) = X(w^i)(0)$ if $\omega = (w_j)_{j\ge 1}$).

We are going to study the fluctuations

$$T_N^f(\omega) = \frac{1}{\sqrt{N}} \sum_1^N (f(X_\cdot^{i,N}(\omega)) - E_{\bar{P}}[f]) \tag{2.2}$$

for $f$ a smooth $(C_b^1(C([0,1], \mathbb{R}^d))$ function).

<u>Notation</u>    $C$ will be the space $C([0,1],\mathbb{R}^d)$, for a continuous process $(H_s)_{0\le s\le 1}$ we set $H_v^* = \sup_{s\le v} |H_s|$, $H$ will be the Banach space of continuous adapted process $(X_u)_{0\le u\le 1}$ such that $E\left[\sup_{0\le u\le 1} X_u^2\right]^{1/2} = \|X\|_H < \infty$, and, in what follows, $K$ will denote a constant independent of time and of $N$ varying from place to place.

Our goal is now to prove that $\{T_N^f, f \in C_b^1(C)\}$ converges in the finite distribution sense towards a gaussian random field. First, we define on $(W \times C, P \otimes \bar{P})$ endowed with the product filtration, the process $U_t(w,x)$ $0 \le t \le 1$ satisfying

$$dU_t(w,x) = \sum_{\ell=0}^m \left[\left[\int_W \frac{\partial Y^\ell}{\partial a} (\bar{X}_s(w), \bar{X}_s(\tilde{w})).U_s(w,x)\right.\right.$$

$$+ \frac{\partial Y^\ell}{\partial b} \, (\overline{X}_s(w), \, \overline{X}_s(\tilde{w})) . U_s(\tilde{w},x) \, dP(\tilde{w}) \qquad (2.3)$$

$$+ \, Y^\ell(\overline{X}_s(w),x_s) \Big] \, dB^\ell_s$$

$$U_o(w,x) \;\; = \; 0.$$

(It is easy to check the existence and uniqueness of $(U_.)$ in $H(W \times C)$ with the usual iteration technique.)

We are going to choose a "nice" version of $U$; more precisely, we have:

<u>Lemma 2.1.</u>   It is possible to choose a version of $U_.$ such that:

(a) for every $x \in C$, $U_.(w,x)$ is a solution on $(W,P)$ of

$$dU_t(w,x) = \sum_0^m \Bigg[ \int \frac{\partial Y^\ell}{\partial a} \, (\overline{X}_s(w), \, \overline{X}_s(\tilde{w})) . U_s(w,x)$$

$$+ \frac{\partial Y^\ell}{\partial b} \, (\overline{X}_s(w), \, \overline{X}_s(\tilde{w})) . U_s(\tilde{w},x) \, dP(\tilde{w}) \qquad (2.4)$$

$$+ \, Y^\ell(\overline{X}_s(w), \, x_s) \Bigg] \, dB^\ell_s .$$

$$U_o = 0.$$

(b) If $(\overline{X}_.^{\epsilon,x}) \; x \in C$ denotes the solution on $(W,P)$ of $(1.1)$ with

$$Y_s^{\ell,\epsilon}(a,b) = Y^\ell(a,b) + \epsilon Y^\ell(a,x_s) \text{ on } (W,P), \text{ then:}$$

$$U_t(w,x) = \lim_{\epsilon \downarrow 0} \frac{1}{\epsilon} \, (\overline{X}_.^{\epsilon,x} - \overline{X}_.) \text{ in } H(w), \quad \forall x \in C. \qquad (2.5)$$

(c) $Z_t(w_1,w_2) = U_t(w_1,\overline{X}_s(w_2))$ is the solution on $(W \times W, \, P \otimes P)$ of

$$dZ_t(w_1,w_2) = \sum_0^m \Bigg[ \int_W \frac{\partial Y^\ell}{\partial a} \, (\overline{X}_s(w_1), \, \overline{X}_s(\tilde{w})) . Z_s(w_1,w_2)$$

$$+ \frac{\partial Y^\ell}{\partial b} \, (\overline{X}_s(w_1), \, \overline{X}_s(\tilde{w})) . Z_s(\tilde{w},w_2) \, dP(\tilde{w}) \qquad (2.6)$$

$$+ \, Y^\ell(\overline{X}_s(w_1), \, \overline{X}_s(w_2)) \Bigg] \, dB^\ell_s$$

$$Z_o = 0.$$

$\underline{\text{Proof.}}$  (a) Choose a version of $(U_.)$ satisfying (2.3) continuous in $t$ and measurable in $(w,x)$.

Using the results of Doleans-Dade [4], it is possible to choose a right continuous $H_t(w,x)$, such that $H_t(w,x)$ is a version of

$$\sum_0^m \int_0^t \left[ \frac{\partial Y^\ell}{\partial a}(\overline{X}_s(w), \overline{X}_s(\tilde{w})) \cdot U_s(w,x) + \frac{\partial Y^\ell}{\partial b}(\overline{X}_s(w), \overline{X}_s(\tilde{w})) \cdot U_s(\tilde{w},x) \, dP(\tilde{w}) \right.$$
$$\left. + Y^\ell(\overline{X}_s(w), x_s) \right] dB_s^\ell$$

both on $(W \times C, P \otimes \overline{P})$ and on $(W,P)$ for each $x \in C$.  As a consequence, there exists $N$ with $\overline{P}(N) = 0$ such that, for $x \notin N$, $H_t(w,x)$ and $U_t(w,x)$ are indistinguishable on $(W,P)$, so for $x \notin N$, $U_t(w,x)$ is a solution of (2.4). Now redefining $U_t(w,x)$ for $x \in N$ by a continuous solution of (2.4), we have built a version satisfying (a).

(b) Define $R^\varepsilon = \frac{1}{\varepsilon}(\overline{X}^{\varepsilon,x} - \overline{X}) - U_.(w,x)$ for $x \in C$   ($R^\varepsilon$ is an element of $H(W)$). We have

$$R_t^\varepsilon(w) = \sum_{\ell=0}^m \int_0^t \left[ \int_W \frac{1}{\varepsilon}(Y^\ell(\overline{X}_s^{\varepsilon,x}(w), \overline{X}_s^{\varepsilon,x}(\tilde{w})) - Y^\ell(\overline{X}_s(w), \overline{X}_s(\tilde{w}))) \right.$$
$$- \frac{\partial Y^\ell}{\partial a}(\overline{X}_s(w), \overline{X}_s(\tilde{w})) \cdot U_s(w,x) + \frac{\partial Y^\ell}{\partial b}(\overline{X}_s(w), \overline{X}_s(\tilde{w})) U_s(\tilde{w},x) \, dP(\tilde{w})$$
$$\left. + (Y^\ell(\overline{X}_s^{,x}(w), x_s) - Y^\ell(\overline{X}_s, x_s)) \right] dB_s^\ell(w).$$

By the $C_b^2$ assumption on $Y^\ell(.,.)$ we can write

$$Y^\ell(a',b') - Y^\ell(a,b) - \frac{\partial Y^\ell}{\partial a}(a,b) \cdot (a'-a) - \frac{\partial Y^\ell}{\partial b}(a,b)(b'-b) =$$
$$= R(a,b,a',b') \quad (|a-a'| + |b-b'|) \quad (2.7)$$

with $|R(a,b,a',b')| \leq K(1 \wedge (|a-a'| + |b-b'|))$.

So we have

$$E\left[(R^\varepsilon)_t^{*2}\right] \leq K \int_0^1 \left[ E\left[ R_s^{\varepsilon 2} + E_{P \otimes P}\left[ |\overline{X}_s^{\varepsilon,x}(w) - \overline{X}_s(w)|^2 \times \right. \right. \right.$$
$$\quad (2.8)$$
$$\left. \left. \left. (1 + \frac{1}{\varepsilon^2}(1 \wedge |\overline{X}_s^{\varepsilon,x}(w) - \overline{X}_s(w)|^2) + \frac{1}{\varepsilon^2}(1 \wedge |\overline{X}_s^{\varepsilon,x}(\tilde{w}) - \overline{X}_s(\tilde{w})|^2) \right] \right] \right] ds.$$

From Gronwall's lemma, it is easy to check that $E\left[(X^{\varepsilon,x}-X)_1^{*4}\right] \leq K\varepsilon^4$, then Cauchy Schwarz's inequality and Gronwall's lemma applied in (2.8) allow us

150

to conclude that $\| R^{\varepsilon} \|_{H(W)} = 0(\varepsilon)$, which proves (b).

(c) follows obviously from the following easy result: if
$(w^i, F^i, (F^i_t), P^i)$, $i = 1,2$, are two filtered probability spaces, $\phi : w^1 \to w^2$
is such that $\phi \, P^1 = P^2$, $\phi^{-1}(F^2_t) \subseteq F^1_t$, if $B^i_t$ are $(F^i_t)$-Brownian motions on $W^i$

such that $B^2 \, \phi = B^1$, $f(s,w_2)$ is predictable and $E^2\left[\int_0^1 f^2(s,w_2) \, ds\right] < \infty$. If

$\int_0^t f(s,w_1) dB^2_s \equiv G(v,w^2)$, $G(t,\phi(w_1))$ is a version of $\int_0^t f(s,\phi(w_1)) dB'_s$.

Define now on $(\Omega,Q)$, for $i,j \geq 1$ distinct,

$$Z^{i,j}_t(\omega) = Z_t(w_i,w_j) - \int U_t(w^i,x) \, d\overline{P}(x) \tag{2.9}$$

$$Z^{i,N}_t = \frac{1}{N} \sum_{\substack{j \neq i \\ j=1}}^{N} Z^{i,j}_t . \tag{2.10}$$

<u>Notation.</u>  In the following, we will omit the superscript N and usually
write $X^i(.)$, $Z^i(.)$ for $X^{i,N}$, $Z^{i,N}$; we will denote by $\overline{X}^i(\omega)$ the process
$\overline{X}(w^i)$.

Our next goal is to try to prove that $X^i = \overline{X}^i + Z^i + o\left(\frac{1}{\sqrt{N}}\right)$.  We have:

<u>Proposition 2.2.</u>    (a)

$$\text{for } i \geq 1, \quad \sup_N N^2 \, E\left[(X^i - \overline{X}^i)^{*4}_1\right] < \infty \tag{2.11}$$

$$\sup_N N \, E\left[\sum_1^N (X^i - \overline{X}^i)^{*4}_1\right] < \infty$$

(b)  If  $R^i = X^i - \overline{X}^i - Z^i$, $i \geq 1$,

$$N \| R^i \|^2_{H(\Omega)} = N \, \| R^1 \|^2_{H(\Omega)} \leq \frac{K}{\sqrt{N}} .$$

<u>Proof.</u>    (a) take $i = 1$,

$$X_t^i - \bar{X}_t^i = \sum_{\ell=0}^{m} \int_0^t \frac{1}{N} \sum_{j=1}^{N} \left[ Y^\ell(X_s^1, X_s^j) - Y^\ell(\bar{X}_s^1, X_s^j) + Y^\ell(\bar{X}_s^1 - X_s^j) - Y(\bar{X}_s^1, \bar{X}_s^j) \right.$$

$$\left. + Y^\ell(\bar{X}_s^1, \bar{X}_s^j) - \int_W Y^\ell(\bar{X}_s^1, \bar{X}_s(\tilde{w})) \, dP(\tilde{w}) \right] dB_s^{\ell,1}.$$

So we have

$$E\left[ (X^1 - \bar{X}^1)_t^{*4} \right] \leq K \int_0^t E\left| (X^1 - \bar{X}^1)_s^{*4} \right| + \frac{1}{N} \sum_{j=1}^{N} E\left[ (X^j - \bar{X}^j)_s^{*4} \right] ds \tag{2.13}$$

$$+ K \cdot \sum_\ell \int_0^t E\left[ \left| \frac{1}{N} \sum_j Y_s^\ell(\bar{X}_s^1, \bar{X}_s^j) \right|^{4*} ds \right]$$

with

$$Y_s^\ell(a,b) = Y^\ell(a,b) - \int_W Y^\ell(a, \bar{X}_s(w)) \, dP(w). \tag{2.14}$$

Writing inequalities similar to (2.13) for $1 \leq i \leq N$, and summing them, we find

$$N \, E\left[ \sum_{i=1}^{N} (X^i - \bar{X}^i)_t^{*4} \right] \leq K \int_0^t N \, E\left[ \sum_{1}^{N} (X^i - \bar{X}^i)_s^{*4} \right] ds$$

$$+ K \sum_\ell E\left[ \frac{1}{N^3} \int_0^t \sum_{i_1,i_2,i_3,i_4,i_5} Y_s^\ell(\bar{X}_s^{i_1}, \bar{X}_s^{i_2}) \cdot Y_s^\ell(\bar{X}_s^{i_1}, \bar{X}_s^{i_3}) \times \right.$$

$$\left. Y_s^\ell(\bar{X}_s^{i_1}, \bar{X}_s^{i_4}) \cdot Y_s^\ell(\bar{X}_s^{i_1}, \bar{X}_s^{i_5}) ds \right]$$

but whenever $\{\text{card } i_1, i_2, i_3, i_4, i_5\} \geq 4$, the term

$$E\left[ (Y_s^\ell(\bar{X}_s^{i_1}, \bar{X}_s^{i_2}) \cdot Y_s^\ell(\bar{X}_s^{i_1}, \bar{X}_s^{i_3})) \times (Y_s^\ell(\bar{X}_s^{i_1}, \bar{X}_s^{i_4}) \cdot Y_s^\ell(\bar{X}_s^{i_1}, \bar{X}_s^{i_5})) \right] = 0.$$

This implies that the last term in the previous inequality is bounded independently of N, and, using Gronwall's lemma, we can conclude that

$$\sup_N N \cdot E\left[ \sum_{j=1}^{N} (X^i - \bar{X}^i)_1^{*4} \right] < \infty .$$

Now with the help of (2.13) and using the same method, we see that

$$\sup_N N^2 \, E\left[ (X^i - \bar{X}^i)_s^{*4} \right] < \infty.$$

This proves (a).

(b) Take $i = 1$, we have

$$R_t^1 = \sum_{\ell=0}^{m} \int_0^t \Biggl[ \int_W Y^\ell(X_s^1, \bar{X}_s(\tilde{w})) - Y^\ell(\bar{X}_s^1, \bar{X}_s(\tilde{w})) - \frac{\partial Y^\ell}{\partial a}(\bar{X}_s^1, \bar{X}_s(\tilde{w})) . Z^1(\omega) \, dP(\tilde{w})$$

$$+ \frac{1}{N} \sum_j Y^\ell(\bar{X}_s^1, X_s^j) - Y^\ell(\bar{X}_s^1, \bar{X}_s^j) - \frac{\partial Y^\ell}{\partial b}(\bar{X}_s^1, \bar{X}_s^j) . Z_s^j$$

$$+ \frac{1}{N} \sum_j Y^\ell(X_s^1, X_s^j) - Y^\ell(X_s^1, \bar{X}_s^j) - Y^\ell(\bar{X}_s^1, X_s^j) + Y^\ell(\bar{X}_s^1, \bar{X}_s^j)$$

$$+ \frac{1}{N} \sum_j Y^\ell(X_s^1, \bar{X}_s^j) - Y^\ell(\bar{X}_s^1, \bar{X}_s^j) - \int_W (Y^\ell(X_s^1, \bar{X}_s(\tilde{w})) - Y^\ell(\bar{X}_s^1, \bar{X}_s(\tilde{w})) ) \, dP(\tilde{w})$$

$$+ \frac{1}{N} \sum_j \frac{\partial Y^\ell}{\partial b}(\bar{X}_s^1, \bar{X}_s^j) . Z_s^j - \int_W \frac{\partial Y^\ell}{\partial b}(\bar{X}_s^1, \bar{X}(\tilde{w})) . Z^1(\tilde{w}, \omega_1) \, dP(\tilde{w}) \Biggr] dB_s^{\ell,1}$$

where we write $\omega = (\tilde{w}, \omega_1) \in \Omega = W^{IN^*}$. Because of (2.7), we find

$$E\left[N(R^1)_t^{*2}\right] \le K \int_0^t N \, E\left[(R^1)_s^{*2}\right] + \sum_{j=1}^{N} E\left[(R^j)_s^{*2}\right] + N \, E\left[|X^1 - \bar{X}^1|_s^{*4}\right]$$

$$+ E\left[\sum_{j=1}^{N} (X^j - \bar{X}^j)_s^{*4}\right] ds + M_t^N, \tag{2.15}$$

where

$$M_t^N = K \, E\Biggl[ \sum_0^m \int_0^t \frac{1}{N} (\sum_j Y^\ell(X_s^1, X_s^j) - Y^\ell(X_s^1, \bar{X}_s^j) - Y^\ell(\bar{X}_s^1, X_s^j) + Y^\ell(\bar{X}_s^1, \bar{X}_s^j))^2$$

$$+ \frac{1}{N} (\sum_j Y_s^\ell(X_s^1, \bar{X}_s^j) - Y_s^\ell(\bar{X}_s^1, \bar{X}_s^j))^2$$

$$+ \frac{1}{N} (\sum_j \frac{\partial Y^\ell}{\partial b}(\bar{X}_s^1, \bar{X}_s^j) . Z_s^j - \int_W \frac{\partial Y^\ell}{\partial b}(\bar{X}_s^1, \bar{X}(\tilde{w})) . Z_s^1(\tilde{w}, \omega_1) \, dP(\tilde{w}))^2 ds \Biggr]$$

$$= I_t^1 + I_t^2 + I_t^3$$

with the notation

$$Y_s^\ell(a,b) = Y^\ell(a,b) - \int_W Y^\ell(a, \bar{X}_s(w)) \, dP(w).$$

__Bound for__ $I_t^1$ :

$$\frac{1}{N}\left[\sum_j Y^\ell(X_s^1,X_s^j) - Y^\ell(X_s^1,\bar{X}_s^j) - Y^\ell(\bar{X}_s^1,X_s^j) + Y^\ell(\bar{X}_s^1,\bar{X}_s^j)\right]^2 =$$

$$\frac{1}{N}\left[\sum_j \int_0^1 (\frac{\partial Y^\ell}{\partial a}(\bar{X}_s^1+v(X_s^1-\bar{X}_s^1),X_s^j) - \frac{\partial Y^\ell}{\partial a}(\bar{X}_s^1+v(X_s^1-\bar{X}_s^1), \bar{X}_s^j)).(X_s^1-\bar{X}_s^1)dv\right]^2$$

$$\leq K(\sum_j |X_s^j-\bar{X}_s^j|^2).|X_s^1-\bar{X}_s^1|^2$$

so $I_t^1 \leq \frac{K}{N}$ (using Cauchy Schwarz's inequality and (a)).

__Bound for__ $I_t^2$ : We have

$$E\left[\frac{1}{N}(\sum_j Y_s^\ell(X_s^1,\bar{X}_s^j) - Y_s^\ell(\bar{X}_s^1,\bar{X}_s^j))^2\right] \leq$$

$$K.E\left[|X_s^1-\bar{X}_s^1|^2\right] + \frac{1}{N} E\left[(\sum_{j>1} Y_s^\ell(X_s^1,\bar{X}_s^j) - Y_s^\ell(\bar{X}_s^1,\bar{X}_s^j))^2\right] \leq$$

$$\frac{K}{N} + \frac{1}{N} E\left[\sum_{\substack{j\neq k \\ >1}} (Y_s^\ell(X_s^1,\bar{X}_s^j) - Y_s^\ell(\bar{X}_s^1,\bar{X}_s^j)).(Y_s^\ell(X_s^1,\bar{X}_s^k) - Y_s^\ell(\bar{X}_s^1,\bar{X}_s^k))\right].$$

Now fix $j \neq k > 1$; in order to study this last term, introduce the processes $(\tilde{X}^i)$, $1 \leq i \leq N$, $i \neq j$, $i \neq k$, solving

$$\begin{cases} d\,\tilde{X}_t^i = \frac{1}{N} \sum_\ell \sum_{\substack{i'\neq j,1'\geq 1 \\ i'\neq k}}^N Y^\ell(\tilde{X}_t^i,\tilde{X}_t^{i'})dB_t^{\ell,i} \\[2ex] \tilde{X}_o^i = X^i(0). \end{cases} \tag{2.16}$$

From this we deduce that for, $i \neq j,k$,

$$E\left[|X^i-\tilde{X}^i|_t^{*2}\right] \leq K \int_0^t \frac{1}{N} \sum_{\substack{i'\neq j \\ i'\neq k}} E\left[(X^{i'}-\tilde{X}^{i'})_s^{*2}\right]ds + \frac{K}{N^2} \tag{2.17}$$

and consequently

$$\frac{1}{N} \sum_{\substack{i'\neq j \\ i'\neq k}}^N E\left[(X^{i'}-\tilde{X}^{i'})_t^{*2}\right] \leq K \int_0^t \frac{1}{N} \sum_{i'\neq j,k} E\left[(X^i-\tilde{X}^i)_s^{*2}\right]ds + \frac{K}{N^2} .$$

Using Gronwall's lemma, this implies that

$$\frac{1}{N} \sum_{i' \neq j,k} E\left[(X^{i'} - \overset{\sim}{X}{}^{i'})^{*2}_t\right] \leq \frac{K}{N^2}$$

and, with (2.17), we get

$$E\left[(X^i - \overset{\sim}{X}{}^i)^{*2}_1\right] \leq \frac{K}{N^2} \ , \quad i \neq j,k. \tag{2.18}$$

We have

$$J_s = E\left[(Y^\ell_s(X^1_s, \overline{X}^j_s) - Y^\ell_s(\overline{X}^1_s, \overline{X}^j_s)) \cdot (Y^\ell_s(X^1_s, \overline{X}^k_s) - Y^\ell_s(\overline{X}^1_s, \overline{X}^k_s))\right]$$

$$= E\left[(Y^\ell_s(X^1_s, \overline{X}^j_s) - Y^\ell_s(\overset{\sim}{X}{}^1_s, \overline{X}^j_s) + A^{\ell,j}) \cdot (Y^\ell_s(X^1_s, \overline{X}^k_s) - Y^\ell_s(\overset{\sim}{X}{}^1_s, \overline{X}^k_s) + A^{\ell,k})\right]$$

where $A^{\ell,j} = Y^\ell_s(\overset{\sim}{X}{}^1_s, \overline{X}^j_s) - Y^\ell_s(\overline{X}^1_s, \overline{X}^j_s)$ and $A^{\ell,k}$ has a similar definition. Using

Cauchy-Schwarz's inequality and the fact that $N \, E\left[(\overset{\sim}{X}{}^1_s - \overline{X}^1_s)^{*2}\right] \leq K$, we find

$$|J_s| \leq \frac{K}{N^{3/2}} + \left|E\left[A^{\ell,j} A^{\ell,k}\right]\right| \ .$$

Now $\overset{\sim}{X}{}^1$ and $\overline{X}^1$ are independent of $(\overline{X}^j, \overline{X}^k)$ by construction and, because of the
definition of $Y^\ell_s$, we find

$$E\left[A^{\ell,j} A^{\ell,k}\right] = 0.$$

As a consequence we have $\quad I^2_t \leq \frac{K}{\sqrt{N}}$ .


Bound for $\quad I^3_t$ : We have

$$\frac{1}{N}(\sum_j \frac{\partial Y^\ell}{\partial b}(\overline{X}^1_s, \overline{X}^j_s) . Z^j_s - \int_W \frac{\partial Y^\ell}{\partial b}(\overline{X}^1_s, \overline{X}(\overset{\sim}{w})) . Z^1(\overset{\sim}{w}, \omega_1) dP(\overset{\sim}{w}))^2$$

$$= \frac{1}{N^3}( \sum_{i \neq j} \frac{\partial Y^\ell}{\partial b}(\overline{X}^1_s, \overline{X}^j_s) . Z_s(w_i, w_j) - \int_W \frac{\partial Y^\ell}{\partial b}(\overline{X}^1_s, \overline{X}_s(\overset{\sim}{w})) . Z_s(\overset{\sim}{w}, w_j) \, dP(\overset{\sim}{w}))^2 .$$


If we define on $W^3$

$$F_s^\ell(w_1,w_2,w_3) = \frac{\partial Y^\ell}{\partial b}(\overline{X}_s(w_1),\overline{X}_s(w_2)).Z_s(w_2,w_3)$$

$$- \int \frac{\partial Y^\ell}{\partial b}(\overline{X}_s(w_1),\overline{X}_s(w_2)).Z_s(w_2,w_3)dP(w_2)$$

we see that

$$\int F_s^\ell(w_1,w_2,w_3)dP(w_2) = 0 \qquad\qquad (2.19)$$

$$\int F_s^\ell(w_1,w_2,w_3)dP(w_3) = 0$$

and
$$I_t^3 \leq \frac{K}{N} + K \sum_\ell \int_0^t E\left[\frac{1}{N^3}\left(\sum_{\substack{i,j>1\\i\neq j}} F_s(w_1,w_i,w_j)\right)^2\right]ds;$$

but
$$\frac{1}{N^3} E\left[\left(\sum_{\substack{i,j>1\\i\neq j}} F_s(w_1,w_i,w_j)\right)^2\right]$$

$$\leq \frac{1}{N^3} \cdot \sum_{\substack{i\neq j\\i,j>1}} E\left[F_s(w_1,w_i,w_j)^2 + F_s(w_1,w_i,w_j).F_s(w_1,w_j,w_i)\right] \leq \frac{K}{N}.$$

So we have $I^3(t) \leq K/N$.

Now going back to (2.15), we see that

$$E\left[N(R^1)_t^{*2}\right] \leq K \int_0^t N\, E\left[(R^1)_s^{*2}\right] + \sum_1^N E\left[(R^j)_s^{*2}\right]ds + \frac{1}{\sqrt{N}}$$

which implies, when we write similar inequalities for $i \in [\overline{1,N}]$, that

$$E\left[\sum_1^N (R^j)_t^{*2}\right] \leq K \int_0^t E\left[\sum_1^N (R^j)_s^{*2}\right]ds + \frac{K}{\sqrt{N}}$$

so, by Gronwall's lemma,

$$E\left[\sum_1^N (R^j)_t^{*2}\right] \leq K/\sqrt{N}.$$

Now with (2.19) we obtain $E\left[N(R^1)_t^{*2}\right] \leq \frac{K}{\sqrt{N}}$ which proves the proposition.

We now introduce an operator $\Pi$ which maps $C_b^1(C)$ into $L^2(C,\overline{P})$, by setting

$$(\Pi f)(x) = \int_W f'(\overline{X}_.(w)).U(w,x)dP(w) \tag{2.21}$$

(see Lemma 2.1 for the definition of $U_.$) and we define

$$\overline{\Pi}f = \Pi f - E_{\overline{P}}[\Pi f]. \tag{2.22}$$

Let $C_{b,o}^1 = \{f \in C_b^1(C)/E_{\overline{P}}[f] = 0\}$; we can construct on some probability space (see Neveu [8]) a centered gaussian random field $\{T^f, f \in C_{b,o}^1\}$ with covariance

$$\mathrm{cov}(T^f,T^g) = E_{\overline{P}}[(I+\overline{\Pi})f.(I+\overline{\Pi})g]. \quad \text{We have}$$

__Theorem 2.3.__  The system $T_N^f = \dfrac{1}{\sqrt{N}} \sum_1^N f(X^i(\omega))$, $f \in C_{b,o}^1$ (on $(\Omega,Q)$) converges in the finite distribution sense towards the centered gaussian random field with covariance (2.23).

__Proof.__  Using characteristic functions, it is clearly sufficient to show that, for $f \in C_{b,o}^1(C)$, $T_N^f$ converges in law towards $N(0,E_{\overline{P}}[((I+\overline{\Pi})f)^2])$.  Now

$$T_N^f = \frac{1}{\sqrt{N}} \times \sum_1^N f(X^i(\omega)) = \frac{1}{\sqrt{N}} \sum_1^N (f(X^i(\omega)) - f(\overline{\overline{X}}(w^i)))$$

$$+ \frac{1}{\sqrt{N}} \sum_1^N f(\overline{X}(w^i)).$$

Since $f \in C_{b,o}^1$,

$$f(w+h) = f(w) + f'(w).h + R(w,h)\,\|h\| \tag{2.24}$$

$$\text{with } \|R(w,h)\| \leq K \text{ and } \lim_{h\downarrow 0} R(w,h) = 0.$$

By Proposition 2.2 $X^i = \overline{X}^i + Z^i + R^i$ with

$$E\left[N(R^i)_1^{*2}\right] \leq \frac{K}{\sqrt{N}}.$$

So we have, for $i \geq 1$,

$$f(X^i) - f(\overline{X}^i) = f'(\overline{X}^i(\omega)).(Z^i + R^i) + R(\overline{X}^i, X^i - \overline{X}^i) \parallel X^i - \overline{X}^i \parallel_C$$

and

$$\frac{1}{\sqrt{N}} \sum_i (f(X^i) - f(\overline{X}^i)) = \frac{1}{\sqrt{N}} \cdot \sum_i f'(\overline{X}^i(\omega)).Z^i$$

$$+ \frac{1}{N} \sum_i f'(\overline{X}^i).\sqrt{N} \, R^i + \frac{1}{\sqrt{N}} \sum_i R(\overline{X}^i, X^i - \overline{X}^i) \parallel X^i - \overline{X}^i \parallel_C .$$

But the last two terms converge towards zero in $L^1$, since for the last term we have

$$E\left[ \frac{1}{\sqrt{N}} \sum_i \parallel R(\overline{X}^i, X^i - \overline{X}^i) \parallel_C \parallel X^i - \overline{X}^i \parallel_C \right] \leq$$

$$\frac{1}{\sqrt{N}} \sum_1^N E\left[ R(\overline{X}^i, X^i - \overline{X}^i)_1^{*2} \right]^{1/2} \times \frac{K}{\sqrt{N}} = K \, E\left[ R(\overline{X}^1, X^1 - \overline{X}^1)_1^{*2} \right]^{1/2} .$$

If we then take $p \geq 1$, and set $A_p$ and $\alpha_p$ such that

$$\overline{P}(A_p)^C \leq 1/p \quad \text{and} \quad \forall x \in A_p \quad \forall u \in C \quad \parallel u \parallel \leq \alpha_p \Rightarrow \parallel R(x,u) \parallel_C^2 \leq \frac{1}{p} ,$$

we can find $N_o \geq 1$ such that, for $N \geq N_o$,

$$Q((X^{1,N} - \overline{X}^1)_1^* > \alpha_p) \leq \frac{1}{p} .$$

So $E_Q\left[ (R(\overline{X}^1(\omega),\ X^{1,N} - \overline{X}^1))_1^{*2} \right] \leq \frac{1}{p} + \frac{K}{p}$ for $N \geq N_o$. This proves that $E_Q\left[ (R(\overline{X}^1(\omega),\ X^{1,N} - \overline{X}^1))_1^{*2} \right]^{1/2}$ goes to zero when $N$ goes to infinity.

We are now going to study

$$I_N = \frac{1}{\sqrt{N}} \cdot \sum_1^N f'(\overline{X}(w^i)).Z^i$$

$$= \frac{1}{N^{3/2}} \sum_{i \neq j} f'(\overline{X}(w^i)).Z^{i,j} . \tag{2.25}$$

Define, on $W \times W$, $G(w_1, w_2) = f'(\overline{X}(w_1)).Z^{1,2}(w_1, w_2) \in L^2(W \times W)$, we have

$$\int G(w_1, w_2) \, dP(w_2) = 0 \quad P \text{ a.s.}$$

If we define $\overline{G}(w_2) = \int G(w_1, w_2) \, dP(w_1)$ and $\overset{\sim}{G}(w_1, w_2) = G(w_1, w_2) - \overline{G}(w_2)$, find

158

$$I_N = \frac{1}{\sqrt{N}} \sum_j \overline{G}(w_j) + \frac{1}{N^{3/2}} \sum_{i \neq j} \widetilde{G}(w_i, w_j) - \frac{1}{N^{3/2}} \sum_j \overline{G}(w_j).$$

But the last two terms go to zero in $L^2$ when $N$ goes to infinity. As a consequence,

$$T_N^f = \frac{1}{\sqrt{N}} \left( f(\overline{X}(w^i)) + \overline{G}(w^i) \right) + R_N$$

where $R_N$ converges towards zero in probability; but

$$\overline{G}(w) = \int f'(\overline{X}(w_1)) . U(w_1, X_.(w)) dP(w_1)$$

$$- \int f'(\overline{X}(w_1)) . U(w_1, x) d\overline{P}(x) dP(w_1)$$

$$= (\overline{\Pi} f)(X_.(w)) P \quad \text{a.s.}$$

So we find that $T_N^f$ converges in law towards

$$N\left(0, E_P\left[((I+\overline{\Pi})f^2\right]\right).$$

This proves the theorem.

REFERENCES


[1] Braun, W. and Hepp, K.   The Vlasov dynamics and its fluctuation in the 1/N limit of interacting particles. *Comm. Math. Phys. 56*, 101–113 (1977).

[2] Dawson, D.A.   Critical dynamics and fluctuations for a mean field model of cooperative behaviour. *J.Statist.Phys.31* (1983), 29–85.

[3] Dobrusin R.L.   Vlasov equations. *Funct. Anal. and Appl., 13*, 115 (1979).

[4] Doleans-Dade, C.   *Intégrales stochastiques par rapport à une famille de probabilité. Séminaire de Probabilités V*, Springer, Lecture Note 191, Berlin, Heidelberg, New York (1971).

[5] Kusuoka, Tamura   The convergence of Gibbs measures associated with mean field potentials. *J.Fac.Sci.Univ.Tokyo Sect.1A Math.*, 31 (1984), 223–245.

[6]  McKean, H.P.  Propagation of chaos for a class of nonlinear
         parabolic equations. *Lecture Series in Diff. Equ.*, *7*,
         Catholic Univ., 41-57 (1967), Washington D.C.

[7]  Marchioro, C. and Pulvirenti, M.  Hydrodynamics in two dimensions
         and vortex theory. *Comm. Math. Phys.*, *84*, 483-503 (1982).

[8]  Neveu, J.  *Processes aléatoires gaussiens*.  Presses Univ. de
         Montréal (1968).

[9]  Sznitman, A.S.  Equations de type de Boltzmann spatialement homogènes.
         *Z. Wahrscheinlichkeitsth. Geb. 66* (1984) 559-592.

[10]  Sznitman, A.S.  An example of nonlinear diffusion process with
         normal reflecting boundary conditions and some related limit
         theorems. *J. Funct. Anal. 56* (3) , (1984), 311-336.

[11]  Tanaka, H.  Some probabilistic problems in the spatially homogeneous
         Boltzmann equation. *Proc. of IFIP-TSI Conf. on Theory and
         Applications of Random Fields, Bangalore* (Jan.1982).

[12]  Tanaka, H.  Limit theorems for certain diffusion process with
         interaction. *Taniguchi Symp. SA* , Kataka 1982, 469-488.

[13]  Tanaka, H. and Hitsuda, M.  Central limit theorems for a simple
         diffusion model of interacting particles. *Hiroshima Math. J., ll*,
         415-423 (1981).


A.S. Sznitman,
Laboratoire de Probabilités,
Tour 56,
associé au CNRS L.A. 224.

J WILLMS

# Convergence of infinite particle systems to the Poisson process under the action of the free dynamic

## 1. INTRODUCTION

This paper originates from a result recently obtained by Dobrushin and Sukhov [1]. Let $\rho$ be a Radon measure on $X = \mathbb{R}^d \times \mathbb{R}^d$ satisfying $\rho(\mathbb{R}^d \times (\{v\} \times \mathbb{R}^{d-1})) = 0$ for all $v \in \mathbb{R}$. $X$ is interpreted as the direct product of the position and velocity space of a classical d-dimensional particle. As in Theorem 3.4, [1], we want to concentrate on the question as to whether the asymptotic behaviour of infinite-dimensional systems of particles in $\mathbb{R}^d$ under the action of the free dynamic is already determined by the asymptotic behaviour of the corresponding first moment measures $\nu^1_{P_t}$. Under weak conditions on the second correlation measure and on the correlation decay of the initial state P, the answer is affirmative. In fact we show that under these conditions the time evolution $P_t$ of the point process P under the action of the free dynamic converges weakly to the Poisson process $P_\rho$ if and only if $\nu^1_{P_t}$ converges vaguely to $\rho$ (which is the first moment measure of $P_\rho$). In particular, this result contains Theorem 3.4, [1].

In contrast to Dobrushin and Sukhov we make use of the fact that the limit process $P_\rho$ is simple and therefore we can apply a theorem due to Kallenberg [2]. To show that $P_t$ converges weakly to $P_\rho$, Dobrushin et al. had to prove that the n-dimensional distributions $P_{t,A_1,\ldots,A_n}$ converge to $P_{\rho,A_1,\ldots,A_n}$, whereas we only have to verify that the probability $P_t(\zeta_A = 0)$ that there is no particle in $A \subseteq \mathbb{R}^d$ at time t converges as $t \to \infty$ to $\exp(-\rho(A))$.

To prove this, we consider a sufficiently fine partition $\{Z_m\}_{m \in J}$ of A. The condition on the correlation decay of P implies that $P_t(\cap_{m \in J} \{\zeta_{Z_m} = 0\})$ (which equals $P_t(\zeta_A = 0)$) converges as $t \to \infty$ to the product $\prod_{m \in J} P_t(\zeta_{Z_m} = 0)$. The condition on the second correlation measure of P guarantees that this product can be approximated by $\prod_{m \in J} (1 - \nu^1_{P_t}(Z_m))$ which tends as $t \to \infty$ to $\prod_{m \in J} (1 - \rho(Z_m))$ (since, by assumption, $\nu^1_{P_t}$ converges vaguely to $\rho$) and therefore can be

approximated by $\exp(-\rho(A))$. Note that in the first version of our result
the form of correlation decay chosen depends on the dynamic, whereas in the
following corollary the corresponding condition does not refer to it.

2. <u>PRELIMINARIES</u>

Let $X := \mathbb{R}^d \times \mathbb{R}^d$ be the phase space of a single-particle system in $\mathbb{R}^d$.
Points of the space X are denoted by $(q,v)$, $(q',v')$, etc., where $q \in \mathbb{R}^d$ is
interpreted as the position and $v \in \mathbb{R}^d$ as the velocity vector of a classical
d-dimensional particle.

$K(X)$ denotes the space of all continuous functions f with compact support
$S(f)$, $B(X^k)$ (resp. $B_o(X^k)$) the class of all Borel (resp. relatively compact
Borel) subsets of $X^k$ and $M(X)$ the class of all positive Radon measures on
$(X, B(X))$. A measure $\mu \in M(X)$ is called a point measure if $\mu(A) \in \mathbb{N}_o$ for all
$A \in B_o(X)$. A point measure $\mu$ is simple if $\mu(\{x\}) \in \{0,1\}$ for all $x \in X$.
Let $M^{..}(X)$ (resp. $M^{.}(X)$) be the subclass of all point measures (resp. simple
point measures). $M^{..}(X)$ consists exactly of all elements of the form
$\sum_{m \geq 1} n_m \, \varepsilon_{(q_m,v_m)}$ where $\varepsilon_{(q_m,v_m)}$ is the unit mass (Dirac measure) at
$(q_m,v_m) \in X$, $n_m \in \mathbb{N}$, and $(q_1,v_1)$, $(q_2,v_2)$, ... are distinct elements of X
with only a finite number of $(q,v)$'s in any $A \in B_o(X)$. Therefore $M^{..}(X)$ can
be used for describing models which deal with an infinite number of particles
in X and where only finitely many particles are allowed in a bounded region.
If $\mu \in M^{.}(X)$, multiple occupancy is excluded, i.e. there cannot be more than
one particle at the same point of X. $\mu(A)$ can be interpreted as the
number of particles in A. For $f \in K(X)$ (resp. $A \in B_o(X)$), $\zeta_f$ (resp. $\zeta_A$)
denotes the map $M^{..} \to \mathbb{R}$ defined by $\zeta_f(\mu) := \mu(f) = \int f(x)\,\mu(dx)$ (resp.
$\zeta_A(\mu) := \mu(A)$). Let $B(M^{..})$ be the $\sigma$-algebra in $M^{..}$ generated by the mappings
$\zeta_f$, $f \in K(X)$. $B(M^{..})$ coincides with the $\sigma$-algebra generated by the vague
topology in $M^{..}$ (cf. Kallenberg $[2]$, Lemma 4.1).

A point process P is a probability measure on $(M^{..}, B(M^{..}))$. P is called
simple if $P(M^{.}) = 1$.

The k-th moment measure $\nu_P^k$ of a point process P in X is a measure on
$(X^k, B(X^k))$ defined by

$$\nu_P^k(A_1 \times \ldots \times A_k) := \int \mu(A_1) \, \ldots \, \mu(A_k) \, P(d_\mu), \quad A_1, \ldots, A_k \in B(X).$$

The point process P is called of k-th order if $\nu_P^k$ is a Radon measure on

$(X^k, B(X^k))$.

The $k$-th correlation measure $\overset{\backprime}{\nu}{}^k_P$ of a point process $P$ in $X$ is a measure on $(X^k, B(X^k))$ defined by $\overset{\backprime}{\nu}{}^k_P(A) := \nu^k_P(A \cap \overset{\backprime}{X}{}^k)$, $A \in B(X^k)$, where

$$\overset{\backprime}{X}{}^k := \{(x_1,\ldots,x_k) \in X^k : x_i \neq x_j \text{ for } i \neq j\}.$$

The free motion of particles is characterized by the family $\{T_t\}_{t \in \mathbb{R}}$ of homeomorphisms on $X$ defined by $T_t(q,v) := (q+vt,v)$. $\mu_t(f) := \mu(f \circ T_t)$ for $f \in K(X)$ defines for all $t \in \mathbb{R}$ a measurable bijection on $M^{\cdot\cdot}(X)$, denoted again by $T_t$. For a point process $P$ in $X$, let $P_t$ be the image of $P$ under $T_t$. $P_t$ is a point process of the same order as $P$ with $\nu^k_{P_t}(A_1 \ldots A_k) =$

$$= \nu^k_P(T_{-t}(A_1) \times \ldots \times T_{-t}(A_k)), \quad A_1,\ldots,A_k \in B(X),$$ and $P_t$ is simple if $P$ is.

An important example of a point process in $X$ is the Poisson process $P_\rho$ with first moment measure $\rho$ (e.g. Krickeberg [3], Mecke [4]). $P_\rho$ is simple if and only if $\rho$ is diffuse, that is, if $\rho(\{x\}) = 0$ for any $x \in X$.

## 3. CONVERGENCE TO THE POISSON PROCESS

We want to reduce the problem of weak convergence $P_t \Rightarrow P_\rho$ (as $t \to \pm\infty$) to the problem of vague convergence of the corresponding first moment measure $\nu^1_{P_t} \overset{\textasciitilde}{\to} \rho$, where $\overset{\textasciitilde}{\to}$ denotes the vague convergence as $t \to \pm\infty$ defined by

$$\lim_{t \to \pm\infty} \nu^1_{P_t}(f) = \rho(f) \text{ for all } f \in K(X).$$ $P_t \Rightarrow P_\rho$ implies $\rho = T_t(\rho)$ (but this does not mean that $\rho$ is of the form $\rho = m \times Q$ where $m$ denotes the Lebesgue measure on $\mathbb{R}^d$ and $Q$ an arbitrary Radon measure on $\mathbb{R}^d$).

For $r > 0$ and $y \in X$, $D(r,y)$ denotes the cube in $X$ with edge length $r$ and centre at the point $y$. Furthermore, let $\Pi_2 : X \to \mathbb{R}$ be the projection defined by $\Pi_2((q,v)) := v_1$ where $v = (v_1,\ldots,v_d)$.

We assume that the initial state $P$ satisfies the following conditions.

I.  The second correlation measure $\overset{\backprime}{\nu}{}^2_P$ is absolutely continuous with respect to the measure $\overline{K}$ on $(X^2, B(X^2))$ defined by

$$\overline{K}(A) = \int_{X^2} 1_A((q,v), (q',v')) \, K(dq \times dq') \, \eta(dv) \, \eta(dv'), \quad A \in B(X^2)$$

where $K$ is some measure on $(\mathbb{R}^{2d}, B(\mathbb{R}^{2d}))$ with

$$H := \sup_{q,p \in \mathbb{R}^d} K(D(1,q) \times D(1,p)) < +\infty,$$

and where $\eta$ is a Radon measure on $\mathbb{R}^d$ with $\eta(\{v\} \times \mathbb{R} \times \ldots \times \mathbb{R}) = 0$ for all $v \in \mathbb{R}$.

Furthermore, for $\overline{K}$-almost all $(x,y) \in X^2$

$$\frac{d\mathring{v}_P^2}{d\overline{K}} (x,y) \leq 1 \quad \text{is assumed.}$$

For a Radon measure $\rho$ on X define $\mathcal{C}(\rho):=\{A \in \mathcal{B}_o(X): \rho(\partial A) = 0\}$. Let us first consider the following condition on the correlation decay.

II($\rho$).  $\lim\limits_{t \to \pm\infty} \left| P_t(\zeta_A=0) \cdot P_t(\zeta_B=0) - P_t(\{\zeta_A=0\} \cap \{\zeta_B=0\}) \right| = 0$

whenever A and B are non-empty sets, $A,B \in \mathcal{C}(\rho)$ and
$\sup\Pi_2(A) < \inf\Pi_2(B)$.

Note that if $B,C \in \mathcal{B}_o(\mathbb{R}^d)$ and $A=C\times B$ then condition I gives us

$$\mathring{v}_{P_t}^2 (A\times A) = \mathring{v}_P^2(T_{-t}(A)\times T_{-t}(A))$$
$$\leq \int 1_{B\times B}(v,v') \cdot 1_{C\times C}(q'+vt,q'+v't) \, K(dq\times dq')\eta(dv)\eta(dv')$$
$$\leq \int 1_{B\times B}(v,v') \, K(\{(q,q') \in \mathbb{R}^{2d} : q+vt \in C, \ q'+v't \in C\})\eta(dv)\eta(dv')$$
$$\leq H\cdot(\text{diam}C+1)^{2d}\cdot(\eta(B))^2 < +\infty \tag{1}$$

for all $t \in \mathbb{R}$.

<u>Theorem.</u>  Let $\rho$ be a Radon measure on X with $\rho(\Pi_2^{-1}(\{v\})= 0$ for all $v \in \mathbb{R}$ and let P be a point process in X of first order which satisfies Conditions I and II($\rho$).  Then

$$v_{P_t}^1 \longrightarrow \rho \leftrightarrow P_t \Rightarrow P_\rho$$

<u>Proof.</u>  We only consider the case $t \to +\infty$; all arguments also apply for $t \to -\infty$.

Let us first assume that $v_{P_t}^1 \longrightarrow \rho$.  $P_\rho$ is simple since $\rho$ is diffuse.

$v_{P_t}^1 \longrightarrow \rho$ implies $\lim\limits_{t\to\infty} v_{P_t}^1 (A) = \rho(A)$ for all $A \in \mathcal{C}(\rho)$.  According to Theorem 4.7, Kallenberg [2] we therefore only have to show

164

$$\lim_{t\to\infty} P_t(\zeta_A = 0) = \exp(-\rho(A)) \quad \text{for all } A \in \mathcal{C}(\rho).$$

For $A \in \mathcal{C}(\rho)$ with $\rho(A) = 0$, there follows

$$\lim_{t\to\infty} \left| P_t(\zeta_A=0) - \exp(-\rho(A)) \right| = \lim_{t\to\infty} P_t(\zeta_A \geq 1) \leq \lim_{t\to\infty} \nu_{P_t}^1(A) = \rho(A) = 0.$$

Consider now $A \in \mathcal{C}(\rho)$ with $\rho(A) > 0$. Let $B$ be a measurable subset of $A$ with $\rho(B) = \rho(A)$ and consider a partition $\{Z_j\}_{j\in J}$ of $B$. We get

$$\left| P_t(\zeta_A=0) - \exp(-\rho(A)) \right|$$

$$\leq \left| P_t(\zeta_A=0) - P_t(\zeta_B=0) \right|$$

$$+ \left| P_t(\bigcap_{j\in J} \{\zeta_{Z_j}=0\}) - \prod_{j\in J} P_t(\zeta_{Z_j}=0) \right|$$

$$+ \left| \prod_{j\in J} (1 - P_t(\zeta_{Z_j}\geq 1)) - \prod_{j\in J} (1-\nu_{P_t}^1(Z_j)) \right| \tag{2}$$

$$+ \left| \prod_{j\in J} (1 - \nu_{P_t}^1(Z_j)) - \prod_{j\in J} (1 - \rho(Z_j)) \right|$$

$$+ \left| \prod_{j\in J} (1 - \rho(Z_j)) - \exp(-\rho(A)) \right|.$$

Given $\varepsilon > 0$, we shall show that we can find a partition $\{Z_j\}_{j\in J}$ of $B$ such that each term on the right-hand side of (2) is, for sufficiently large $t$, smallen than $\varepsilon$. This will prove the first part of the theorem.

To construct a suitable partition we consider $\Pi_2(A)$ which is a bounded set in $\mathbb{R}$. Hence we can find $a,b \in \mathbb{R}$ such that $\Pi_2(A) \subseteq \,]a,b]$. For all $m \in \mathbb{N}$ and $0 \leq j \leq m-1$, put

$$I_j^{(m)} := \,]a + \frac{j\cdot(b-a)}{m}, \; a + \frac{(j+1)\cdot(b-a)}{m}] \quad \text{and}$$

$$Z_j^{(m)} := A \cap \Pi_2^{-1}(I_j^{(m)}).$$

Since $\partial Z_j^{(m)} \subseteq \partial A \cup \Pi_2^{-1}(\partial I_j^{(m)})$ and since $\Pi_2(\rho)$ is diffuse, we have $Z_j^{(m)} \in \mathcal{C}(\rho)$. Let $J_m$ denote the set of all $j \in \{1,2,\ldots,m\}$ for which $\rho(Z_j^{(m)}) > 0$ and let us put $A_m := \bigcup_{j\in J_m} Z_j^{(m)}$. Clearly, $A_m$ is a measurable subset of $A$ with

$\rho(A_m) = \rho(B)$ and $\{Z_j\}_{j \in J_m}$ is a partition of $A_m$.

Since $\Pi_2(\rho)$ is a diffuse Radon measure on $\mathbb{R}$ (and thus the distribution function of $\Pi_2(\rho)$ is uniformly continuous on $[a,b]$), it is not difficult to show that $\lim_{m \to \infty} \max_{j \in J_m} \rho(Z_j^{(m)}) = 0$. Furthermore, we have $\sum_{j \in J_m} \rho(Z_j^{(m)}) = \rho(A)$ for all $m \in \mathbb{N}$. Hence, given $\varepsilon > 0$, there exists $m_o \in \mathbb{N}$ such that, for all $m \geq m_o$,

$$\left| \exp(-\rho(A)) - \prod_{j \in J_m} (1-\rho(Z_j^{(m)})) \right| \leq \varepsilon. \tag{3}$$

Since $A$ is a bounded set in $X$, we can find $C \in \mathcal{B}_o(\mathbb{R}^d)$ and $B \in \mathcal{B}_o(\mathbb{R}^{d-1})$ such that $A \subseteq C \times ([a,b] \times B)$. Hence (1) gives us

$$\nu^2_{P_t} (Z_j^{(m)} \times Z_j^{(m)}) \leq H \cdot (\operatorname{diam}C+1)^{2d} \cdot (\eta(I_j^{(m)} \times B))^2 < +\infty.$$

Again we have that $\lim_{m \to \infty} \max_{j \in J_m} \eta(I_j^{(m)} \times B) = 0$ and thus

$$0 \leq \lim_{m \to \infty} \sum_{j \in J_m} \nu^2_{P_t} (Z_j^{(m)} \times Z_j^{(m)}) \leq H \cdot (\operatorname{diam}C+1)^{2d} \cdot \eta([a,b] \times B) \lim_{m \to \infty} \max_{j \in J_m} \eta(I_j^{(m)} \times B)$$

$$= 0 \quad \text{for all } t \in \mathbb{R}.$$

Hence we can find $n \in \mathbb{N}$ such that $n \geq m_o$, $\max_{j \in J_n} \rho(Z_j^{(n)}) \leq \frac{1}{2}$, $\frac{\varepsilon}{n} \leq \frac{1}{2}$ and

$$\sum_{j \in J_n} \nu^2_{P_t} (Z_j^{(n)} \times Z_j^{(n)}) \leq \varepsilon \quad \text{for all } t \in \mathbb{R}. \tag{4}$$

Write $Z_j$ for $Z_j^{(n)}$, $B$ for $A_n$ and $J$ for $J_n$. In the following we shall see that $\{Z_j\}_{j \in J}$ is a suitable partition.

Next we want to show that the second term of the right-hand side of (2) tends to 0 as $t \to \infty$. This is a direct consequence of the assumed correlation decay. For $B \in \mathcal{B}(X)$, abbreviate $\{\zeta_B = 0\}$ to $\hat{B}$. We claim that

$$\lim_{t \to \infty} \left| P_t(\hat{Z}_j) \cdot P_t(\hat{Z}_i) - P_t(\hat{Z}_j \cap \hat{Z}_i) \right| = 0 \quad \text{for all } i,j \in J \text{ with } i > j. \tag{5}$$

This follows directly from condition $II(\rho)$ if $\sup \Pi_2(Z_j) < \inf \Pi_2(Z_i)$. If $\sup \Pi_2(Z_j) \geq \inf \Pi_2(Z_i)$ then $i = j+1$ (and equality holds). Given $\delta > 0$, we can find non-empty sets $V, W \in \mathcal{C}(\rho)$ such that $V \cup W = Z_j$, $V \cap W = \emptyset$, $\rho(W) \leq \frac{\delta}{3}$ and

$\sup\Pi_2(V) = \inf\Pi_2(W) < \sup\Pi_2(W) = \inf\Pi_2(Z_{j+1})$ (since $\rho(Z_j) > 0$).

Therefore

$$\left| P_t(\hat{Z}_j) \cdot P_t(\hat{Z}_i) - P_t(\hat{Z}_j \cap \hat{Z}_i) \right|$$

$$\leq \left| P_t(\hat{Z}_j) \cdot P_t(\hat{Z}_i) - P_t(\hat{V}) \cdot P_t(\hat{Z}_i) \right|$$

$$+ \left| P_t(\hat{V}) \cdot P_t(\hat{Z}_i) - P_t(\hat{V} \cap \hat{Z}_i) \right|$$

$$+ \left| P_t(\hat{V} \cap \hat{Z}_i) - P_t(\hat{Z}_j \cap \hat{Z}_i) \right|.$$

Since both the first and the third term of the right-hand side of the above
inequality are smaller than $P_t(\zeta_W \geq 1)$ and therefore smaller than $\nu^1_{P_t}(W)$, it
follows from condition II($\rho$) and $\nu^1_{P_t} \longrightarrow \rho$ that, for sufficiently large $t$, the
right-hand side of the above inequality is smaller than $\delta$. Thus (5) is
proved. Now it is not difficult to show that by induction we can find
$t_o \in \mathbb{R}$ such that, for all $t \geq t_o$,

$$\left| P_t( \bigcap_{j \in J} \{\zeta_{Z_j} = 0\}) - \prod_{j \in J} P_t(\zeta_{Z_j} = 0) \right| \leq \varepsilon. \tag{6}$$

Finally, let us consider the remaining terms of the right-hand side of (2).
Since $\nu^1_{P_t} \longrightarrow \rho$, we can find $t_1 \geq t_o$ such that, for all $t \geq t_1$, we have

$$\left| \nu^1_{P_t}(Z_j) - \rho(Z_j) \right| \leq \frac{\varepsilon}{m} \quad \text{for all } j \in J \tag{7}$$

$$\nu^1_{P_t}(A \backslash B) \leq \varepsilon. \tag{8}$$

Hence we have, for all $t \geq t_1$,

$$\left| P_t(\zeta_A = 0) - P_t(\zeta_B = 0) \right| \leq P_t(\zeta_{A \backslash B} \geq 1) \leq \nu^1_{P_t}(A \backslash B) \leq \varepsilon. \tag{9}$$

Since, for all $n \geq 2$ and all $D \in \mathcal{B}_o(X)$,

$$P_t(\zeta_D \geq n) \leq \int_{\{\zeta_D \geq n\}} \zeta_D \, dP_t$$

$$= \int_{\{\zeta_D \geq n\}} \frac{1}{\zeta_D(\mu) - 1} \cdot \mu^2(D \times D \cap \hat{X}^2) \, P_t(d\mu) \leq \frac{1}{n-1} \nu^2_{P_t}(D \times D) \tag{10}$$

and since, for all $t \geq t_1$ and all $j \in J$, $P_t(\zeta_{Z_j} \geq 1) \leq \nu^1_{P_t}(Z_j) \leq \rho(Z_j) + \frac{\varepsilon}{n} \leq 1$, it follows from (4) that, for all $t \in \mathbb{R}$,

$$\prod_{j \in J} (1 - P_t(\zeta_{Z_j} \geq 1)) - \prod_{j \in J} (1 - \nu^1_{P_t}(Z_j))$$

$$\leq \sum_{j \in J} |P_t(\zeta_{Z_j} \geq 1) - \nu^1_{P_t}(Z_j)| \tag{11}$$

$$\leq \sum_{j \in J} \int_{\{\zeta_{Z_j} \geq 2\}} \zeta_{Z_j} \, dP_t$$

$$\leq \sum_{j \in J} \nu^2_{P_t}(Z_j \times Z_j) \leq \varepsilon$$

and from (7) that, for all $t \geq t_1$,

$$\left| \prod_{j \in J} (1 - \nu^1_{P_t}(Z_j)) - \prod_{j \in J} (1 - \rho(Z_j)) \right|$$

$$\tag{12}$$

$$\leq \sum_{j \in J} |\nu^1_{P_t}(Z_j) - \rho(Z_j)| \leq \varepsilon.$$

Inequalities (9), (6), (11), (12) and (3) show that, for all $t \geq t_1$,

$$|P_t(\zeta_A = 0) - \exp(-\rho(A))| \leq 5\varepsilon,$$

which proves the first part of the theorem.

Let us now assume that $P_t \Rightarrow P_\rho$. Consider $f \in K(X)$ with support $S(f) \subseteq A = C \times B \in \mathcal{B}_o(X)$. Since $\zeta_f$ is continuous with respect to the vague topology, $\zeta_f(P_t) \Rightarrow \zeta_f(P_\rho)$ follows.

Therefore to prove that $\lim_{t \to \infty} \int \zeta_f \, dP_t = \int \zeta_f \, dP_\rho$ it is sufficient to show that

$$\lim_{n \to \infty} \sup_{t \in \mathbb{R}} \int_{\{|\zeta_f| \geq n\}} |\zeta_f| \, dP_t = 0.$$

But this follows immediately from (10), (1) and

$$\int_{\{|\zeta_f| \geq n\}} |\zeta_f| \, dP_t \leq c \cdot \int_{\{\zeta_A \geq n/c\}} \zeta_A \, dP_t$$

168

for all $t \in \mathbb{R}$ where $c := \sup\{|f(x)|: x \in X\}$.

Finally we consider a form of correlation decay of the initial state $P$ which does not refer to the dynamic.

Let $\Pi_1: X \to \mathbb{R}$ be the projection defined by $\Pi_1(q,v) := q_1$ $(q=(q_1,\ldots,q_d))$. For $C_1, C_2 \in \mathcal{B}(\mathbb{R})$ and $r,s \in \mathbb{R}_+ := \{x \in \mathbb{R}: x > 0\}$, define

$$\tilde{\alpha}_P(C_1, C_2) := \sup\{|P(\{\zeta_{A_1}=0\} \cap \{\zeta_{A_2}=0\}) - P(\zeta_{A_1}=0)\cdot P(\zeta_{A_2}=0)|:$$

$$A_i \in \mathcal{B}(X) \text{ with } \Pi_1(A_i) \subseteq C_i, i = 1,2\},$$

$$\alpha_P(r,s) := \sup_{y \in \mathbb{R}} \tilde{\alpha}_P(I(r,y), \mathbb{R}\setminus I(r+s,y))$$

where $I(r,y) := \left[ y - \frac{r}{2}, y + \frac{r}{2}\right[$.

III.  $\lim_{s \to \infty} \alpha_P(s, a\cdot s) = 0$  for all  $a \in \mathbb{R}_+$.

<u>Corollary.</u>  Let $\rho$ be a Radon measure on $X$ with $\rho(\Pi_2^{-1}(\{v\})) = 0$ for all $v \in \mathbb{R}$ and let $P$ be a point process in $X$ of first order which satisfies Conditions I and III. Then

$$\nu^1_{P_t} \longrightarrow \rho \quad \Leftrightarrow \quad P_t \Rightarrow P.$$

<u>Proof.</u>  It only remains to show that Condition III implies Condition II($\rho$).

Consider $A,B \in \mathcal{B}_0(X)$ with $\sup\Pi_2(A) < \inf\Pi_2(B)$. There exist $x,y \in \mathbb{R}$ and $s,r,\tilde{r} \in \mathbb{R}_+$ such that $\tilde{r} - r =: \delta > 0$, $\Pi_1(A) \subseteq I(s,x)$, $\Pi_1(B) \subseteq I(s,x)$, $\Pi_2(A) \subseteq I(r,y)$ and $\Pi_2(B) \subseteq \mathbb{R}\setminus I(\tilde{r},y)$. Note that $\Pi_1(T_{-t}(A)) \subseteq I(s+rt, x-yt)$ and that, for $t > 0$, $\sup(\Pi_1(T_{-t}(B)) \leq (x + \frac{s}{2}) - (y + \frac{\tilde{r}}{2})t = x-yt - \frac{1}{2}(\tilde{r}t-s)$. This gives us

$$\Pi_1(T_{-t}(B)) \subseteq \mathbb{R}\setminus I(\tilde{r}t-s, x-yt) \text{ for all } t > \frac{s}{\tilde{r}}.$$

Since $\lim_{t \to \infty} \frac{\tilde{r}t-s}{rt+s} = 1 + \frac{\delta}{r} > 1$, there exists $t_o > \frac{s}{\tilde{r}}$ such that, for all $t \geq t_o$, $\tilde{r}t - s \geq (1+a)\cdot(rt+s)$ where $a := \frac{\delta}{2r} > 0$. Thus we get, for all $t \geq t_o$,

$$\Pi_1(T_{-t}(B)) \subseteq \mathbb{R}\setminus I((1+a)\cdot(rt+s), x-yt).$$

169

Therefore Condition III implies

$$\lim_{t \to \infty} |P_t(\{\zeta_A=0\} \cap \{\zeta_B=0\}) - P_t(\zeta_A=0) \cdot P_t(\zeta_B=0)|$$

$$\leq \lim_{t \to \infty} \alpha_P(s+rt, a \cdot (s+rt)) = 0.$$

Both condition I and Condition III are weaker than the corresponding assumptions in [1]. Contrary to Dobrushin et al., we have to make no further assumption about the first correlation measure $\nu_P^1$ (as in Condition 3.II in [1]). Therefore our result contains Theorem 3.4, Dobrushin et al. [1]. Note that in Theorem 3.4 [1] the probability measure Q describing the distribution of the velocities is absolutely continuous with respect to the Lebesgue measure on $\mathbb{R}^d$.

## REFERENCES


[1] Dobrushin, R.L. and Sukhov, Yu.M. (Boldrighini, C.) Time asymptotics for some degenerate models of evolution of systems with an infinite number of particles. *Journal of Soviet Mathematics, 16, 4*, 1277-1340 (1981). (Translated from *Itogi Nauki i Tekhniki, Seriya Sovremennye Problemy Matematiki, 14*, 147-254 (1979).)

[2] Kallenberg, O. *Random Measures*. Akademie-Verlag, Berlin (1976).

[3] Krickeberg, K. The Cox Process. *Sympos. Math. IX, Calcolo Prob., teor. Turbulenza*, 151-167 (1972).

[4] Mecke, J. Stationäre zufällige Masse auf lokalkompakten Abelschen Gruppen, Z. *Wahrscheinlichkeitstheorie verw. Gebiete 9*, 36-58 (1967).


## ACKNOWLEDGEMENT


I am particularly grateful to Dr. Hans Zessin who introduced me to the theory of point processes and provided valuable advice throughout my work on this subject.



Jürgen Willms,
Fakultät für Mathematik,
Universitätsstrasse,
D-4800 Bielefeld,
West Germany.

H ZESSIN

# BBGKY hierarchy equations for Newtonian and gradient systems

<u>INTRODUCTION</u>

This paper originated in my attempt to understand the results of Lang [9] on the infinite one-dimensional gradient dynamics from the point of view of the BBGKY (Bogoljubov-Born-Green-Kirkwood-Yvon) hierarchy which had been developed for the infinite one-dimensional Newtonian dynamics by Gallavotti, Lanford and Lebowitz [4] (See also Sinai and Suhov [18] and Takahashi [19]). We consider the time development of certain infinite particle systems. These systems have the following two features in common: (1) at time zero the particles will be distributed in $\mathbb{R}^\nu$, $\nu \geq 1$, according to some suitable initial state; (2) the motion of the particles after time zero is completely deterministic in nature. For a particle with initial condition $z(0,a,\mu) = a \in \mathbb{R}^\nu$ moving in an external field A and interacting via a force B with other particles having initial conditions distributed over a measure $\mu$, the motion is formally described by

$$\frac{d}{dt} z(t,a,\mu) = A(z(t,a,\mu)) + \int_{\mathbb{R}^\nu} B(z(t,a,\mu) - z(t,b,\mu))(\mu - \varepsilon_a)(db). \quad (*)$$

Here A and B are suitable functions $A,B : \mathbb{R}^\nu \to \mathbb{R}^\nu$. This framework contains the canonical dynamics of finitely or infinitely many mass points (i.e. the Newtonian dynamics) as well as gradient systems in the sense of Lang [9] (take $A \equiv 0$ in (*)).

There are three basic problems in the analysis of such models: (1) to show that the system (*) has a unique solution for sufficiently many and regular enough initial conditions a and $\mu$ (a theorem of this type will be called an *evolution theorem*); (2) to find all *equilibrium states*, i.e. to find all states which are invariant under the motion; (3) for each equilibrium state $\bar{P}$, to find all initial states which in the course of time weakly converge to $\bar{P}$.

In this paper we will be concerned only with problems (2) and (3), and we take for granted that an evolution theorem for the underlying dynamic is true. The treatment of these problems will require the method of BBGKY hierarchy equations, which has been developed in statistical mechanics to describe the time evolution $\{P_t;\ t \in \mathbb{R}\}$ of the state $P$ of an infinite particle system. Here $P_t$ is obtained from $P_o = P$ by translation along a solution $\mu \to \mu_t$ of the equations of motion (*). These hierarchy equations can be regarded as a system of differential equations for the moment measures of the time evolving states $P_t$.

<u>Description of the results.</u>  First, in Theorem 2 the existence of weak solutions of the BBGKY hierarchy for local perturbations of equilibrium states (= states which are invariant under the motion) is shown to be a simple consequence of the boundedness of all moments of the number of particles in bounded regions over compact intervals of time (compare (1.11)) combined with a disintegration of the compound Campbell measure of equilibrium states (see Theorem 1). For this we have to assume that an evolution theorem is true such that the solutions $t \to \mu_t$ are continuous (which implies that the number of particles in bounded regions remains bounded over compact intervals of time) (see (2.6)). This general result is applied to the case where the underlying dynamics is given by the Newtonian equations and the initial state is given by a locally perturbed Gibbs state. Using the evolution theorem of Marchioro, Pellegrinotti and Presutti, Theorem 3 gives the existence of weak solutions of the BBGKY hierarchy for this basic example in any dimension. This generalizes a theorem of Gallavotti, Lanford and Lebowitz [4]. We remark that this has been proved in another way in [22].

We are not able to give (even partial) answers to Problems (2) and (3) for this example. For this reason we study these questions for gradient systems in any dimension with spatially homogeneous initial states as Lang [9] did it for $\nu = 1$; i.e., we ignore the first term on the right-hand side of (*) and want to control the second term on the right-hand side. The reason for considering gradient systems is that they are more accessible to mathematical treatment, firstly because one knows Ljapunov functions, and secondly because it is possible to exhibit equilibrium states explicitly.

For these systems we get most of Lang's results in any dimension by a detailed analysis of the associated BBGKY hierarchy, for which a new method is proposed. This consists in reducing the hierarchy equations to a new hierarchy which corresponds to the so-called reduced moment measures of the time-evolving states $P_t$. This in turn amounts to a system of hierarchy equations corresponding to the moment measures of the so-called Palm measures $P_t^o$ of the

172

time-evolving states $P_t$. This reduction is the main idea of Section 4 of this paper and enables us to use properties, in particular invariance properties of Palm measures, which play an essential role in the analysis of the reduced hierarchy. It gives a solution to Problem (2) and a partial solution to Problem (3). In Theorem 4, combined with Proposition 3, the equilibrium states for the gradient dynamic are characterized as the so-called rigid states. Furthermore, we show that equilibrium states in the variational sense, ground states and lattice type states are rigid states. Finally, in Theorem 5 we prove that the evolution $\{P_t;\ t \in \mathbb{R}\}$ converges weakly to the set of rigid states, if the initial state P is spatially homogeneous such that, for any time t, the specific energy and the mean square velocity of the particle, which at time t = 0 started in 0, considered as a function in time, satisfy some natural regularity conditions.

## 1. DISINTEGRATION OF CAMPBELL MEASURES

To make this paper self-contained we repeat first the general framework already formulated in $\lfloor 22 \rfloor$.

Let X be a locally compact space with a countable base; denote by $K(X)$ the set of all continuous functions f of X with compact carrier. We denote by $\mathcal{B}(X)$, resp. $\mathcal{B}_o(X)$, the class of all Borel, resp. relatively compact (i.e. bounded) Borel, subsets of X. Let $M = M(X)$ denote the set of all positive Radon measures on $\mathcal{B}(X)$. A measure $\mu \in M(X)$ is called a point measure if $\mu(A) \in \mathbb{N}_o$ for each $A \in \mathcal{B}_o(X)$; it is called simple if it is a point measure such that $\mu\{x\} \in \{0,1\}$ for each $x \in X$. Let $M^{\cdot\cdot} = M^{\cdot\cdot}(X)$ denote the set of all point measures and $M^{\cdot} = M^{\cdot}(X)$ the set of all simple point measures in M. It is well known that $M^{\cdot\cdot}$ is vaguely closed in M and $M^{\cdot}$ is a $G_\delta$-set in $M^{\cdot\cdot}$. A point process, resp. simple point process, P in X is a probability measure on the $\sigma$-algebra $\mathcal{B}(M)$ of all Borel subsets of M such that $P(M^{\cdot\cdot}) = 1$, resp. $P(M^{\cdot}) = 1$. A sequence $P_n$ of point processes converges weakly to a point process P, and we write $P_n \Rightarrow P$, if $P_n(\varphi) \to P(\varphi)$ for each bounded continuous real function $\varphi$ on M, i.e. for each $\varphi \in C^b(M)$. If P is a point process in X and $G \in \mathcal{B}(X)$, then $P_G$ denotes the image of P with respect to the mapping $\mu \to \mu_G$, where $\mu_G = 1_G \cdot \mu$. $M_f^{\cdot\cdot}$ denotes the set of all $\mu \in M^{\cdot\cdot}$ with finite support, and $N(G)$, $G \in \mathcal{B}(X)$, is defined by $N(G)(\mu) = \mu(G)$, $\mu \in M(X)$.

Consider the topological sum $X = \sum_{n \geq 0} X^n$ of the powers $X^n$ of X, where $X^o$

is the set whose sole element is the 'empty sequence' denoted by $\emptyset$. $X$ is a locally compact second countable topological space. Let $K(X)$ denote the space of real continuous functions with compact support on $X$, i.e. the direct sum of the spaces $K(X^n)$ of real continuous functions with compact support on $X^n$.

Let $k$ be a positive integer and let $\mu^k$ be the $k$-th power of $\mu$, i.e. $\mu^k$ is the measure in $X^k$ defined by $\mu^k(f_1 \otimes \ldots \otimes f_k) = \mu(f_1) \cdot \ldots \cdot \mu(f_k)$, $f_1,\ldots,f_k \in K(X)$. A point process $P$ in $X$ is called of $k$-th order if

$$\nu_P^k(f) = \int_{M(X)} \mu^k(f) \; P(d\mu) \tag{1.1}$$

exists and is finite for each $f \in K(X^k)$. In this case $\nu_P^k$ defines a Radon measure in $X^k$, called $k$-th moment measure of $P$. A necessary and sufficient condition for this to happen is

$$\nu_P^k(B^k) < +\infty, \quad \text{for each } B \in \mathcal{B}_o(X). \tag{1.2}$$

$P$ is called of infinite order if this is true for each $k$.

Now let $P$ be a point process in $X$ of infinite order and $f = (f^n)_{n \geq 0} \in K(X)$. Then

$$K_P(f) = \Sigma_{k \geq 0} \; \nu_P^k(f^k) \tag{1.3}$$

defines a Radon measure in $X$. Denote by $\check{\nu}_P^k$ the measure $D \to \nu_P^k(D \cap \check{X}^k)$, $D \in \mathcal{B}(X^k)$, where $\check{X}^k = \{(x_1,\ldots,x_k) \in X^k : (j_1 \neq j_2 \Rightarrow x_{j_1} \neq x_{j_2})\}$. The measure $\check{K}_P = \Sigma_{k \geq 0} \; \check{\nu}_P^k$ is called correlation measure of $P$.

We also need the so-called reduced Campbell measure of a point process $P$ given for rectangles $B \times N$, $B \in \mathcal{B}(X)$, $N \in \mathcal{B}(M^{..}(X))$ by

$$C_P^!(B \times N) = \int_{M^{..}(X)} \int_X 1_B(x) \cdot 1_N(\mu - \varepsilon_x) \; \mu(dx) P(d\mu). \tag{1.4}$$

Given a Radon measure $\rho$ in $X$, a point process $P$ in $X$ is said to satisfy $(\Sigma_\rho)$ if

$$(\Sigma_\rho) \qquad C_P^! \ll \rho \otimes P.$$

We denote the corresponding derivative by $f$. Condition $(\Sigma_\rho)$ is equivalent to saying: for each nonnegative, measurable function $g$ on $X \times M^{..}(X)$,

$$\int_{M^{..}(X)} \int_X g(x,\mu) \; \mu(dx) P(d\mu) = \int_{M^{..}(X)} \int_X g(x,\mu+\varepsilon_x) f(x,\mu) \rho(dx) P(d\mu).$$

$$(1.5)$$

Define, for $n = 1,2,\ldots$, nonnegative, measurable functions $e_n$ on $X^n \times M^{..}(X)$ by

$$e_n(x_1,\ldots,x_n;\mu) = f(x_1,\mu) \cdot f(x_2,\mu+\varepsilon_{x_1}) \cdot \ldots \cdot f(x_n,\mu+\varepsilon_{x_1}+\ldots+\varepsilon_{x_{n-1}}). \quad (1.6)$$

For $n \geq 2$ these functions $e_n$ are symmetric in $x_1,\ldots,x_n$ $\rho^n \otimes P$-a.e. Therefore we can define $L_\rho \otimes P$-a.e. a nonnegative, measurable function $e$ on $M^{..}_f(X) \times M^{..}(X)$ by $e(0,\mu) = 1$, $e(\varepsilon_{x_1}+\ldots+\varepsilon_{x_n},\mu) = e_n(x_1,\ldots,x_n;\mu)$ $(n \geq 1)$. Here $L_\rho$ denotes the image of the measure $\Sigma_{n\geq 0} \frac{\rho^n}{n!}$ with respect to $X \ni (x_1,\ldots,x_n) = x \to \varepsilon_{x_1}+\ldots+\varepsilon_{x_n} = \nu_x$.

The equations appearing in the following proposition are the equilibrium equations due to Ruelle $[17]$.

<u>Proposition 1</u>  (Matthes et al. $[7,12]$, Nguyen and Zessin $[7]$). If $\rho$ is a Radon measure in X and P is a point process in X satisfying $(\Sigma_\rho)$ then, for each $G \in \mathcal{B}(X)$ such that $\mu(G) < +\infty$ P-a.e. $[\mu]$ and measurable $\varphi : M^{..}(X) \to [0,+\infty]$,

$$P(\varphi) = \int_{M^{..}(G^c)} \int_{M^{..}(G)} \varphi(\nu+\mu) \cdot e(\nu,\mu) \, L_{\rho,G}(d\nu) P(d\mu). \qquad (1.7)$$

It follows that $P_G << L_{\rho,G}$ for each $G \in \mathcal{B}_0(X)$ with density

$$\sigma_G(\nu) = \int_{M^{..}(G^c)} e(\nu,\mu) \, P(d\mu). \qquad (1.8)$$

The following theorem gives a disintegration for the compound Campbell measure of point processes P of infinite order satisfying $(\Sigma_\rho)$, which serves as one of the main tools in the following.

<u>Theorem 1.</u>  Let $\rho$ be a Radon measure and let P be a point process in X of infinite order satisfying $(\Sigma_\rho)$. Then, for each measurable, nonnegative function h on $X \times M^{..}(X)$,

$$\int_{M^{..}(X)} \int_X h(x,\mu) \; \Lambda_\mu(dx) P(d\mu) = \int_{M^{..}(X)} \int_X h(x,\mu+\nu_x) e(\nu_x,\mu) \; \Lambda_\rho(dx) P(d\mu)$$

$$(1.9)$$

175

where $\overset{\lambda}{X} = U_{n\geq 0} X^n$ and $\wedge_\mu = \Sigma_{n\geq 0} 1_{\overset{\lambda}{X}n} \cdot \mu^n$.

<u>Proof.</u>   1.  We first prove (1.9) for functions h of the form
$h(x,\mu) = 1_N(\mu) \cdot 1_H(x)$, $N \in B(M^{..}(X))$, $H \in B_o(X)$.  From $(\Sigma_\rho)$ we get

$$\int_N \wedge_\mu(H) P(d\mu) = \int_N \underset{n\geq 0}{\Sigma} \int_{\overset{\lambda}{X}n} 1_H(x_1,\ldots,x_n)\mu(dx_1)\ldots\mu(dx_n)P(d\mu)$$

$$= \underset{n\geq 0}{\Sigma} \int_{M^{..}(X)} \int_X \int_{X^{n-1}} 1_N(\mu+\varepsilon_{x_n}) \cdot 1_{\overset{\lambda}{X}n_{\wedge H}}(x_1,\ldots,x_n)\mu(dx_1)\ldots\mu(dx_{n-1})$$

$$\times f(x_n,\mu)\rho(dx_n)P(d\mu).$$

Repeating this step several times, this equals

$$\underset{n\geq 0}{\Sigma} \int_{M^{..}(X)} \int_X \cdots \int_X 1_N(\mu+\varepsilon_{x_n}+\ldots+\varepsilon_{x_1}) \cdot 1_{\overset{\lambda}{X}n_{\wedge H}}(x_1,\ldots,x_n) \; e_n(x_1,\ldots,x_n;\mu)$$

$$\times \rho(dx_1)\ldots\rho(dx_n)P(d\mu)$$

$$= \int_{M^{..}(X)} \int_{\overset{\lambda}{X}} 1_N(\mu+\nu_x) \cdot 1_H(x)e(\nu_x,\mu)\wedge_\rho(dx)P(d\mu).$$

2.  Denote by $E$ the collection of rectangles $H \times N$, $H \in B_o(X)$, $N \in B(M^{..}(X))$.
$E$ generates the $\sigma$-algebra $B(X) \times B(M^{..}(X))$ and is closed with respect to
finite intersections.  Furthermore, there exists an increasing sequence $(E_n)$
in $E$ such that $E_n \nearrow X \times M^{..}(X)$.  Now the right-hand and left-hand sides of
(1.9) define measures on $B(X) \times B(M^{..}(X))$ which coincide on $E$ (according to 1.
and which are finite there (since P is of infinite order).  Consequently they
coincide on $B(X) \times B(M^{..}(X))$.


The following corollary in a general context answers the question as to
whether, for a certain class of initial states which can be described by
their correlation functions, the time-evolving states are still described by
their correlation functions.  Its proof is contained in the proof of Theorem
3 of [22].

<u>Corollary.</u>   Let $\overline{P}$ be a point process in X of infinite order satisfying $(\Sigma_\rho)$
for some $\rho$.  If $\Psi \in L^p(\overline{P})$ for some $p > 1$, $\Psi \geq 0$ with $\overline{P}(\Psi) = 1$, and if
$(T_t)_{t\in\mathbb{R}}$ is a continuous flow on $(M^{..},\overline{P})$, then each $P_t = (\Psi\cdot\overline{P}) \circ T_{-t}^{-1}$, $t \in \mathbb{R}$,

176

is of infinite order, the correlation functions $\rho_{P_t} = \dfrac{d\grave{K}_{P_t}}{d\wedge_\rho}$ of each $P_t$ exist and are given by

$$\rho_{P_t}(x) = \int_{M^{..}} \cdot \Psi(T_t(\mu+\nu_x)) \cdot e(\nu_x,\mu)\overline{P}(d\mu), \quad x \in X, \ t \in \mathbb{R}. \tag{1.10}$$

For later use we recall here the basic inequality which shows that each $P_t$ is of infinite order. For each $A \in \mathcal{B}_o(X)$, $n,k \in \mathbb{N}$,

$$\sup_{t \in \mathbb{R}} \ \nu_{P_t}^n (A^k) \leq \left[\int_{M^{..}(X)} \mu(B)^{q\cdot n} \ \overline{P}(d\mu)\right]^{1/q} \cdot \left[\int_{M^{..}(X)} \Psi(\mu)^P \ \overline{P}(d\mu)\right]^{1/P} < +\infty.$$

This follows from the Cauchy–Schwarz inequality combined with $\overline{P}$ being invariant and of infinite order.

## 2. THE BBGKY HIERARCHY

We first describe a class of flows which arises in the context of classical statistical mechanics (see, e.g., Dobrushin [2], Neunzert [14] ). Let $X$ be the d-dimensional Euclidean space and let $A,B : X \to X$ be continuous mappings where $B$ has also compact support. We consider the equation

$$\frac{d}{dt} z(t,a,\mu) = A(z(t,a,\mu)) + \int_X B(z(t,a,\mu) - z(t,b,\mu))(\mu-\varepsilon_a)(db) \tag{2.1}$$

$(t \in \mathbb{R})$ for a particle with initial condition

$$z(0,a,\mu) = a \in X \tag{2.2}$$

moving in an 'external field' $A$ and interacting via a 'two body for force' $B$ with other particles having initial conditions distributed over a measure $\mu \in M^{..}(X)$. This framework contains the free motion ($A(q,p) = (p,0)$, $B \equiv 0$), the Newtonian dynamics (Example 1 below) and also gradient systems in the sense of Lang [9] (example 2 below).

Example 1 (Newtonian equations). Let $\mu$ be a point measure in $X = \mathbb{R}^\nu \times \mathbb{R}^\nu$ and let $\Phi : \mathbb{R}^\nu \to \mathbb{R}$ be a measurable function such that $\phi(q) = \phi(-q)$ (such a $\phi$ we call pair potential). Assume that $\phi$ is twice continuously differentiable, has finite range and that it is the sum of a stable pair potential $\phi'$ and a nonnegative continuous pair potential $\phi''$ with $\phi''(0) > 0$. (In particular, $\phi$ is superstable in the sense of Ruelle [17].) If we set

$$A((q,p)) = (p,0),$$
$$B(q,p) = (0,F(q)) \text{ with } F(q) = -\text{grad } \phi(q) = -F(-q),$$
$$z(t,(q,p),\mu) = (x(t,(q,p),\mu),\ y(t,(q,p),\mu)),$$

then Equation (2.1) corresponds to

$$\frac{d}{dx} x(t,a,\mu) = y(t,a,\mu)$$

$$\frac{d}{dx} y(t,a,\mu) = -\sum_{a \neq b \in \mu} \text{grad } \phi\ (x(t,a,\mu) - x(t,b,\mu)) \qquad (a=(q,p))$$

with initial condition

$$x(0,a,\mu) = q,\ y(0,a,\mu) = p.$$

In this case we get the Newtonian equation for a particle with initial position $(q,p)$ interacting via a pair potential $\phi$ with the particles of $\mu$.

<u>Example 2</u> (Gradient systems). If $\mu$ is a point measure in $\mathbb{R}^\nu$ ($\nu \geq 1$) and if $A \equiv 0$ and $B = F = -\text{grad } \phi$ for certain continuously differentiable pair potentials $\phi$ with finite range, then Equation (2.1) corresponds to

$$\frac{d}{dt} z(t,a,\mu) = -\sum_{a \neq b \in \mu} \text{grad } \phi\ (z(t,a,\mu) - z(t,b,\mu))$$

with initial condition $z(0,a,\mu) = a$.

To introduce some notation, let $x = (a_1,\ldots,a_n) \in X^n$ and define

$$A(x) = (A(a_1),\ldots,A(a_n)),$$
$$B(x) = (B(a_1),\ldots,B(a_n)).$$

Given $\mu \in \overset{..}{M}(X)$ we define $B_\mu : X \to X$ by

$$B_\mu(a) = (\mu - \varepsilon_a)(B(a-\cdot)),\ \text{if } a \in \mu;\ B_\mu(a) = \mu(B(a-\cdot)),\ \text{if } a \notin \mu$$

and $B_\mu : X^n \to X^n$ by

$$B_\mu(x) = (B_\mu(a_1),\ldots,B_\mu(a_n)),\ x = (a_1,\ldots,a_n) \in X^n.$$

We denote by $Df$ the gradient of $f$. If $x = (x_1,x_2) \in \mathbb{R}^{\nu_1} \times \mathbb{R}^{\nu_2}$, $\nu_1 + \nu_2 = \nu$, $D_{x_1} f$ and $D_{x_2} f$ will mean gradients with respect to the corresponding group of variables, so that $Df = (D_{x_1} f, D_{x_2} f)$. With this notation for $x = (a_1,\ldots,a_n) \in X^n$ and suitable $f$, the usual inner product of $A + B_\mu$ and $Df$ is

178

$$\langle (A+B_\mu), Df\rangle \, (x) = \sum_{j=1}^{n} \langle D_{a_j} f(x), A(a_j)\rangle + \sum_{j=1}^{n} \langle D_{a_j} f(x), B_\mu(a_j)\rangle. \quad (2.3)$$

We now make the assumption that the system of equations (2.1) is solvable for sufficiently many initial conditions $\mu \in M^{..}(X)$.* To be more precise we assume that there exists a measurable subset $\hat{M} \subset M^{..}(X)$ with the following properties:

for each $\mu \in \hat{M}$ and each $a \in \mu$ the system of equations (2.1) has a solution $t \to z(t,a,\mu)$ with $z(0,a,\mu) = a$; $\qquad (2.4)$

the mapping
$$T_t : \mu \to \sum_{a \in \mu} \varepsilon_{z(t,a,\mu)}, \quad t \in \mathbb{R}, \quad \mu \in \hat{M} \qquad (2.5)$$

is a measurable transformation on $\hat{M}$ such that $T_o = id_{\hat{M}}$, $T_{t+s} = T_t \, T_s$, $s,t \in \mathbb{R}$;

for each $\mu \in \hat{M}$ the mapping $t \to \mu_t$ is continuous with respect to $\qquad (2.6)$ vague topology.

Here $\mu_t = T_t \mu$. Note that $\mu_t$ is the image of $\mu$ with respect to $z(t,\cdot,\mu)$, and that for each $k \in \mathbb{N}$, $t \to \mu_t^k$ is continuous if (2.6) is true. These assumptions have been verified for the Newtonian equations of Example 1 (see Lanford [8], Marchioro, Pellegrinotti and Presutti [10]).

Lemma 1. Let $n \in \mathbb{N}$ and let P be a point process in X satisfying

$$P(\hat{M}) = 1; \qquad (2.7)$$

for each function f which is continuously differentiable and of $\qquad (2.8)$ compact support on $X^n$, the functions $t \to \nu_{P_t}^n (f)$ are well defined,

finite valued, differentiable and the interchange of differentiation and expectation is allowed.

Then, for each function f which is continuously differentiable and of compact support on $X^n$,

---

*We remark that existence and uniqueness of solutions of (2.1) usually do not hold for all initial data, and therefore the time evolution may not be defined for all initial states. The problem of getting a good evolution theorem consists in choosing the largest possible subset $\hat{M}$, on which one can define $(T_t)$.

$$\frac{d}{dt}\, \nu^n_{P_t}(f) = \nu^n_{P_t}(<A+B_\nu.,Df>) + \nu^{n+1}_{P_t}(f_1), \quad t \in \mathbb{R}, \qquad \text{(BBGKY)}$$

where $f_1(a_1,\ldots,a_n,a_{n+1}) = <B((a_1,\ldots,a_n) - (a_{n+1})),Df(a_1,\ldots,a_n)>$,
$(a_{n+1}) = (a_{n+1},\ldots,a_{n+1})$.

<u>Proof.</u>  1.  Let $f$ be a continuously differentiable function on $X^n$ with compact support and $\mu \in \hat{M}$.  For $x = (a_1,\ldots,a_n) \in \hat{X}^n$ with $a_i \in \mu$ we set $z(t,x,\mu)_{Df} = (z(t,a_1,\mu),\ldots,z(t,a_n,\mu))$, $t \in \mathbb{R}$.  Then

$$\frac{d}{dt}\, f(z(t,x,\mu)) = <Df(z(t,x,\mu)), \frac{d}{dt}\, z(t,x,\mu)>$$

$$= \sum_{j=1}^{n} <D_{a_j} f(z(t,x,\mu)), (A(z(t,a_j,\mu))$$

$$+ \int_X B(z(t,a_j,\mu) - z(t,b,\mu))\, (\mu-\varepsilon_{a_j})(db))>.$$

The right-hand side equals $<A+B_{\mu_t}, Df>\,(z(t,x,\mu))$, due to (2.3), so that

$$\frac{d}{dt}\, f(z(t,x,\mu)) = <A+B_{\mu_t}, Df>\,(z(t,x,\mu)). \qquad (2.9)$$

2.  Consequently

$$\frac{d}{dt}\, \nu^n_{P_t}(f) = \int_{\hat{M}} \int_{\hat{X}^n} \frac{d}{dt}\, f(z(t,x,\mu))\, \mu^n(dx)P(d\mu)$$

$$= \int_{\hat{M}} \int_{\hat{X}^n} <A+B_{\mu_t}, Df>\,(z(t,x,\mu))\, \mu^n(dx)P(d\mu)$$

$$= \int_{\hat{M}} \int_{\hat{X}^n} <A+B_{\mu}, Df>\,(x)\, \mu^n(dx)P_t(d\mu).$$

On the other hand,

$$<A+B_\mu, Df>\,(x) = <A+B_{\nu_x}, Df>\,(x) + <B_{\mu-\nu_x}, Df>\,(x),$$

$$B_{\mu-\nu_x}(x) = \int_X B(x-(b))(\mu-\nu_x)(db), \qquad (2.10)$$

and $\displaystyle \int_{\hat{X}^n} \int_X <Df(x), B(x-(b))>\,(\mu-\nu_x)(db)\, \mu^n(dx)$

$$= \int_{\hat{X}^{n+1}} <Df((a_1,\ldots,a_n)), B((a_1,\ldots,a_n)-(a_{n+1}))>\mu^{n+1}(d(a_1,\ldots,a_{n+1}))$$

Making use of these relations completes the proof.

<u>Theorem 2.</u>   Let $\rho$ be a Radon measure in X and $\overline{P}$ a point process in X satisfying $(\Sigma_\rho)$ which is concentrated on $\hat{M}$ and invariant under $(T_t)_{t \in \mathbb{R}}$. We further assume that $\overline{P}$ is of infinite order. Let $\Psi \in L^p(\overline{P})$ for some $p > 1$, nonnegative with $E_{\overline{P}}(\Psi) = 1$, and consider the probability measures $P = \Psi\overline{P}$ and $P_t = P \circ T_{-t}^{-1}$. Then, for each n and each function f which is continuously differentiable and of compact support on $X^n$, the function $t \to \nu_{P_t}^n(f)$ is well defined, finite valued and differentiable. Furthermore, the correlation functions $\rho_{P_t}$ of $P_t$ exist, are given by (1.10) and satisfy the BBGKY hierarchy in the sense

$$\frac{d}{dt}\, \hat{\rho}^n(f \cdot \rho_{P_t}) = \hat{\rho}^n(<A+B_\nu,\ Df> \cdot \rho_{P_t}) + \hat{\rho}^{n+1}(f_1 \cdot \rho_{P_t})\ . \tag{2.11}$$

<u>Proof.</u>   That each $P_t$ is of infinite order has been shown above (see 1.11)). Therefore the function $t \to \nu_{P_t}^n(f)$ is well defined for each $f \in K(X^n)$ and has finite values.   Consider

$$\nu_{P_t}^n(f) = \int_{\hat{M}} \int_{\hat{X}^n} f(z(t,x,\mu))\ \mu^n(dx)P(d\mu),\quad t \in \mathbb{R}.$$

On account of (2.10) combined with (2.6) and Dini's theorem, for each $T > 0$

$$\frac{d}{dt} \int_{\hat{X}^n} f(z(t,x,\mu))\ \mu^n(dx)$$

$$= \int_{\hat{X}^n} <A+B_{\nu_x},\ Df>(x)\ \mu_t^n(dx) + \int_{\hat{X}^{n+1}} \Xi_1(x)\ \mu_t^{n+1}(dx),\quad |t| < T.$$

The integrands on the right-hand side are functions in $K(X^n)$, resp. $K(X^{n+1})$. Each $P_t$ is of infinite order.   Therefore

$$\int_{\hat{M}} \left| \frac{d}{dt} \int_{\hat{X}^n} f(z(t,x,\mu))\ \mu^n(dx) \right| P(d\mu) < +\infty,$$

so that the right-hand side in (BBGKY) makes sense. Furthermore, due to (1.11), for each $T > 0$

$$\left\{ \frac{d}{dt} \int_{\hat{X}^n} f(z(t,x,\mu))\ \mu^n(dx) \right\}_{|t| < T}$$

is uniformly integrable as a bounded subset of $L^2(P)$ by the theorem of

181

la Vallée-Poussin. Therefore $t \to \tilde{\nu}^n_{P_t}(f)$ is differentiable, differentiation with respect to t and integration with respect to P can be interchanged and the derivative of $t \to \tilde{\nu}^n_{P_t}(f)$ is given by (BBGKY).

Combining this with the corollary of Theorem 1 completes the proof.

## 3.  BBGKY HIERARCHY FOR NEWTONIAN DYNAMICS

We now give Theorem 2 a concrete content in describing a class of point processes $\overline{P}$ in $X = \mathbb{R}^\nu \times \mathbb{R}^\nu$ ($\nu \geq 1$) which satisfy the assumptions of Theorem 2 for the Newtonian dynamics $(T_t)_{t \in \mathbb{R}}$ of Example 1. Let $\phi : \mathbb{R}^\nu \to \mathbb{R}$ be a pair potential as in Example 1. For each $q \in \mathbb{R}^\nu$ and $\nu \in M^{..}(\mathbb{R}^\nu)$ let

$$U_\phi(q,\nu) = \int_{\mathbb{R}^\nu} \phi(q-q') \; \nu(dq'). \tag{3.1}$$

We now consider point processes $\overline{P}$ in $X$ satisfying $(\Sigma_\rho)$, where $\rho = z \cdot \lambda \otimes (2\pi)^{-\nu/2} \cdot \lambda$ ($z > 0$, $\lambda$ Lebesgue measure in $\mathbb{R}^\nu$) and where the corresponding density f is given in terms of $\phi$ in the following way:

$$f((q,p),\mu) = \exp(-U_\phi(q,\pi(\mu)) - \frac{|p|^2}{2}), \quad (q,p) \in X, \; \mu \in M^{..}(X). \tag{3.2}$$

Here $\pi(\mu)$ is the image of $\mu$ with respect to the projection $\pi((q,p)) = q$. In particular $\overline{P}$ satisfies the equilibrium equations (1.7). We call $\overline{P}$ a *Gibbs state with potential $\phi$ and Maxwellian velocity distribution* if the image of $\overline{P}$ with respect to $\pi$ is tempered in the sense of Ruelle [17]. The equilibrium equation (1.7) has important consequences for $\overline{P}$:

<u>Lemma</u>.   Let $\overline{P}$ be a Gibbs state with potential $\phi$ and Maxwellian velocity distribution.

(1)   (Ruelle [17]).  $\overline{P}$ is of infinite order.

(2)   (Marchioro, Pellegrinotti and Presutti [10]).  $\overline{P}$ is invariant under the Newtonian dynamics $(T_t)$ of Example 1.

Combining this with Theorem 2 we have shown the existence of a weak solution of the BBGKY hierarchy for the Newtonian dynamics of Example 1 for each initial state of the form $\Psi \cdot \overline{P}$ in any dimension in the following sense:

182

$\underline{\text{Theorem 3.}}$  Let $X = \mathbb{R}^\nu \times \mathbb{R}^\nu$, $\rho = z \cdot \lambda \otimes \tau$ with $z > 0$ and $\tau$ given by (3.4), and let $\phi$ be a pair potential in $\mathbb{R}^\nu$ twice continuously differentiable, of finite range, superstable in the sense above (see the second example), and lower regular. If $\overline{P}$ is a Gibbs state with potential $\phi$ having a Maxwellian velocity distribution and if $\Psi \in L^p(\overline{P})$ for some $p > 1$, nonnegative with $E_{\overline{P}}(\Psi) = 1$, then the correlation functions $\rho_{P_t}$ of the time-evolved states $P_t = (\Psi\overline{P}) \circ T_{-t}^{-1}$ exist for all $t$, are given by (1.17), are differentiable in the sense that $t \to \hat{\rho}^n(f \cdot \rho_{P_t})$ is differentiable for each $f$ continuously differentiable and satisfy the BBGKY hierarchy in the sense of (2.11).

## 4. DETAILED ANALYSIS OF THE FIRST THREE BBGKY HIERARCHY EQUATIONS IN THE CASE OF GRADIENT SYSTEMS

We now analyse the first three hierarchy equations in the case of gradient systems for spatially homogeneous initial states $P$ in $X = \mathbb{R}^\nu$, $\nu \geq 1$, and consider the problem of describing the weak limit points of the time-evolution $\{P_t;\ t \in \mathbb{R}\}$ of $P$.

Let $M_o^{..}(X)$ be the set of all $\mu \in M^{..}(X)$ such that $0 \in \mu$. Fix some $R > \delta \geq 0$ and denote by $M_\delta^{..}(X)$, resp. $M_\delta^o(X)$, the set of all configurations $\mu \in M^{..}(X)$, resp. $M_o^{..}(X)$, such that, for each $x,y \in \mu$, $x \neq y$, we have $|x-y| > \delta$. Let $K_\delta(0)$ be the closed ball in $X$ with centre $0$ and radius $\delta$, let $X_\delta = X \setminus K_\delta(0)$ and $X_{\delta,R} = \{x \in X: \delta < |x| < R\}$.

We consider pair potentials of the following form: let $\phi : X \to \mathbb{R} \cup \{+\infty\}$ be even, with finite range $R$, finite valued and continuously differentiable on $X_\delta$. We assume also that $\phi$ is superstable and lower regular in the sense of Ruelle [17]. Finally, we assume that an evolution theorem is true in the following sense. There exists a measurable subset $\hat{M} \subset M_\delta^{..}(X)$ having the following properties:

(1) for each $\mu \in \hat{M}$ and each $x \in \mu$ the system of equations

$$\frac{d}{dt} z(t,x,\mu) = -\sum_{x \neq y \in \mu} D\phi(z(t,x,\mu)-z(t,y,\mu)) \tag{4.1}$$

has a unique solution $t \to z(t,x,\mu)$ with $z(0,x,\mu) = x$;

(2) the mapping $T_t : \mu \to \sum_{x \in \mu} \varepsilon_{z(t,x,\mu)}$, $t \in \mathbb{R}$, $\mu \in \hat{M}$, is a measurable transformation on $\hat{M}$ such that $T_o = \text{id}$, $T_{t+s} = T_t T_s$ $(s,t \in \mathbb{R})$.

In the one-dimensional case, evolution theorems of this type have been
obtained by Lang [9] and Fichtner and Freudenberg [3]. For higher dimensions,
no such result is available.

For each $x \in X = \mathbb{R}^\nu$ we define an automorphism $T_x$ on $M^{..}(X)$ by

$$(T_x \mu)(B) = \mu(B+x), \quad \mu \in M^{..}(X), \ B \in \mathcal{B}(X). \tag{4.2}$$

Thus, for each measurable function $f : X \to \mathbb{R}_+ \cup \{+\infty\}$,

$$\int_X f(y) \ (T_x \mu)(dy) = \int_X f(y-x) \ \mu(dy). \tag{4.2'}$$

Instead of $T_x \mu$ we shall write $\mu-x$. We call a point process P in X *spatially
homogeneous* if, for each measurable, nonnegative function u on $M^{..}(X)$ and
each $x \in X$,

$$\int_{M^{..}(X)} u(\mu-x)P(d\mu) = \int_{M^{..}(X)} u(\mu)P(d\mu). \tag{4.3}$$

In the case of spatially homogeneous point processes P in X with finite
nontrivial intensity measure, we can reduce the moment measures of P in the
following way. Let $f : X^n \to \mathbb{R}$ be continuous with compact support. Then

$$\nu_P^n(f) = \int_{M^{..}(X)} \int_X h(x,\mu) \ \mu(dx)P(d\mu),$$

where $h(x,\mu) = \int_{X^{n-1}} f(x,y) \ \mu^{n-1}(dy)$. Recall that the Palm measure $P^o$
corresponding to P is given by

$$P^o(v) = \frac{1}{z(P)} \cdot \int_{M^{..}(X)} \int_X g(x) \cdot v(\mu-x) \ \mu(dx)P(d\mu),$$

where v is a nonnegative, measurable function on $M^{..}(X)$ and g is a nonnegative,
measurable function on X with $\lambda(g) = 1$, and the so-called *intensity* z(P) of
P is determined by $\nu_P^1 = z(P) \cdot \lambda$. Here $\lambda$ denotes the $\nu$-dimensional Lebesgue
measure. Using well-known properties of Palm measures (see Mecke [13]), we
get

$$\nu_P^n(f) = z(P) \cdot \int_{M_o^{..}(X)} \int_X h(x,\mu+x) \ \lambda(dx)P^o(d\mu)$$

$$= z(P) \cdot \int_X \int_{M_o^{..}(X)} \int_{X^{n-1}} f(x,(x)+u) \ \mu^{n-1}(du)P^o(d\mu) \ \lambda(dx).$$

184

Here $(x) = (x, \ldots, x)$. To summarize, we have, for spatially homogeneous point process P in $X$ with finite, nontrivial intensity, that

$$\nu_P^n(f) = z(P) \cdot \int_X \int_{X^{n-1}} f(x, (x)+u) \, \nu_{P^o}^{n-1}(du) \, \lambda(dx), \quad f \in K(X^n). \qquad (4.4)$$

In particular, if $g \in K(X)$, $\ell \in K(X^{n-1})$ and $f(x,u) = g(x) \cdot \ell(u-(x))$, $x \in X$, $u \in X^{n-1}$, we get

$$\nu_P^n(f) = z(P) \cdot \lambda(g) \cdot \nu_{P^o}^{n-1}(\ell). \qquad (4.4')$$

Given a spatially homogeneous initial state P with $P(\hat{M}) = 1$, the study of the behaviour of the functions $t \to \nu_{P_t}^n(f)$ therefore reduces to the analysis of $t \to z(P_t)$ and $t \to \nu_{(P_t)^o}^{n-1}(\ell)$. * (For this observe that $P_t$ is spatially homogeneous if P has this property, since each $T_x$ commutes with each $T_t$.) But the intensity $z(P_t)$ does not depend on t, since, for each $B \in B(X)$,

$$\nu_{P_t}^1(B) = \int\int 1_B(z(t,0,\mu-x)+x) \, \mu(dx) P(d\mu)$$

$$= z(P) \cdot \int\int 1_B(z(t,0,\mu)+x) P^o(d\mu) \, \lambda(dx)$$

$$= z(P) \cdot \lambda(B).$$

Consequently $\nu_{P_t}^1 = \nu_P^1$ for each t. Here we have used that

$$z(t,y-x,\mu-x)+x = z(t,y,\mu) \quad \text{for each } x,y \in \mu \in \hat{M}, \qquad (4.5)$$

which can be seen in the following way. This equation is true for $t = 0$, and both sides satisfy the same differential equation on account of $B_\mu(y) = B_{\mu-x}(y-x)$. Therefore the intensity is a constant of the motion. On the other hand, combining (4.5) with well-known properties of Palm measures, we get an interesting equation due to Harris [5] which will be important in the following:

---

*Note that $z(P_t)$, resp. $\nu_{(P_t)^o}^1(\phi)$, can be interpreted as the particle density, resp. the specific energy, of $P_t$ (see Nguyen and Zessin [15]).

$\underline{\text{Lemma 2.}}$   If P is a spatially homogeneous point process in X with finite, nontrivial intensity and concentrated on $\hat{M}$, then

$$\int_{M_o^{\cdot\cdot}(X)} \int_X f(x,\mu)\ \mu(dx)(P_t)^o(d\mu) = \int_{M_o^{\cdot\cdot}(X)} \int_X f(x,\mu)\ \mu(dx)(P^o)_t(d\mu),$$

$$\tag{4.6}$$

where $(P^o)_t(u)_{Df} = \int_{M_o^{\cdot\cdot}(X)} u(\mu_t - z(t,o,\mu))P^o(d\mu)$. Here f and u are nonnegative, measurable functions.

$\underline{\text{Proof.}}$   Let g be a nonnegative, measurable function on X with $\lambda(g) = 1$. Then

$$\iint f(x,\mu)\ \mu(dx)(P_t)^o(d\mu)$$

$$= \frac{1}{z(P_t)} \cdot \iint\left[\int g(z(t,x,\mu))f(z(t,y,\mu)-z(t,x,\mu),\mu_t-z(t,x,\mu))\mu(dy)\right] \times$$

$$\times\ \mu(dx)P(d\mu).$$

Combining (4.5) with a well-known characterization of Palm measures (see Mecke $\boxed{13}$, Satz 2.3) this equals

$$\iint g(z(t,o,\mu)+x)f(z(t,u,\mu)-z(t,o,\mu),\mu_t-z(t,o,\mu))\ \mu(du)P^o(d\mu)\ \lambda(dx).$$

Here we have used also that $z(P_t) = z(P)$. The last integral equals the right-hand side of (4.6) because $\lambda(g) = 1$.

$\underline{\text{Remarks.}}$   1.  Particularly, the lemma implies that $(P_t)^o = (P^o)_t$ (take $f(x,\mu) = 1_{\{o\}}(x) \cdot f'(\mu)$ in (4.6)). Therefore in the following we write simply $P_t^o$.

2.  Another consequence of the lemma are the equations

$$v_{P_t^o}^n(f) = \int_{M_o^{\cdot\cdot}(X)} \int_{X^n} f(z(t,x,\mu)-z(t,o,\mu))\ \mu^n(dx)P^o(d\mu)$$

where f is a nonnegative, measurable function on $X^n$.

In the following lemma we derive a system of differential equations for the reduced moment measures of $P_t$, considered outside $o$, which we denote by $\tilde{v}_{P_t^o}^n$.

More precisely,

$$\hat{\nu}^n_{P^o_t}(f) = \int_{M^{\cdot\cdot}_o(X)} \int_{X^n} f(x)\,(\mu-\varepsilon_o)^n(dx)P^o_=(d\mu)$$

$$\tag{4.6'}$$

$$= \int_{M^{\cdot\cdot}_o(X)} \int_{X^n} f(z(t,x,\mu)-z(t,0,\mu))(\mu-\varepsilon_o)^n(dx)P^o(d\mu),$$

where $f$ is a nonnegative, measurable function on $X^n$.

<u>Lemma 3.</u>   For $n \in \mathbb{N}$ let $f : X^n \to \mathbb{R} \cup \{+\infty\}$ be a function of compact support, which is finite valued and differentiable on $X^n_\delta$. Let $P$ be a spatially homogeneous point process in $X$ satisfying

$$P(\hat{M}) = 1; \tag{4.7}$$

the function $t \to \hat{\nu}^n_{P^o_t}(f)$ is well defined, finite valued, differentiable,

and the interchange of differentiation and integration in (4.6') is

allowed. $\tag{4.8}$

Then the function $t \to \hat{\nu}^n_{P^o_t}(f)$ satisfies the n-th *reduced BBGKY hierarchy*

*equation*

$$\frac{d}{dt}\,\hat{\nu}^n_{P^o_t}(f) = \int_{M^{\cdot\cdot}_o(X)} \int_{X^n} \sum_{i=1}^n \langle D_{x_i} f(x),\, B_\mu(x_i) - B_\mu(0)\rangle(\mu-\varepsilon_o)^n(dx)P^o_t(d\mu).$$

$$\tag{4.9}$$

<u>Proof.</u>   Let $\mu \in \hat{M} \cap M^{\cdot\cdot}_o(X)$ and $x = (x_1,\ldots,x_n) \in X^n$ with $x_i \in \mu-\varepsilon_o$ for $i = 1,\ldots,n$. Then

$$\frac{d}{dt}\,f(z(t,x,\mu)-z(t,0,\mu))$$

$$= \sum_{i=1}^n \langle D_{x_i} f(z(t,x,\mu)-z(t,0,\mu)),\, B_{\mu_t}(z(t,x_i,\mu))-B_{\mu_t}(z(t,0,\mu))\rangle,$$

so that, on account of (4.6) combined with $B_\mu(x) = B_{\mu-y}(x-y)$, we get Equation (4.9).

We now return to the problem of describing the solutions of the reduced hierarchy equations.  Let $P^O(M)$ denote the set of all spatially homogeneous point processes in X with nontrivial, finite intensity which are concentrated on $\hat{M}$.

__Theorem 4.__   Let $P \in P^O(M)$ and let $\phi$ be a pair potential satisfying the assumption given above.  We furthermore assume that the functions $t \to \hat{\nu}^1_{P^O_t}(\phi)$ and $t \to \hat{\nu}^2_{P^O_t}(D\phi \otimes D\phi)$ are well defined, finite valued, differentiable, and the interchange of differentiation and integration is allowed.  Consider the following properties of P:

(1) P is an *equilibrium state* for the gradient dynamic (4.1), i.e.  P is invariant under $(T_t)_{t \in \mathbb{R}}$ ;

(2) P is a *rigid state* with potential $\phi$ in the sense of Lang, i.e.

$$\int_X D\phi(x)(\mu - \varepsilon_o)(dx) = 0 \quad P^O - a.s. \; [\mu] \; ;$$

(3) $\quad \int_X D\phi(x-y)(\mu - \varepsilon_y)(dx) = 0$ for each $y \in \mu \quad P^O - a.s. \; [\mu] \; ;$

(4) P is an *equilibrium state in the variational sense* with potential $\phi$, i.e.

$$\hat{\nu}^1_{P^O}(\phi) = \min \{\hat{\nu}^1_{Q^O}(\phi) \; : \; Q \in P^O(\hat{M})\};$$

(5) P is a *ground state* with potential $\phi$, i.e.

$$e(\mu) = \inf \{e(\nu); \; \nu \in \hat{M}\} \quad P - a.e. \; [\mu] \; .$$

Here e denotes the specific energy, defined by

$$e(\mu) = \lim_{n \to \infty} \frac{1}{|G_n|} \cdot \frac{1}{2} \cdot \Sigma\{\phi(x-y); \{x,y\} \in \mu, \; x \neq y, \; \{x,y\} \wedge G_n \neq \emptyset\}.$$

(Note that the limit, taken for suitable sequences $G_n \nearrow X$, exists P - a.s. on account of $\hat{\nu}^1_{P^O}(\phi) < + \infty$; see Nguyen and Zessin [15], Theorem (7.15).)

(6) P is a *lattice type state*, i.e. P is concentrated on a discrete subgroup of X.

Then the following relations are true:

188

$$(1) \Rightarrow (2) \Leftrightarrow (3) \Leftarrow (4) \Leftarrow (5)$$
$$\Uparrow$$
$$(6)$$

If, in addition to the assumptions above, $\phi$ is also convex and twice continuously differentiable on $X_\delta$, then $(2) \Rightarrow (1)$.

Proof.    $(1) \Rightarrow (2)$ : Combining the last lemma in the case $n = 1$ and $f = \phi$ with the $(T_t)$ - invariance of $P$ yields

$$\int_{\overset{..}{M}_o(X)} <B_\mu(0), \int_X (D\phi(-x)-D\phi(x))(\mu-\varepsilon_o)(dx)> P^o(d\mu) = 0,$$

if we take into account the basic invariance property of Palm measures (Mecke [13], Satz 2.5) that the Campbell measure of $P^o$ is invariant under $(x,\mu) \to (-x,\mu-x)$.

But $\int_X (D\phi(-x)-D\phi(x))(\mu-\varepsilon_o)(dx) = -2 \cdot B_\mu(0)$ ,

since $D\phi(-x) = -D\phi(x)$, and this implies $(2)$.

$(2) \Leftrightarrow (3)$ : Using Mecke's Theorem 2.5 again, we have

$$\int_{\overset{..}{M}_o(X)} \int_X |\int_X D\phi(x)(\mu-\varepsilon_o)(dx)|^2 \mu(dy) P^o(d\mu)$$

$$= \int_{\overset{..}{M}_o(X)} \int_X |\int_X D\phi(x-y)(\mu-\varepsilon_y)(dx)|^2 \mu(dy) P^o(d\mu).$$

Therefore $(2)$ is equivalent to $(3)$.

$(4) \Rightarrow (2)$ : If $P$ is not a rigid state with potential $\phi$, then $P^o(|B_.(0)|^2) > 0$. On the other hand, we have assumed that the function $t \to \hat{v}^2_{P^o_t}(D\phi\otimes D\phi) = P^o_t(|B_.(0)|^2)$ is continuous; therefore there exists $\varepsilon > 0$ such that $P^o_t(|B_.(0)|^2) > 0$ for each $t \in (-\varepsilon,+\varepsilon)$. Furthermore, Lemma 3 yields

$$\frac{d}{dt} \hat{v}^1_{P^o_t}(\phi) = -2 \, P^o_t(|B_.(0)|^2),$$

so that $t \to \hat{v}^1_{P^o_t}(\phi)$ is strictly decreasing in $(-\varepsilon,+\varepsilon)$; in particular,

$\hat{v}^1_{P^o}(\phi) > \hat{v}^1_{P^o_t}(\phi)$ for $0 < t < \varepsilon$. Therefore P cannot be an equilibrium state

in the variational sense with potential $\phi$.

$(5) \Rightarrow (4)$ : Let P be a ground state for $\phi$. Then e equals $\frac{1}{2} \hat{v}^1_{P^o}(\phi)$ P - a.s.
This follows from the ergodic theorem for the energy mentioned above. This
theorem also implies that, for each $Q \in P^o(\hat{M})$ which is ergodic with respect
to $(T_x)_{x \in X}$, we have $\frac{1}{2} \hat{v}^1_{Q^o}(\phi) = e$ $Q$ - a.s. Consequently, $\hat{v}^1_{P^o}(\phi) \leq \hat{v}^1_{Q^o}(\phi)$
for each ergodic $Q \in P^o(\hat{M})$. This implies (4), using the fact that each
spatially homogeneous point process with finite intensity can be decomposed
into ergodic components.

$(6) \Rightarrow (2)$ is obvious, and $(2) \Rightarrow (1)$ will be postponed (see Prop. 3).

<u>Remarks.</u>    1. In the one-dimensional case, Lang has shown for point processes
$P \in P^o(\hat{M})$, which are concentrated on configurations with a constant density
$\frac{1}{r}$, $r > 0$, and for pair potentials, which essentially are convex and
repulsive with $\delta < r \leq R$, that $(3) \Rightarrow (6)$ and $(6) \Rightarrow (5)$ (see Lang [9]).
Combining these results with Proposition 3 below yields that all conditions
of Theorem 4 are equivalent.*  The problem of finding the ground states with
potential   in higher dimensions is connected with the unsolved problem of
the closest packing of balls in X. We note also that Theorem 1 of Ventevogel
([20]) combined with the ergodic theorem for the energy mentioned above
gives a short direct proof of $(6) \Rightarrow (5)$ in the one-dimensional case.

2. Theorem 4 shows that rigid states are intimately related to one
invariant of the motion, namely the specific energy:  if P is an initial
state which is spatially homogeneous with finite non-trivial intensity,
concentrated on $\hat{M}$ for which the specific energy $t \to \hat{v}^1_{P^o_t}(\phi)$ is an invariant of

the motion, then P is a rigid state.

---

*As has been remarked by Ventevogel [20], a theorem $(6) \Rightarrow (5)$ has been
conjectured by Wigner [21] forty years ago.  See also Born [1].

190

Assuming also convexity for the potential $\phi$ (and therefore also $\phi \geq 0$), and combining the first reduced hierarchy equations with some ideas Lang used in the one-dimensional case, we get the following more dimensional version of an ergodic theorem due to Lang ([9]) and Fichtner and Freudenberg ([3]).

<u>Theorem 5.</u>  Let $P \in \mathcal{P}^o(\hat{M})$ and, in addition to the assumptions in Theorem 4, let $\phi$ be also convex and twice continuously differentiable on $X_\delta$. Then, for each sequence of real numbers $t_n$ tending to infinity, there exists a subsequence $t_{n_k}$ and a spatially homogeneous point process $\overline{P}$ in X such that $P_{t_{n_k}}$ admit the weak limit $\overline{P}$. Furthermore, $\overline{P}$ is a rigid state with potential $\phi$ having the same intensity as the initial state P.

<u>Proof.</u>  1.  It follows immediately from Proposition 3.2.8 in Matthes, Kerstan and Mecke [11] that the time-evolution $\{P_t;\ t \in \mathbb{R}\}$ is relatively compact in the space of all point processes in X endowed with the weak topology in ovserving that for each $n \in \mathbb{N}$, $K \in \mathcal{B}_o(X)$

$$\sup_{t \in \mathbb{R}} P_t\{N(K) \geq n\} \leq \frac{1}{n} \sup_t \nu^1_{P_t}(K) = \frac{1}{n} \nu^1_P(K) \ .$$

Consequently, for each sequence $t_n \to \infty$ there exists a point process $\overline{P}$ in X and a subsequence $t_{n_k} \to \infty$ such that $P_{t_{n_k}} \Rightarrow \overline{P}$ if $k \to \infty$. $\overline{P}$ is spatially homogeneous because the set of all spatially homogeneous point processes in X is weakly closed  (see Matthes, Kerstan and Mecke [11], 10.3.4).

2. To see that $\overline{P}$ is a rigid state with potential $\phi$ having the same intensity as P we proceed as follows.  We know from the proof of Theorem 4 that

$$\frac{d}{dt} \nu^1_{P^o_t}(\phi) = -2 \cdot \int_{M^{\cdot\cdot}_o(X)} |B_\mu(0)|^2\, P^o_t(d\mu) . \tag{a}$$

Consequently the specific energy $t \div \nu^1_{P^o_t}(\phi)$ is decreasing.

3. Since the integral on the right-hand side of (a) equals $\nu^2_{P^o_t}(D\phi \otimes D\phi)$, we can consider the second reduced hierarchy equation evaluated for $f = D\phi \otimes D\phi$. In this case, for each $x_1, x_2 \in X \setminus \{0\}$

$$D_{x_1} f(x_1,x_2) = D^2\phi(x_1) D\phi(x_2) = D_{x_2} f(x_2,x_1),$$

and the second reduced hierarchy equation amounts to

$$\frac{d}{dt} \hat{\nu}^2_{P^o_t}(D\phi\otimes D\phi) = 2\cdot\iint <B_\mu(0),D^2\phi(x_1)(B_\mu(x_1)-B_\mu(0))>(\mu-\varepsilon_o)(dx_1)P^o_t(d\mu)$$

$$= -\iint <B_\mu(0),D^2\phi(x_1)B_\mu(0)>+<B_\mu(x_1),D^2\phi(x_1)B_\mu(x_1)>$$

$$-2\cdot<B_\mu(0),D^2\phi(x_1)B_\mu(x_1)>(\mu-\varepsilon_o)(dx_1)P^o_t(d\mu).$$

Here we have used again the invariance property of Palm measures (see Mecke [13], Satz 2.5). Since $\phi$ is convex, $D^2\phi(x_1)$ is positive semidefinite; combining this with the symmetry of $D^2\phi(x_1)$, the last integral equals

$$-\iint (B_\mu(0)-B_\mu(x_1))D^2\phi(x_1)(B_\mu(0)-B_\mu(x_1))(\mu-\varepsilon_o)(dx_1)P^o_t(d\mu)$$

and is non-positive. Consequently the specific energy $t \to \hat{\nu}^1_{P^o_t}(\phi)$ is convex; since it is also decreasing and nonnegative, we get

$$\lim_{t\to\infty} \frac{d}{dt} \hat{\nu}^1_{P^o_t}(\phi) = 0. \tag{b}$$

4. We now show that $\lim_{n\to\infty} P_{t_n} = \overline{P}$ implies $z(P) = z(\overline{P})$. To see this we first note that the superstability of the potential $\phi$ implies that there exist $A > 0$, $B \geq 0$ such that

$$\sup_{t\in\mathbb{R}} \nu^2_{P_t}(F_o \times F_o) \leq \frac{1}{2A} \hat{\nu}^1_{P^o}(\phi) + \frac{B}{A} \cdot z(P) < +\infty, \tag{c}$$

where $F_o = \{(x_1,\dots,x_\nu) : -\frac{1}{2} \leq x_i < \frac{1}{2}, i = 1,\dots,\nu\}$.

This can be seen in the following way: superstability of $\phi$ means that there exist $A > 0$, $B \geq 0$ such that, for each $G \in \mathcal{B}_o(X)$ and $\mu \in M^{\cdot\cdot}(X)$,

$$\frac{1}{2} \cdot \sum_{\substack{x,y\in\mu\cap\overline{G} \\ x\neq y}} \phi(x-y) \geq \sum_{u\in G\cap Z^\nu} \left[ A \cdot N(F_o+u)(\mu)^2 - B \cdot N(F_o+u)(\mu) \right],$$

where

$$\overline{G} = \cup \ \{F_o + u : (F_o+u) \cap G\neq 0, u \in Z^\nu\}.$$

192

Consequently, for each $G \in B_o(X)$ and $t \in \mathbb{R}$,

$$\nu^2_{P_t}(F_o \times F_o) \leq \frac{1}{2A} \cdot \frac{1}{|\bar{G}|} \cdot \int_{\hat{M}} \sum_{\substack{x,y \in \mu \cap \bar{G} \\ x \neq y}} \phi(x-y) P_t(d\mu) + \frac{B}{A} \cdot z(P).$$

Going to the limit taken for suitable sequences $G \nearrow X$, the ergodic theorem for the energy (see Nguyen and Zessin [15], Theorem (7.15)) implies

$$\nu^2_{P_t}(F_o \times F_o) \leq \frac{1}{2A} \hat{\nu}^1_{P_t^o}(\phi) + \frac{B}{A} \cdot z(P), \quad t \in \mathbb{R}.$$

Since $t \to \hat{\nu}^1_{P_t^o}(\phi)$ is decreasing with $\hat{\nu}^1_{P^o}(\phi) < +\infty$, we get (c).

To show that (c) implies $z(P) = z(\bar{P})$, we use the Cauchy-Tschebyscheff inequality to see that

$$\sup_{t \in \mathbb{R}} \int_{N(F_o)(T_t) \geq n} \mu_t(F_o) P(d\mu) \leq \frac{1}{n} \sup_{t} \nu^2_{P_t}(F_o \times F_o).$$

Therefore (c) implies that the set $\{N(F_o)(T_t); t \in \mathbb{R}\}$ is uniformly integrable with respect to P, and this yields $z(P) = z(\bar{P})$ due to a well-known theorem (see Matthes et. al. [11], 10.1.1).

5. Now consider a sequence $t_n \to \infty$ such that $\lim_{n \to \infty} P_{t_n} = \bar{P}$. We know from 4. above that in this case $z(P) = z(\bar{P})$. But this in turn implies $\lim_{n \to \infty} P^o_{t_n} = \bar{P}^o$ (see Kallenberg [6], Theorem 6.2). Therefore, since each $\mu \to |B_\mu(0)|^2 \wedge N$, $N \in \mathbb{N}$, is bounded and continuous on $M_o^{\cdot\cdot}(X)$,

$$\lim_{n \to \infty} P^o_{t_n}(|B_\mu(0)|^2 \wedge N) = \bar{P}^o(|B_\mu(0)|^2 \wedge N)$$

for each N. Consequently, from (a) and (b),

$$\bar{P}^o(|B_\mu(0)|^2 \wedge N) = 0 \quad \text{for each } N$$

and finally $\bar{P}^o(|B_\mu(0)|^2) = 0$. This proves the theorem.

<u>Remarks.</u>  1.  Looking once more at the proof of the theorem, we see that the convexity of $\phi$ implies that the mean square velocity of the particle which at time 0 started in 0 is decreasing in such a way that

$$P^O(|B_{\mu_t}(z(t,0,\mu)))|^2) \leq \frac{1}{2t} \cdot (\mathring{v}^1_{P^O}(\phi) - \mathring{v}^1_{P^O_t}(\phi)).$$

2. Combining the theorem with the fact that the time-evolution $\{P_t ; t \in \mathbb{R}\}$ is relative compact, we immediately get the following version of the theorem. Under the assumptions of the theorem, $P_t$ converges weakly to the non-empty set $R^O_{z(P)}(\phi)$ of all spatially homogeneous rigid states with potential $\phi$ having the same intensity as P, i.e. for each weakly open subset U which contains $R^O_{z(P)}(\phi)$ $P_t \in U$ for all t sufficiently large.

In the situation of the last theorem, we can see now that rigid states are equilibrium states. Particularly, equilibrium states in the variational sense, ground states and lattice type states are invariant under the gradient dynamic.

<u>Proposition 3.</u>   If $P \in P^O(\hat{M})$ and $\phi$ is a potential satisfying the assumptions of Theorem 5, then P is a rigid state with potential $\phi$ if and only if P is an equilibrium state for the gradient dynamic (4.1).

<u>Proof.</u>   We know already from Theorem 4 that equilibrium states are rigid states. Conversely, if P is a rigid state with potential $\phi$, then for each $t \in \mathbb{R}$ $P_t$ is also a rigid state with potential $\phi$. This can be seen in the following way. We know from the proof of Theorem 5 that, for each $Q \in P^O(\hat{M})$, the function

$$\Psi_Q : t \to -2 \cdot \int_{M^O_\delta(X)} |B_\mu(0)|^2 Q^O_t(d\mu)$$

is non-decreasing and non-positive. Therefore, because $\Psi_P(0) = 0$, $P_t$ is a rigid state for each $t \geq 0$.

Combining Lemma 2 and Theorem 4, this implies that, for each t,

$$0 = B_{\mu_t}(z(t,x,\mu)) = \frac{d}{dt} z(t,x,\mu) \text{ for each } x \in \mu \quad P^O - \text{a.e. } [\mu] .$$

But this immediately implies, for each $t \geq 0$, that

$$z(t,x,\mu) = x \quad \text{for each } x \in \mu \quad P^O - \text{a.e. } [\mu] ,$$

i.e., $T_t\mu = \mu$ $P^O$- a.e. for each $t \geq 0$ and thus for each $t \in \mathbb{R}$ (since each $T_t$ is invertible). Therefore $P^O_t = P^O$ for each $t \in \mathbb{R}$, and finally $P_t = P$ for each $t \in \mathbb{R}$ because the first moment measures of P and each $P_t$ coincide.

194

<u>ACKNOWLEDGEMENT</u>

Section 4 of this paper owes much to oral and written discussions with Reinhard Lang.

<u>REFERENCES</u>

[1]  Born, M.  On the stability of cristal lattices. I. *Proc. Camb. Soc. 36*, 160-172 (1940).

[2]  Dobrushin, R.L.  Vlasov's equation. *Funct. Anal. Applications 13*, 115-123 (1979).

[3]  Fichtner, K.-H. and Freudenberg,W.  Asymptotic behaviour of time evolutions of infinite particle systems. *Z. Wahrscheinlichkeitstheorie verw. Gebiete 54*, 141-159 (1980).

[4]  Gallavotti, G., Lanford, O.E. and Lebowitz, J.L. Thermodynamic limit of time-dependent correlation functions for one-dimensional systems. *J. Math. Phys. 11*, 2898-2905 (1970).

[5]  Harris, T.E.  Random measures and motions of point processes. *Z. Wahrscheinlichkeitstheorie verw. Gebiete 18*, 85-115 (1971).

[6]  Kallenberg, O.  Characterization and convergence of random measures and point processes. *Z. Wahrscheinlichkeitstheorie verw. Gebiete 27*, 9-21 (1973).

[7]  Kerstan, J., Matthes, K., Mecke, J. *Infinitely Divisible Point Processes* (in Russian). Moscow.  Nauka (1982).

[8]  Lanford, O.E.  Classical mechanics of one-dimensional systems of infinitely many particles. *Commun. math. Phys. 9*, 169-181; and *11*, 257-292 (1969).

[9]  Lang, R.  On the asymptotic behaviour of infinite gradient systems. *Commun. math. Phys. 65*, 129-149 (1979).

[10]  Marchioro, C., Pellegrinotti, A. and Presutti, E.  Existence of time evolution of $\nu$-dimensional statistical mechanics. *Commun. math. Phys. 40*, 175-185 (1975).

[11]  Matthes, K., Kerstan, J. and Mecke, J. *Infinitely divisible point processes*. Chichester, New York, Brisbane, Toronto. Wiley (1978).

[12]  Matthes, K., Mecke, J. and Warmuth, W.  Bemerkungen zu einer Arbeit
      von Nguyen, X.X. and Zessin, H. *Math. Nachr. 88*, 117-127 (1979).

[13]  Mecke, J. Stationäre zufällige Masse auf lokalkompakten Abelschen
      Gruppen. *Z. Wahrscheinlichkeitstheorie verw. Gebiete 9*, 36-58
      (1967).

[14]  Neunzert, H. Neuere qualitative und numerische Methoden in der Plasma-
      physik, *Gastvorlesung im Rahmen des Paderborner Ferienkurses in
      Angewandter Mathematik* (1975).

[15]  Nguyen, X.X. and Zessin, H.  Ergodic theorems for spatial processes.
      *Z. Wahrscheinlichkeitstheorie verw. Gebiete 48*, 133-158 (1979).

[16]  Nguyen, X.X. and Zessin, H.  Integral and differential characterization
      of the Gibbs process, *Math. Nachr. 88*, 105-115 (1979).

[17]  Ruelle, D.  Superstable interactions in classical statistical
      mechanics. *Commun. math. Phys. 18*, 127-159 (1970).

[18]  Sinai, Ya.G. and Suhov, Yu.M.  Existence theorems for solutions of the
      Bogolyubov equations. *Teoreticheskaya i Matematicheskaya Fizika
      19*, 344-363 (1974).

[19]  Takahashi, Y.  On a class of Bogoliubov equations and the time
      evolution in classical statistical mechanics. In *Random Fields,*
      Vol.II (ed. Fritz,J., Lebowitz, J.L., Szász, D.J., 1033-1056.
      Amsterdam-Oxford-New York, North Holland (1979).

[20]  Ventevogel, W.J.  On the configuration of a one-dimensional system of
      interacting particles with minimum potential energy per particle.
      *Physica, 92 A*, 343-361 (1978).

[21]  Wigner, E.P.  *Phys. Rev. 46*, 1002 (1934); *Trans. Faraday Soc. 34*,
      678 (1938).

[22]  Zessin, H.  Equilibrium time evolutions and BBGKY hierarchy equations
      of large classical systems. *Z.Wahrscheinlichkeitsth. verw.
      Gebiete 68*, 461-472 (1985).

Hans Zessin,
Fakultät für Mathematik,
Universitätsstrasse,
D-4800 Bielefeld 1,
West Germany.

S ALBEVERIO, W KARWOWSKI, M RÖCKNER & L STREIT

# Capacity, Green's functions and Schrödinger operators

## 1. INTRODUCTION

The theory of general Markov processes with one-dimensional state space has been developed both analytically and probabilistically in fundamental work by Feller, Dynkin and Ito-McKean. A natural extension of this theory to the case of a more dimensional state space is provided, for symmetric processes, by the Beurling-Deny-Fukushima-Silverstein theory of Dirichlet forms [1] - [4]. The latter theory permits in particular the development of a general stochastic calculus in the case of generators with coefficients which are singular, and in particular which need not be measurable functions. Therefore, in this sense the theory goes beyond both Ito-Stratonovič's and Stroock-Varadhan's theories of stochastic differential equations, see e.g. [1], [2], [5], [6]. It is also well known that in quantum mechanics there is a direct need to handle differential operators with singularities due to the nature of the forces involved. In quantum field theory additional singularities arise from constraints due to invariance properties and the necessity of considering infinitely many degrees of freedom.

The theory of Dirichlet forms thus appears to be a natural tool for handling problems of quantum theory and this has been exploited in a series of publications, see e.g. [2], [5] - [15] and references therein (this approach to quantum theory is termed "quantum theory in terms of ground states" or "Dirichlet quantum theory"). As far as quantum mechanics of finitely many particles is concerned, the basic observation is that the Hamiltonian of the theory can be realized, by unitary equivalence, as the infinitesimal generator of a symmetric Markov process associated with a particular Dirichlet form (also called "energy form") obtained (in the absence of electromagnetic fields) by closure from $\frac{1}{2} \int |\nabla f|^2 d\mu$ on $C_o(\mathbb{R}^d)$-functions, with $\mu$ a Radon measure on $\mathbb{R}^d$, the form being considered in $L^2(\mathbb{R}^d, d\mu)$ (d = number of degrees of freedom). A detailed study of such forms, in particular in relation to applications to quantum mechanics, has been provided especially in [2], [5], [6], [8], and references therein. In particular it has been shown that there is a relation between the concepts

197

of vanishing capacity of a set and of quantum mechanical tunneling through
the set [11], [15].

In this paper we first recall (in Section 2) the relation between the
concepts of capacity and equilibrium potential in the theory of Dirichlet
forms and the corresponding notions of classical potential theory. We
present some basic concepts and illustrate them by examples, using, in
particular, illustrations from our previous work. We do this because we need
such concepts and relations in Section 3 and also because we think that
these relations are perhaps not well known even to workers in the inter-
disciplinary field of applications of probability theory to problems of
mathematical physics.

In Section 3 we point out a relation between capacity and Green function,
concerning in particular capacity zero sets, which we think may be of some
interest in view of the above-mentioned quantum mechanical interpretation of
capacity. The relation is not given in its natural full generality, the
point here being just to indicate, also with the help of examples, a possible
further line of investigation.


## 2.  SOME CONCEPTS OF POTENTIAL THEORY FOR DIRICHLET SPACES

We shall here briefly recall some basic notions of the theory of Dirichlet
forms on locally compact separable Hausdorff spaces.

Let $X$ be a locally compact Hausdorff space with a countable base for the
topology. Let $m$ be a positive Radon measure on $X$ with supp $[m]$ = $X$. We
write $L^2(X;m)$ for the real $L^2$-space with inner product $(u,v) \equiv \int_X u(x)v(x)m(dx)$

We consider symmetric bilinear positive forms $\varepsilon$ on $L^2(X;m)$, densely
defined, with definition domain $D(\varepsilon)$. Hence $\varepsilon$ is a real-valued symmetric
bilinear map on $D(\varepsilon) \times D(\varepsilon)$ such that $\varepsilon(u,u) \geq 0$ for all $u \in D(\varepsilon)$. $\varepsilon$ is
closed if $D(\varepsilon)$ is a complete normed space with respect to a norm $\| \ \|_\alpha$ for
some $\alpha > 0$ (and then automatically for all $\alpha > 0$) given by $\|u\|_\alpha \equiv \varepsilon_\alpha(u,u)$,
with $\varepsilon_\alpha(u,v)$ the symmetric bilinear positive form given by
$\varepsilon_\alpha(u,v) \equiv \varepsilon(u,v) + \alpha(u,v)$ for all $u,v \in D(\varepsilon)$. $\varepsilon$ is closable if it has a
closed extension.

A symmetric bilinear positive densely defined form $\varepsilon$ on $L^2(X;m)$ is said
to be a *Dirichlet form* if it is closed and has the contraction property
$\varepsilon(u^{\#},u^{\#}) \leq \varepsilon(u,u)$ for all $u \in D(\varepsilon)$, with $u^{\#} \equiv (u \vee 0) \wedge 1$ (with $\vee$ for supremum
and $\wedge$ for infimum). For all concepts related to Dirichlet forms we refer

to $[1]$. In this paper we shall always consider Dirichlet forms which are *regular* in the sense that $D(\varepsilon) \cap C_c(X)$ is dense in $D(\varepsilon)$ in the $\| \ \|_\alpha$-norm and in $C_c(X)$ in the uniform norm, where $C_c(X)$ denotes the space of continuous functions on X which have compact support (see $[1]$).

For any open subset $A \subset X$ let $L_A \equiv \{u \in D(\varepsilon) \mid u \geq 1 \text{ m-a.e. on } A\}$. We define the $((\varepsilon,\alpha) - )$ *capacity of A* by

$$\text{Cap}_\alpha (A) = \begin{cases} \inf\limits_{u \in L_A} \varepsilon_\alpha(u,u) & \text{if } L_A \neq \emptyset \\ \\ \infty & \text{if } L_A = \emptyset. \end{cases}$$

For any subset A of X we put

$$\text{Cap}_\alpha (A) = \inf_{\substack{B \supset A \\ B \text{ open}}} \text{Cap}_\alpha (B)$$

The following theorem is proved in $[1]$ (Lemma 3.1.1):

<u>Theorem 1.</u>   (i) For each open subset $A \subset X$ of finite capacity there is a unique element $e_\alpha (A) \in L_A$ such that $\varepsilon_\alpha(e_\alpha(A), e_\alpha(A)) = \text{Cap}_\alpha(A)$.

(ii) $0 \leq e_\alpha(A) \leq 1$ m-a.e., $e_\alpha(A) = 1$ m-a.e. on A.

(iii) $e_\alpha(A)$ is the unique element of $D(\varepsilon)$ satisfying $e_\alpha(A) = 1$ m-a.e. on A and $\varepsilon_\alpha(e_\alpha(A), v) \geq 0$ for all $v \in D(\varepsilon)$ with $v \geq 0$ m-a.e. on A.

(iv) $v \in D(\varepsilon)$, $v = 1$ m-a.e. on A imply $\varepsilon_\alpha(e_\alpha(A), v) = \text{Cap}_\alpha(A)$.

(v) If $A \subset B$ are open and s.t. $L_A$ and $L_B$ are non-void, then
$$e_A \leq e_B \text{ m-a.e.}$$

$e_\alpha(A)$ is called the $\alpha$-*equilibrium potential of A.*

It was shown by Deny and Fukushima that $\text{Cap}_\alpha$ is a Choquet capacity, see $[1]$, Th. 3.1.1.

A statement depending on x in subset A of X is said to hold *quasi-everywhere* (q.e.) if there exists a set $N \subset A$ with $\text{Cap}_\alpha(N) = 0$ for some $\alpha > 0$ such that the statement holds for every $x \in A - N$. The definition is independent of $\alpha$, as is seen using $(1 \wedge \alpha)\varepsilon_1(u,u) \leq \varepsilon_\alpha(u,u) \leq (1 \vee \alpha)\varepsilon_1(u,u)$.

Let v be a function defined q.e. on X. v is called *quasi-continuous* if there exists for any $\varepsilon > 0$ an open subset $U \subset X$ such that $\text{Cap}_\alpha(U) < \varepsilon$ for

199

some $\alpha > 0$ and $u \upharpoonright X - U$ is continuous. If $u \upharpoonright X_\Delta - U$, with $X_\Delta \equiv X \cup \Delta$ the one point compactification of X, is continuous we say that u is *quasi-continuous in the restricted sense*. For any function f on X we define $\tilde{f}$ as a modification (with respect to m) of f which is quasi-continuous in the restricted sense, if it exists. Since $\varepsilon$ is regular it follows by [1], Theorem 3.1.3 that every $u \in D(\varepsilon)$ has a quasi-continuous modification in the restricted sense. According to [1] one defines the $\alpha$-*equilibrium potential* $e_\alpha(B)$ for any Borel subset B of X with $Cap_\alpha(B) < \infty$ as the unique element in $D(\varepsilon)$ satisfying $\tilde{e}_\alpha(B) = 1$ quasi-everywhere on B and $\varepsilon_\alpha(e_\alpha(B),v) \geq 0$ for all $v \in D(\varepsilon)$, with $\tilde{v} \geq 0$ quasi-everywhere on B.

Then $e_\alpha(B)$ so defined reduces to the previously defined one when B is open, on the basis of above theorem and the fact that for open sets U and u quasi-continuous on U we have that $u \geq 0$ m-a.e. on $U \Rightarrow u \geq 0$ q.e. on U, cf. [1], Lemma 3.1.4. It is also proved in [1], Th. 3.3.1, that for any Borel set B with $Cap_\alpha(B) < \infty$ we have $Cap_\alpha(B) = \varepsilon_\alpha(e_\alpha(B), e_\alpha(B))$ and then Theorem 1 extends to these Borel sets. Moreover [1] (proof of Lemma 3.1.1, and p.75), $e_\alpha(B)$ is the unique element in $L_B \equiv \{u \in D(\varepsilon) \mid u \geq 1$ q.e. on B$\}$ such that $\varepsilon_\alpha(e_\alpha(B), e_\alpha(B)) = \inf_{u \in L} \varepsilon_\alpha(u,u)$.

Remark. Actually Theorem 3.3.1 in [1] and also Theorem 1 above hold even for a more general type of sets than the Borel sets, namely for analytical sets. This is because, according to the well-known theorem on the capacitability of a Choquet capacity, we have for any analytical set $A \subset X$, $Cap_\alpha(A) = \sup_{\substack{K \subset A \\ K\,compact}} Cap_\alpha(A)$. (cf. [1] (3.1.6)). Hence in the following "Borel" may always be replaced by "analytical".

Now we can prove the following:

Proposition 2. Let A be a Borel set such that $Cap(A) < \infty$. Then there exists a decreasing sequence $(A_n)_{n \in \mathbb{N}}$ of open sets such that

$$\lim_{n \to \infty} \varepsilon_\alpha(e_\alpha(A_n) - e_\alpha(A), e_\alpha(A_n) - e_\alpha(A)) = 0.$$

In particular, $Cap_\alpha(A_n) \downarrow Cap_\alpha(A)$.

Proof. Since for any Borel set B with $Cap_\alpha(B) < \infty$ we have $Cap_\alpha(B) = \varepsilon_\alpha(e_\alpha(B), e_\alpha(B))$, we can find a decreasing sequence $(A_n)_{n \in \mathbb{N}}$ of open sets containing

A such that

$$\varepsilon_\alpha(e_\alpha(A), e_\alpha(A)) = \lim_{n\to\infty} \varepsilon_\alpha(e_\alpha(A_n), e_\alpha(A_n)).$$

But by Theorem 1, (ii) and (iv), we obtain

$$\varepsilon_\alpha(e_\alpha(A_n), e_\alpha(A), e_\alpha(A_n) - e_\alpha(A)) = \varepsilon_\alpha(e_\alpha(A_n), e_\alpha(A_n)) - \varepsilon_\alpha(e_\alpha(A), e_\alpha(A)),$$

and the assertion follows.

<u>Corollary</u>. If $(A_n)$ is a decreasing sequence of closed sets and $A \equiv \bigcap_{n\in\mathbb{N}} A_n$, and $\text{Cap}_\alpha(A_1) < \infty$ then

$$\lim_{n\to\infty} \varepsilon_\alpha(e_\alpha(A_n) - e_\alpha(A), e_\alpha(A_n) - e_\alpha(A)) = 0.$$

In particular, $\text{Cap}_\alpha(A_n) \downarrow \text{Cap}(A)$.

<u>Proof.</u> The assertion follows from [20], 3.4, using the fact that $e_\alpha(A_n) = (e_\alpha(A_1))_{A_n}$ and $e_\alpha(A) = (e_\alpha(A_1))_A$, where for $f \in D(\varepsilon)$ and $B \subset X$ we denote by $f_B$ the $\alpha$-reduced function of $f$ on $B$ i.e. the unique element in $D(\varepsilon)$ such that for some quasi-continuous modifications $\tilde{f}_B$ and $\tilde{f}$, $\tilde{f}_B \geq \tilde{f}$ q.e. on B and $f_B$ minimizes $\varepsilon_\alpha(.,.)$ over this class of functions (see [1], p.77).

<u>Remark.</u> If $(A_n)_{n\in\mathbb{N}}$ is a decreasing sequence of Borel sets and $\text{Cap}_\alpha(A_1) < \infty$, one has

$$\text{Cap}_\alpha(A_n) = \text{Cap}_\alpha(\overline{A}_n^f) \downarrow \text{Cap}_\alpha(\bigcap_n \overline{A}_n^f),$$

where $\overline{B}^f$ for any $B \subset X$ denotes the closure of B with respect to the fine topology ([1], p.93).

We now recall the definition *of measures of finite energy integral (relative to a Dirichlet space)*. A positive Random measure $\mu$ on X is said to be of $\alpha$-*finite energy integral* if for $\alpha > 0$ one has

$$\int_X |v|(x)\, \mu(dx) \leq C\left[\varepsilon_\alpha(v,v)\right]^{1/2},$$

for some positive constant C and all $v \in D(\varepsilon) \cap C_c(X)$. Here are some remarks on these concepts, cf. [1]:

(a) A sufficient and necessary condition for a positive Random measure $\mu$ to be of $\alpha$-finite energy integral, for some $\alpha > 0$, is that there is a unique element $U_\alpha \mu \in D(\varepsilon)$ such that

$$\varepsilon_\alpha(U_\alpha\mu,v) = \int_X v(x)\, \mu(dx) \quad \forall v \in D(\varepsilon) \cap C_c(X).$$

The function $U_\alpha\mu$ is called an $\alpha$-*potential of the measure* $\mu$. $E(\mu) \equiv \varepsilon_\alpha(U_\alpha\mu, U_\alpha\mu)$ is called the $\alpha$-*energy integral of* $\mu$.

(b) If $K \subset X$ is a closed set and $u \in D(\varepsilon)$, then $u = U_\alpha\mu$ with $\mu$ of $\alpha$-finite energy integral and supp $\mu \subset K$ iff $\varepsilon_\alpha(u,v) \geq 0$ for any $v \in D(\varepsilon) \cap C_c(X)$ and $v \geq 0$ on K. This is so iff $\varepsilon_\alpha(u,v) \geq 0$ $\forall v \in D(\varepsilon)$, $\tilde{v} \geq 0$ q.e. on K, for some quasi-continuous modification $\tilde{v}$ of $v$ in the restricted sense.

(c) For $u \in D(\varepsilon)$, $a > 0$ the following are equivalent:

(i) $u$ is an $\alpha$-potential

(ii) $u \in L^2(X,m)$, $u \geq 0$, $e^{-\alpha t} P_t u \leq u$ m.a.e. $\forall t > 0$

   (this means that $u$ is $\alpha$-excessive)

(iii) $\varepsilon_\alpha(u,v) \geq 0$ $\quad \forall v \in D(\varepsilon)$, $\tilde{v} \geq 0$ m.a.e.

(iv) $\varepsilon_\alpha(u,v) \geq 0$ $\quad \forall v \in D(\varepsilon) \cap C_c(X)$, $v \geq 0$

   (for these properties see [1], Th. 3.2.1).

(d) For any Borel subset B of X with $\text{Cap}_\alpha(B) < \infty$ we have that the equilibrium potential $e_\alpha(B)$ is an $\alpha$-potential of a unique positive Randon measure $\nu_B$ of finite energy integral supported by $\overline{B}$, thus $U_\alpha\nu_B \in D(\varepsilon)$ and

$$\varepsilon_\alpha(e_\alpha(B),v) = \int v(x)\, \nu_B(dx) \quad \forall v \in D(\varepsilon) \cap C_c(X).$$

$\nu_B$ is called $\alpha$-*equilibrium measure of* B. When B is relatively compact open, $\nu_B(\overline{B}) = \text{Cap}_\alpha(B)$. This follows by the regularity of $\varepsilon$ and Theorem 1(iv).


Remark 1.  We give an example where A is open and $e_\alpha(A) \neq e_\alpha(\overline{A})$. Take $X = \mathbb{R}$, $m(dx) = \phi^2 dx$, with $dx$ Lebesgue measure and $\phi(x) = |x|$ for $x < 0$, $\phi(x) \equiv 1$ for $x \geq 0$ and define, for $f,g \in C_0^\infty(\mathbb{R})$,

$$\varepsilon(f,g) \equiv \int_{-\infty}^{+\infty} \frac{df}{dx}\, \frac{dg}{dx}\, \phi(x)^2 dx.$$

Using, for instance, the criteria in [9] we see that the form is closable.
Writing again $\varepsilon$ for its closure we get from [11] that for any $f \in D(\varepsilon)$ we
have $\theta(x)f(x) \in D(\varepsilon)$, with $\theta(x) \equiv 0$ for $x < 0$, $\theta(x) \equiv 1$ for $x > 0$. Let
$A = (-1,0)$, then we get $e_\alpha(A)(x)$ for $x > -1$ by looking for functions in $L_A$
which minimize $\varepsilon_\alpha$; we have $e_\alpha(A)(x) = 1$ for $-1 < x < 0$, $e_\alpha(A)(x) = 0$ for
$x > 0$.

We shall now compute $e_\alpha(\overline{A})(x)$ for $x > -1$. Let $a < -1$, $b > 0$. The
function $f \in L(a,b)$ that minimizes $\varepsilon_\alpha(f,f)$ satisfies $f(x) \equiv 1$ for $x \in (a,b)$
and $f(x)$ satisfies the Euler–Lagrange equation of the form $f(x) = 0$ with
$f(b) = 1$ and $\int_b f^2 dx < \infty$, hence it is given for $x \geq 0$ by $e^{-\sqrt{\alpha}(x-b)}$.

By the above, $e_\alpha(\overline{A})(x) = \lim\limits_{a \to -1} \lim\limits_{b\,0} e_\alpha(a,b) = 1$ for $-1 < x < 0$ and
$e_\alpha(\overline{A})(x) = \exp(-\sqrt{a}\,x)$ for $x \geq 0$. Thus, in particular, $e_a([-1,0]) \neq e_\alpha((-1,0))$.
This is an example in which $\overline{A}^f \neq \overline{A}$.


<u>Remark 2.</u>   Let $P_t^B$ be a transient symmetric stable semigroup on $\mathbb{R}^d$ of index
$\beta$, $0 < \beta \leq 2$, $\beta < d$. In particular for $\beta = 2$ we have the heat equation
semigroup $P_t^2 = e^{(1/2)\Delta} \equiv P_t$. Let $\rho(p) \equiv c\,|p|^\beta$, for all $p \in \mathbb{R}^d$ and some
constant $c > 0$. Consider the form $\varepsilon(u,v) \equiv \int_{\mathbb{R}^d} \hat{u}(p)\,\hat{v}(p)\rho(p)dp$ in
$L^2(\mathbb{R}^d,dx)$, with domain $D(\varepsilon) \equiv \{u \in L^2(\mathbb{R}^d) \mid \int |\hat{u}(p)|^2\,\rho(p)dp < \infty\}$, where $-$
denotes complex conjugation. $\varepsilon$ is a regular Dirichlet form in $L^2(\mathbb{R}^2,dx)$,
cf. [1], p.30. For $\beta = 2$, $D(\varepsilon)$ coincides with the Sobolev space of functions
$u \in L^2(\mathbb{R}^d)$ s.t. $\dfrac{\partial u}{\partial x_i} \in L^2(\mathbb{R}^d)$, with $\dfrac{\partial}{\partial x_i}$ u derivatives in the sense of
Schwartz distributions.

The self-adjoint Markov generator $H$ associated with $\varepsilon$ by
$\varepsilon(u,u) = (H^{1/2}u, H^{1/2}u)$, (with $(\ ,\ )$ the $L^2(\mathbb{R}^d,dx)$–scalar product) is such
that $P_t^\beta = e^{-tH}$, $t \geq 0$. For $\beta = 2$ we have $H = -\dfrac{1}{2}\Delta$. In the general case
we have $P_t^\beta(x,A) = \nu_t^\beta(A-x)$, with $P_t^\beta(x,dy)$ the kernel of $P_t^\beta$ and $\nu_t^\beta$ the
convolution semigroup on $\mathbb{R}^d$ with Fourier transform

$$\int_{\mathbb{R}^d} e^{ipx}\,\nu_t^\beta(dx) = e^{-t\rho(p)}.$$

In particular, for $\beta = 2$, $\nu_t^2(dx) = (2\pi t)^{-d/2}\,e^{-|x|^2/2t}\,dx$. $P_t^\beta$ is transient
iff $1/\rho \in L^1_{loc}(\mathbb{R}^d)$ or iff $\int_0^\infty \nu_t^\beta(K)dt < \infty$ for any compact $K$ or iff

203

$$\int_0^\infty (P_t^\beta f,f)\,dt < \infty \quad \text{for all positive } f \text{ of compact support on } \mathbb{R}^d.$$

(In particular, of course $P_t$ is transient.)

The completion $\overline{D(\varepsilon)}$ of $D(\varepsilon)$ in the metric given by $\varepsilon$ gives the transient Dirichlet space $(\overline{D(\varepsilon)},\varepsilon)$. One has $\overline{D(\varepsilon)} = \{u \in L^1_{loc}(\mathbb{R}^d) \mid u$ is a tempered distribution and $\hat{u} \in L^2(\rho\,dp)\} = \{u \in R^{(\beta/2)}f,\ f \in L^2(\mathbb{R}^d)\}$ and

$$\varepsilon(u,v) = \int_{\mathbb{R}^d} \hat{u}(p)\ \overline{\hat{v}(p)}\ \rho(p)\,dp = \int_{\mathbb{R}^d} f(p)\ g(p)\,dp$$

for all $u,v \in \overline{D(\varepsilon)}$, $u = R^{(\beta/2)}f, v \equiv R^{(\beta/2)}g$, with $R^{(\beta)}(x) \equiv \dfrac{\Gamma((d-\beta)/2)}{c2^\beta \pi^{d/2}\Gamma(\beta/2)}|x|^{\beta-d}$, the Riesz kernel of index $\beta$. In fact $\overline{D(\varepsilon)}$ is the set of all Riesz potentials (of index $\beta/2$) of $L^2(\mathbb{R}^d)$-functions. $(\overline{D(\varepsilon)},\varepsilon)$ is embedded continuously in $L^2_{loc}(\mathbb{R}^d)$.

For $\beta = 2$, $c = 1/2$, $R^{(\beta)}(x)$ is the Newtonian potential kernel and is the $\alpha = 0$-order potential associated with $P_t$ in the sense that

$$R^{(2)}(x) = \int_0^\infty (2\pi t)^{-d/2}\ e^{-|x|^2/2t}\,dt \quad [16].$$

A positive Radon measure $\mu$ on $\mathbb{R}^d$ is of finite $\alpha = 0$-energy integral iff

$$\int_{\mathbb{R}^d} \int_{\mathbb{R}^d} R^{(\beta)}(x-y)\mu(dy)\mu(dx) < \infty.$$

This Riesz potential $R^{(\beta)}\mu(x) \equiv \int R^{(\beta)}(x-y)\mu(dy)$ is a quasi-continuous and lower semi-bounded version of the $\alpha = 0$-potential $U_{\alpha=0}\mu$ and

$$E(\mu) = \int_{\mathbb{R}^d} \int_{\mathbb{R}^d} R^{(\beta)}(x-y)\mu(dy)\mu(dx)$$

in this case. For $\beta = 2$, $R^{(\beta)}\mu(x)$ is the Newtonian potential of the measure $\mu$. One defines also for $u$ continuous with compact support in $\mathbb{R}^d$:

$$R^{(\beta)}u(x) \equiv \int_{\mathbb{R}^d} R^{(\beta)}(x-y)\ u(y)\,dy$$

as the Riesz (resp. for $\beta = 2$: Newtonian) potential of $u$. One has $R^{(\beta)}u \in \overline{D(\varepsilon)}$ and $\varepsilon(R^{(\beta)}u,v) = (u,v)$ for all $v \in \overline{D(\varepsilon)}$.

It is shown in $[1]$ (Ex. 3.3.1, p.81) that $Cap_{\alpha=0}(K) = \mu_K(K)$ for any compact $K \subset \mathbb{R}^d$, where $\mu_K$ is the unique element in the class $\{\mu$ positive Radon measure on $\mathbb{R}^d$, supp $\mu \subset K$, $E(\mu) < \infty\}$ minimizing the Gaussian quadratic form $G(\mu) \equiv \int_{\mathbb{R}^d} \int_{\mathbb{R}^d} R^{(\beta)}(x-y)\, \mu(dx)\mu(dy) - 2\mu(K)$.

This then shows that $Cap_{\alpha=0}$ coincides with the classical capacity for Riesz potentials. For $\beta = 2$ one defines the $\alpha$-Newtonian potential kernel by $g_\alpha(x) \equiv \int_0^\infty e^{-\alpha t} (2\pi t)^{-d/2} e^{-|x|^2/2t}\, dt$. One has $g_\alpha \geq 0$, $g_\alpha(0) = \infty$ if $d \geq 2$, $g_\alpha(0) < \infty$ for $d = 1$. $g_\alpha$ is continuous for $x \neq 0$ and bounded away from zero on compacts. Set $g_\alpha(x,y) \equiv g_\alpha(x-y)$. The $\alpha$-Newtonian potential of a measure $\mu$ on $\mathbb{R}^d$ is by definition the lower semicontinuous function $g_\alpha\mu(x) \equiv \int g_a(x,y)\mu(dy)$.

In classical potential theory the $\alpha$-equilibrium measure of a Borel set B is by definition $\mu_B^\alpha(\cdot) \equiv \alpha \int_B h_B^\alpha(y,\cdot)dy$, with $h_B^\alpha(y,A) \equiv E^x(e^{-\alpha\tau}B, \tau_B < \infty$, $X(\tau_B) \in A)$, where $E^x$ is the expectation with respect to the Wiener process $X$ started at x, $\tau_B$ the first hitting time of B. One has $(g_\alpha\mu_B^\alpha)(x) = h_B^\alpha 1(x) = E^x(e^{-\alpha\tau}B)$. $\mu_B^\alpha$ is a Radon measure concentrated on the closure $\bar{B}$ of B (in fact on $\partial B \cap B^\sim$, $B^\sim$ being the points which are regular for B, i.e. $B^\sim = \{x \in \mathbb{R}^d \mid P^x(\tau_B = 0) = 1\}$.

The $\alpha$-potential of the $\alpha$-equilibrium measure $g_\alpha\mu_B^\alpha(x) = E^x(e^{-\alpha\tau}B)$ is *the $\alpha$-equilibrium potential of B* in the sense of classical potential theory.

It is shown in $[1]$ (p.106) for Borel sets B of finite capacity that $g_\alpha\mu_B^\alpha$ is a quasi-continuous version of the $\alpha$-equilibrium potential $e_B(x)$ for the Dirichlet form $\varepsilon(u,u) = \frac{1}{2} \int |\nabla u(x)|^2 dx$ with the above domain. Moreover it is also shown there that $E^x(e^{-\alpha\dot{\tau}}B)$, $\dot{\tau}_B = \inf\{t \geq 0, X_t(\omega) \in B\}$ is also a quasicontinuous version of $e_B(x)$. This gives immediately that $e_B(x) = 1$ for $x \in B^\sim$. The classical $\alpha$-capacity of B is defined by

$$C_\alpha(B) = \mu_B^\alpha(\mathbb{R}^d) = \alpha \int E^y(e^{-\alpha\tau}B)dy = \int_{\mathbb{R}^d} E^y(e^{-\alpha\tau}B)\mu_B^\alpha(dy).$$

It follows by a result of Deny, see $[1]$, that, for open sets, $Cap_\alpha(B)$ coincides with the $\alpha$-capacity associated with the Dirichlet space $(\varepsilon, D(\varepsilon))$, in the case where the latter capacity is finite.

<u>Remark 3.</u>   Let $X = (r_1, r_2)$ be an open interval with $-\infty \leq r_1 < r_2 \leq +\infty$.
Let $(y,k,m)$ a standard triple in the sense of Feller-Dynkin theory, cf. [17],
[18], i.e. $y$ is a strictly monotone continuous function, $k$ is a monotone
function continuous from the right and $m$ a strictly monotone function,
continuous from the right.  Let us denote by $dm$, $dk$, $dy$ the corresponding
Lebesgue-Stieltjes measures.  Let us consider the form

$$\varepsilon(u,v) = \frac{1}{2} \int_X \frac{du}{dy} \frac{dv}{dy} \, dy + \int_X uv \, dk,$$

for all $u,v \in F \equiv \{u \in L^2(X,dm) \mid \; \exists$ version (with respect to m) $\tilde{u}$ of u such
that $\tilde{u}$ is of local bounded variation, $d\tilde{u} << dy$ with $\frac{d\tilde{u}}{dy} \in L^2(X,dy)$, $\tilde{u} \in L^2(X,dk)\}$
It is shown in [17] (Satz 10.5) that $\varepsilon$ with domain $F$ is a local Dirichlet
form in $L^2(X,dm)$ (i.e. $(\varepsilon,F)$ is a Dirichlet space).
$(\varepsilon,F)$ is not necessarily regular.  Let $F_o$ be the closure in the $\varepsilon_1$-norm of
the functions which are $C^\infty$ with respect to $y$ and have compact support on $X$.
It is shown in [17] that $(\varepsilon,F_o)$ is a regular local Dirichlet form and one
has $F_o = \{u \in F \mid r_i$ is regular $\Rightarrow u(r_i) = 0\}$, where $r_1$ is regular iff

$MY+KY < \infty$ and $YM+YK < \infty$, where $MY \equiv \int_{r_1}^c \left[m(c)-m(z)\right] \, dy(z),$

$$KY \equiv \lim_{c \downarrow r_1} \int_{r_1}^c \left[k(c) - k(z)\right] dy(z),$$

$$YM \equiv \int_{r_1}^c \left[(y(c)-y(z)\right] m(dz),$$

$$YK \equiv \int_{r_1}^c \left[y(c)-y(z)\right] dk(z),$$

and with a corresponding definition of regularity for $r_2$.

Moreover it is shown in [17] that, for given $\alpha > 0$, for each $x \in X$ there
exists precisely one $g_\alpha(\cdot,x) \in F_o$ such that $\varepsilon_\alpha(g_\alpha(\cdot,x),v) = v(x)$ for all
$v \in F_o$.   $g_\alpha$ is called a *reproducing kernel*.   One has $0 < g_\alpha(x,x) < \infty$ for
all $x \in X$.  Moreover for the $\alpha$-capacity given by the Dirichlet form $(\varepsilon,F_o)$
we have

206

$$\text{Cap}_\alpha(\{x\}) = \frac{1}{g_\alpha(x,x)} > 0.$$

It is further shown in $[17]$ that the $\alpha$-equilibrium potential $e_\alpha(x - \frac{1}{n}, x + \frac{1}{n})$ converges as $n \to \infty$ to the function $e_\alpha(\{x\})$ in $F_o$, characterized by $e_\alpha(\{x\}) = 1$ and $\varepsilon_\alpha(e_\alpha\{x\},v) \geq 0$ for all $v \in F_o$ with $v \geq 0$. Furthermore one has ($[17]$, Th. 12.6):

$$e_\alpha(\{x\})(y) = \frac{g_\alpha(y,x)}{g_\alpha(x,x)} = u_2(y)/u_2(x) \text{ for } y \leq x, \; u_1(y)/u_1(x) \text{ for } y \geq x,$$

where $u_1(x) \equiv u(x) \int_x^{r_2} u^{-2}(t)dt$, $u_2(x) \equiv u(x) \int_{r_1}^x u^{-2}(t)dt$,

$$u(x) \equiv \sum_{n=0}^\infty u^n(x), \; u^o \equiv 1, \; u^{n+1}(x) \equiv 2\alpha \int_c^x (\int_c^z u^n(t)d(k+m)(t))dy(t).$$

$u,u_1,u_2$ are positive $\alpha$-harmonic functions on X in the sense that they are continuous functions satisfying $\frac{1}{2} d(D_y^+f) - fdk = \alpha fdm$, with $D_y^+f(x) \equiv \lim_{z \downarrow x} \frac{f(z)-f(x)}{y(z)-y(x)}$. Moreover,

$$r_\alpha(x,y) = g_\alpha(x,y) = \frac{2}{W} \begin{cases} u_2(x)u_1(y), & x \leq y \\ u_1(x)u_2(y), & x \geq y \end{cases}$$

with $W \equiv u_2(x)D_y^+u_1(x) - u_1(x)D_y^+u_2(x)$, which is independent of x, and $r_\alpha$ the kernel of the resolvent $R_\alpha$ associated with the triple (y,k,m), i.e., as shown in $[17]$, to the Dirichlet form $(\varepsilon,F_o)$.

In the present case the quasi-continuous functions on X are simply the continuous functions on X. The $\alpha$-potential of a measure $\mu$ is given by

$$U_\alpha\mu(x) = \int_X g_\alpha(x,y)\mu(dy).$$

Remark 4. Let $X = \mathbb{R}$, $m(dx) = dx$ and $\varepsilon(u,u) = \frac{1}{2} \int_\mathbb{R} (\frac{du}{dx})^2 dx$. This is a special case of Remark 3. $\varepsilon$ with domain the Sobolev space $H^1(\mathbb{R})$ (the closure of $C_o^\infty(\mathbb{R})$ in the $\varepsilon_1$-norm) is a local regular Dirichlet form, and coincides with $\varepsilon$ with domain the functions $u \in L^2(\mathbb{R}^d, dx)$, absolutely continuous and such that $\varepsilon(u,u) < \infty$. In this case the $\alpha$-harmonic functions satisfy $\frac{1}{2} \frac{d^2}{dx^2} u = \alpha u$ and form a two-dimensional vector space spanned by

$e^{\sqrt{2\alpha}\;x}$ and $e^{-\sqrt{2\alpha}\;x}$. No non-constant harmonic function belongs to $D(\varepsilon)$. As in Remark 3, for each $x \in \mathbb{R}$ there is a unique $g_\alpha(\cdot,x) \in D(\varepsilon)$ such that $\varepsilon_\alpha(g_\alpha(\cdot,x),v) = v(x)$ for all $v \in D(\varepsilon)$. One has $g_\alpha(x,y) = Ce^{-\sqrt{2\alpha}|x-y|}$, with $C = 1/\sqrt{2\alpha}$. As in Remark 3, $g_\alpha$ is a reproducing kernel, one has $0 < g_\alpha(x,x) < \infty$ for all $x \in \mathbb{R}$, and $\mathrm{Cap}_\alpha(\{x\}) = 1/g_\alpha(x,x) = \sqrt{2\alpha}$. The $\alpha$-equilibrium potential is in this case

$$e_\alpha(\{x\})(y) = e^{-\sqrt{2\alpha}\;|x-y|}.$$

<u>Remark 5.</u> Let $X = \mathbb{R}$, $m$ a Radon measure on $\mathbb{R}$, positive on any open interval and charging no one-point sets. Let $\rho$ be a non-negative function on $\mathbb{R}$. Suppose that the form $\varepsilon$ defined on $C_o^\infty(\mathbb{R})$ by $\varepsilon(u,u) = \displaystyle\int_{\mathbb{R}} (\frac{du}{dx})^2 \rho(x)\,dx$ is closable in $L^2(\mathbb{R};dm)$ (for sufficient conditions see [9]). Then one shows as in [11], that

(i) if there exists a neighbourhood $U_o$ of zero and positive constants $C$ and $\gamma$ with $0 < \gamma < \frac{1}{2}$ so that $\rho(x) \geq C|x|^{2\gamma}$, $\forall x \in U_o$, then $\mathrm{Cap}_\alpha(\{0\}) = 0$;

(ii) if there exists a neighbourhood of zero $U_o$ and a positive constant $C$ such that $\rho(x) \leq C|x|$ for all $x \in U_o$, then $\mathrm{Cap}_\alpha(\{0\}) = 0$.

Corresponding results hold of course for any point $x \in \mathbb{R}$.

As discussed in [11], [15], one obtains also results for $\mathbb{R}^d$, $d \geq 2$. For instance, if $\rho \in L^1_{loc}(\mathbb{R}^d)$, and $\rho$ is bounded away from zero such that $\varepsilon(u,u) = \frac{1}{2}\displaystyle\int (\nabla u)^2 \rho\,dx$ is closable in $L^2(\mathbb{R}^d,m)$ (cf. [9]), then using (ii) we see that, for $d \geq 2$, $\mathrm{Cap}_\alpha(\{0\}) = 0$. Since $\mathrm{Cap}_\alpha(\{0\}) = \varepsilon_\alpha(e_\alpha(\{0\}), e_\alpha(\{0\}))$, we then have also $e_\alpha(\{0\}) = 0$.


## 3.  CAPACITY AND GREEN FUNCTIONS

As in Section 2, let $\varepsilon$ be a regular Dirichlet form on a locally compact Hausdorff space with a countable base for the topology. Let $\alpha > 0$ and $x_o \in X$. By Proposition 2 there exists a decreasing sequence $(A_n)_{n\in\mathbb{N}}$ of open sets such that $\mathrm{Cap}_\alpha(A_n) \downarrow \mathrm{Cap}_\alpha(\{x_o\})$ and $e_\alpha(A_n) \to e_\alpha(\{x_o\})$ in $\|\;\|_n$-norm. Since $\{x_o\}$ is a $G_\delta$-set we may assume that $\bigcap_n A_n = \{x_o\}$ and that every $A_n$ is relatively compact, in particular $\mathrm{Cap}_\alpha(A_n) < \infty$ for every $n \in \mathbb{N}$. We

208

then have

__Theorem 4.__    Let $\varepsilon, \alpha$, $x_o$ and $(A_n)_{n \in \mathbb{N}}$ be as above.  Then for every
$v \in D(\varepsilon) \cap C(X)$ we have

$$\lim_{n \to \infty} \left[ \mathrm{Cap}_\alpha (A_n) \right]^{-1} \varepsilon_\alpha (e_\alpha (A_n), v) = v(x_o).$$

__Proof.__    Let $v \in D(\varepsilon) \cap C(X)$.  Put

$$a_n \equiv \inf_{x \in A_n} v(x), \qquad b_n \equiv \sup_{x \in A_n} v(x).$$

By Theorem 1(iii) we have

$$v - a_n e_\alpha (A_n) \geq 0 \qquad \text{m-a.e.}$$

$$b_n e_\alpha (A_n) - v \geq 0 \qquad \text{m-a.e. on } A_n.$$

Using Theorem 1(iii) we conclude that

$$\varepsilon_\alpha (e_\alpha (A_n), v) \geq a_n \; \varepsilon_\alpha (e_\alpha (A_n), e_\alpha (A_n)) = a_n \mathrm{Cap}_\alpha (A_n)$$

and

$$b_n \mathrm{Cap}_\alpha (A_n) = b_n \, \varepsilon_\alpha (e_\alpha (A_n), e_\alpha (A_n)) \geq \varepsilon_\alpha (e_\alpha (A_n), v) \text{ or}$$

$$a_n \leq \left[ \mathrm{Cap}_\alpha (A_n) \right]^{-1} \varepsilon_\alpha (e_\alpha (A_n), v) \leq b_n.$$

Since $A_n \downarrow \{x_o\}$, the assertion follows by the continuity of $v$.

__Remark.__    Because of the Corollary following Proposition 2, Theorem 4 is also
valid for *every* sequence $(A_n)_{n \in \mathbb{N}}$ of compact sets such that $A_n \downarrow \{x_o\}$.

__Corollary 1.__    Let $\varepsilon, \alpha$, $x_o$, $(A_n)_{n \in \mathbb{N}}$ be as in Theorem 4 or in the preceding
Remark.  For $n \in \mathbb{N}$ let $\mu_n$ be the measure with finite energy integral
corresponding to the $\alpha$-potential $e_\alpha (A_n)$, i.e. $e_\alpha (A_n) = U_\alpha \mu_n$.

Then

$$\left[ \mathrm{Cap}_\alpha (A_n) \right]^{-1} \mu_n \to \varepsilon (x_o)$$

weakly  (where $\varepsilon (x_o)$ is the Dirac measure in $x_o$).

Proof.   Theorem 4 and $[1]$ Lemma 1.4.2 imply vague convergence.   Then weak convergence follows, since $\left[\text{Cap}_\alpha(A_n)\right]^{-1} \mu_n$ is a probability measure, because $\mu_n(X) = \mu_n(\bar{A}_n) = \text{Cap}_\alpha(A_n)$, by d) in Section 2.

Remark.   In the case $\text{Cap}_\alpha(\{x_o\}) > 0$ we have

$$\lim_{n\to\infty} \left[\text{Cap}_\alpha(A_n)\right]^{-1} \varepsilon_\alpha(e_\alpha(A_n),v) = \left[\text{Cap}_\alpha(\{x_o\})\right]^{-1} \varepsilon_\alpha(e_\alpha(\{x_o\}),v)$$

and then Theorem 4 shows that

$$g_\alpha(\cdot,x_o) \equiv \left[\text{Cap}_\alpha(\{x_o\})\right]^{-1} \tilde{e}_\alpha(\{x_o\})(\cdot)$$

is a reproducing kernel with respect to $\varepsilon_\alpha$ in the sense that

$$\varepsilon_\alpha(g_\alpha(\cdot,x_o),v) = v(x_o) \text{ for all } v \in D(\varepsilon) \cap C(X).$$

For $\text{Cap}_\alpha(\{x_o\}) = 0$, however, "$\lim_{n\to\infty} \left[\text{Cap}_\alpha(A_n)\right]^{-1} e_\alpha(A_n)$" is not an element in $D(\varepsilon)$.   This is because this "limit" will turn out to be the Green function corresponding to $\varepsilon_\alpha$ – a potential which is in general not of finite energy. This will be clarified in the following corollary.

Let $(G_\alpha)_{\alpha>0}$ denote the resolvent associated to $(D(\varepsilon),\varepsilon)$.

Corollary 2.   Let $\varepsilon,\alpha$, $x_o$ and $(A_n)_{n\in\mathbb{N}}$ be as in Theorem 4.

(i) Given $f \in L^2(X,m)$ such that $G_\alpha f$ has a continuous $(m-)$ version $\widetilde{G_\alpha f}$, then

$$\widetilde{G_\alpha f}(x_o) = \lim_{n\to\infty} \int \left[\text{Cap}_\alpha(A_n)\right]^{-1} e_\alpha(A_n) \ f \ dm.$$

(ii) If for every $f \in L^2(X,m)$ there exists a continuous $(m-)$ version $\widetilde{G_\alpha f}$ of $G_\alpha f$, then there exists a function $g_\alpha(\cdot,x_o) \in L^2(X;m)$ such that

$$\widetilde{G_\alpha f}(x_o) = \int g_\alpha(\cdot,x_o) \ f \ dm \text{ for every } f \in L^2(X,m), \text{ i.e. } \varepsilon_\alpha \text{ has a Green function.}$$

(iii) If $\varepsilon_\alpha$ has a Green function, i.e. there exists a kernel, $g_\alpha$, for $G_\alpha$ and if in addition, for every $f \in C_c(X)$, $y \to \int g_\alpha(\cdot,y) \ f \ dm$ is continuous, then

$$g_\alpha(\cdot,x_o) \ dm = \lim_{n\to\infty} \left(\left[\text{Cap}_\alpha(A_n)\right]^{-1} e_\alpha(A_n) \ dm\right),$$

where the limit is taken with regard to the vague topology on the

Radon measures.

Proof.    (i) Let f be as in assertion (i), then by Theorem 4 we have

$$\widetilde{G_\alpha f}(x_o) = \lim_{n\to\infty} \epsilon_\alpha \left( \left[ Cap_\alpha(A_n) \right]^{-1} e_\alpha(A_n), G_\alpha f \right)$$

$$= \lim_{n\to\infty} \int \left[ (Cap_\alpha(A_n) \right]^{-1} e_\alpha(A_n) \, f \, dm.$$

(ii) The assertion follows by (i) and by the fact that $L^2(X,m)$ is weakly
   complete, since it is a Hilbert space.

(iii) is an immediate consequence of (i).

Remark 1.    Let us consider a special case of Corollary 2.  Let $X = \mathbb{R}^d$,
$m(dx) = \phi^2 dx$, with $\phi$ such that $\epsilon(u,u) = 1/2 \int |\nabla u|^2 \phi^2 \, dx$, $u \in C_o^\infty(\mathbb{R}^d)$
extends to a regular Dirichlet form (for sufficient conditions see, e.g.,
[9]).  Assume that the condition of Corollary 2(ii) is fulfilled, then
$g_\alpha(x,x_o)$ is the kernel in $L^2(\mathbb{R}^d, m)$ of the resolvent $G_\alpha = (H_\phi + \alpha)^{-1}$ of the
positive self-adjoint operator $H_\phi$ in $L^2(\mathbb{R}^d, m)$ given by $\epsilon$.  The resolvent of
the corresponding Schrödinger operator H in $L^2(\mathbb{R}^d, dx)$ is $\phi(x) \, g_\alpha(x,x_o)$.


Remark 2.    Let us consider the setting of [19].  $\xi,\hat\xi$  are standard processes
on X in duality with respect to a given $\sigma$-finite measure m and such that
they satisfy the additional regularity conditions of Ch. VI in [19].  It is
shown then in [19] (Ch. VI, p.290) that for points $x_o$ such that $g_\alpha(x,x_o)$ is
bounded and continuous at $x = x_o$,

$$Cap_\alpha(\{x_o\}) \, g_\alpha(x,x_o) = e_\alpha(\{x_o\}) \, (x).$$

The condition on $g_\alpha(x_o,x)$ is satisfied, for all $x_o$, e.g. for all stable
processes of index $\beta > 1$ (and for some more general processes with stationary
independent increments).

   We see that Theorem 4 extends this relation to the class of processes
studied in the theory of Dirichlet forms.  Moreover Theorem 4 also gives
information on the case where $Cap_\alpha(\{x_o\}) = 0$.

Corollary 3.   Let $X = \mathbb{R}$, $\varepsilon$ be as a Corollary 2.   Put $\phi^2 = \rho$ and assume $\rho$, $\frac{d\rho}{dx}$ are continuous on $\mathbb{R} - \{x_o\}$.   Then

$$\text{Cap}_\alpha(\{x_o\}) = \lim_{x \uparrow x_o} \frac{de_\alpha}{dx}(x,x_o)\,\rho(x) - \lim_{x \downarrow x_o} \frac{de_\alpha}{dx}(x,x_o)\,\rho(x).$$

Proof.   Let $A = (a,b)$, $x_o \in A$, $v \in C_o^\infty(\mathbb{R})$ such that $v(x_o) \neq 0$.   Then, from the definition of $\varepsilon_\alpha$:

$$\varepsilon_\alpha(e_\alpha(A),v) = \int_\mathbb{R} \frac{de_\alpha}{dx}(A)\,\frac{dv}{dx}\,\rho dx + \alpha \int_\mathbb{R} e_\alpha(A)\,v\rho\,dx. \tag{3.1}$$

But $e_\alpha(A)(x) = 1$ for $x \in A$ and $e_\alpha(A)(x)$ is a solution of the Euler-Lagrange equation

$$\rho\,\frac{d^2}{dx^2}e_\alpha(A) + \frac{d\rho}{dx}\,\frac{de_\alpha}{dx}(A) - \alpha\rho e_\alpha(A) = 0 \text{ for } x \notin A. \tag{3.2}$$

We decompose the integrals on the right-hand side of (3.1) into $\int_{-\infty}^a + \int_a^b + \int_b^\infty$. An integration by parts yields then, using $v \in C_o^\infty(\mathbb{R}^d)$, that the right-hand side of (3.1) is equal to

$$(v\,\frac{d}{dx}\,e_\alpha(A)\rho)\,(x=b) - (v\,\frac{d}{dx}\,e_\alpha(A)\rho)(x=a) + \alpha \int_a^b e_\alpha(A)\,v\rho dx. \tag{3.3}$$

By the theorem, the left-hand side of (3.1) converges as $a \uparrow x_o$, $b \downarrow x_o$ to $v(x_o)\,\text{Cap}_\alpha(\{x_o\})$.   Since (3.3) converges to the right-hand side of the formula in the Corollary, the proof is complete.

Examples.   Let us consider the Dirichlet form $\varepsilon(f,f) = \frac{1}{2}\int_{\mathbb{R}^3} |\nabla f|^2\,\phi^2 dx$ in $L^2(\mathbb{R},m)$, with $\phi$ a function $\phi(r)$ of $r \equiv |x|$ alone.

1.   For $\phi = 1$ the assumptions of Corollary 2 are fulfilled.   In this case we have

$$e_\alpha(K_R)\,(r) = \begin{cases} 1 & r \leq R \\ f(r) & r > R \end{cases}$$

with $K_R$ the sphere of radius $R$ centered at zero,

$f(r) = \frac{a}{r} e^{-\sqrt{\alpha} r}$ the solution of the Euler equation to

$$4\pi \int_0^\infty (\frac{df}{dr})^2 r^2 dr + 4\pi\alpha \int_0^\infty f^2 r^2 dr, \quad \text{namely} \quad \frac{d^2 f}{dr^2} + \frac{2}{r} \frac{df}{dr} - \alpha f = 0.$$

In this case $\mathrm{Cap}_\alpha (K_R) = 4\pi (R + \sqrt{\alpha}\, R^2 + \frac{1}{3} R^5 e^{2\sqrt{\alpha} R} \alpha)$. Hence

$$\lim_{R \downarrow 0} \frac{e_\alpha (K_R)(r)}{\mathrm{Cap}_\alpha (K_R)} = \frac{1}{4\pi r} e^{-\sqrt{\alpha} r} \quad \text{(which implies the corresponding convergence for}$$

the associated measures, since the sequence is increasing) and, of course,

the right-hand side is the Green function of $-\Delta + \alpha$ (i.e.

$$(-\Delta + \alpha) \, (\frac{1}{4\pi r} e^{-\sqrt{\alpha} r}) = \varepsilon(0)(x).$$

2.   $X = \{ x \in \mathbb{R}, \ x > 0, \ \phi(x) = |x| \}$.  The Euler equation $\dfrac{d^2 f}{dx^2} + \dfrac{2}{x} \dfrac{df}{dx} - \alpha f = 0$
has the two independent solutions $\frac{1}{x} e^{\sqrt{\alpha} x}$ and $\frac{1}{x} e^{-\sqrt{\alpha} x}$.

   Set $A_\varepsilon = (x_o - \varepsilon, \ x_o + \varepsilon))$, $x_o > 0$.  We have from the above Euler equation

$$e_\alpha (A_\varepsilon)(x) = \begin{cases} \dfrac{b_\varepsilon}{x} (e^{\sqrt{\alpha} x} - e^{-\sqrt{\alpha} x}), & 0 < x < x_o - \varepsilon \\[2mm] 1 & x_o - \varepsilon < x < x_o + \varepsilon \\[2mm] \dfrac{a_\varepsilon}{x} e^{-\sqrt{\alpha} x} & x > x_o + \varepsilon \end{cases}$$

with $a_\varepsilon \equiv (x_o + \varepsilon) e^{\sqrt{\alpha}(x_o + \varepsilon)}$

$$b_\varepsilon \equiv (x_o - \varepsilon) \left[ e^{\sqrt{\alpha}(x_o - \varepsilon)} - e^{-\sqrt{\alpha}(x_o - \varepsilon)} \right]^{-1}.$$

Using Corollary 3 we get directly

$$\mathrm{Cap}_\alpha (\{x_o\}) = 2 x_o + \frac{2\sqrt{\alpha}\, x_o^2}{1 - e^{2\sqrt{\alpha} x_o}} .$$


ACKNOWLEDGEMENTS

We are very grateful to Professor Raphael Høegh-Krohn for useful discussions and to Dr. Uwe Spönemann for corrections.  The partial financial support of the Deutsche Forschungsgemeinschaft and the Stiftung Volkswagenwerk is also gratefully acknowledged.

## REFERENCES

[1] M. Fukushima, *Dirichlet Forms and Markov Processes*, North Holland/ Kodansha (1980).

[2] M. Fukushima, Energy forms and diffusion processes, ZiF-Preprint 1984 (to appear in ZiF *Lect. Mathematics and Physics*, Ed. L. Streit, World Scient. Publ., Singapore).

[3] M.L. Silverstein, *Symmetric Markov Processes*, Lect. Notes in Math. **426**, Springer-Verlag: Berlin (1974).

[4] M.L. Silverstein, *Boundary Theory for Symmetric Markov Processes*, Lect. Notes in Math. **516**, Springer-Verlag: Berlin (1976).

[5] S. Albeverio, R. Hoegh-Krohn and L. Streit, Energy forms, Hamiltonians, and distorted Brownian paths, *J. Math. Phys.* **18**, 907-917 (1977).

[6] S. Albeverio and R. Høegh-Krohn, Diffusions, quantum fields and fields with values in Lie groups, in *Stochastic Analysis and Applications*, Ed. M. Pinsky, M. Dekker (1984).

[7] S. Albeverio and R. Høegh-Krohn, Dirichlet forms and diffusion processes on rigged Hilbert spaces, *Z. Wahrscheinlichkeitsth. verw. Geb.* **40**, 1-57 (1977).

[8] S. Albeverio, R. Høegh-Krohn and H. Holden, Markov cosurfaces and gauge fields, *Acta Phys. Austr.* Suppl. XXVI, 211-231 (1984).

[9] M. Röckner and N. Wielens, Dirichlet forms - closability and change of speed measure, ZiF-Preprint 1983 (this volume).

[10] S. Albeverio, Ph. Blanchard, F. Gesztesy and L. Streit, Quantum mechanical low energy scattering in terms of diffusion processes, in *Proc. Marseille Conf. 1983*, Stochastic Aspects of Classical and Quantum Systems, Eds. S. Albeverio, Ph. Combe, M. Sirugue-Collin, Lect. Notes in Math. 1109, Springer-Verlag: Berlin (1985).

[11] S. Albeverio, M. Fukushima, W. Karwowski and L. Streit, Capacity and quantum mechanical tunneling, *Commun. Math. Phys.* **81**, 501-513 (1981).

[12] M. Fukushima, On a stochastic calculus related to Dirichlet forms
     and distorted Brownian motion, *Phys. Repts. 77*, 255-262 (1981).

[13] M. Fukushima, On absolute continuity of multidimensional symmetrizable
     diffusion, pp. 146-176 in *Functional Analysis in Markov
     Processes*, Ed. M. Fukushima, Lect. Notes in Math. *923*, Springer-
     Verlag: Berlin (1982).

[14] M. Fukushima, Markov processes and functional analysis, pp. 187-202
     in *Proc. Int. Math. Conf.*, L.H.Y. Chen et al., Eds, North
     Holland (1982).

[15] M. Fukushima, A note on irreducibility and ergodicity of symmetric
     Markov processes, pp. 200-208 in *Stochastic Processes in Quantum
     Theory and Statistical Physics*, Eds, S. Albeverio, Ph. Combe,
     M. Sirugue-Collin, Lect. Notes in Phys. 173, Springer-Verlag:
     Berlin (1982).

[16] S.C. Port and C.J. Stone, *Brownian Motion and Classical Potential
     Theory*, Academic Press, 1978.

[17] K. Rullkötter and U. Spönemann, *Dirichletformen und Diffusionsprozesse*,
     Diplomarbeit, Bielefeld (1983).

[18] U. Spönemann, in preparation.

[19] R.M. Blumenthal and R.K. Getoor, *Markov Processes and Potential Theory*,
     Academic Press, 1968.

[20] M. Röckner, Generalized Markov fields and Dirichlet forms, Bielefeld
     Preprint 1983 (to appear in *Acta Appl. Math.*).

Sergio Albeverio,                    Michael Röckner,
Mathematisches Institut,             Fakultät fur Mathematik,
Ruhr-Universität,                    Universität Bielefeld,
D-4630 Bochum 1                      Mathematics Department,
West Germany.                        Cornell University.

Witold Karwowski,                    Ludwig Streit,
Inst. Theor. Physics,                Fakultät für Physik,
University of Wroclaw,               Universität Bielefeld.
Wroclaw.

S ALBEVERIO, R HØEGH-KROHN & D MERLINI

# Euler flows, associated generalized random fields and Coulomb systems

## 1. INTRODUCTION

In this paper we mention briefly some problems concerning random fields
associated with certain two-dimensional fluids and related Coulomb and plasma
systems.  In Section 2 we give a short review of some recent work on two-
dimensional inviscid fluids (mainly contained in [1] - [7]).  We introduce
the classical equations for a Euler fluid in a general domain of $\mathbb{R}^2$ and
discuss some properties of classical solutions.  We also associate to the
classical equation of motion for this fluid, stationary statistical solutions
defined in terms of formally invariant infinite-dimensional Gaussian
measures P, of "Gibbs type", and associated generalized random fields.

It turns out that the time invariance of the measures is quite subtle.
The measures are, in fact, infinitesimally invariant in the sense that they
are stationary solutions of the Hopf equation of motion.  Moreover, they are
invariant for a suitable finite-dimensional version of the flow ("Galerkin-
type approximation").  They have support on solutions of the Euler equation
having infinite energy.  In the general case, the existence of a limit flow
(in the sense of a continuous group of transformations on the support of P)
defined by a weak limit of solutions of the Euler equation of motion with
initial values on the support of the invariant measures P is proven.

We also show that there exists at least a strongly continuous unitary
group of transformations in $L^2(P)$ with generator coinciding on a dense domain
with that associated with the above flow.  We briefly discuss the uniqueness
question, which remains open.

In Section 3 we set out briefly the basic connection between a class of
statistical solutions of the Euler equation which are generalized random
fields of Poisson type, the vortex model, and two-dimensional Coulomb
systems and plasmas.

We mention a few basic references and recent results in these areas, our
main aim being, however, to indicate a fascinating network of relations
between different areas of physics, which should become even more of a
fertile ground for application of the methods of the theory of random fields

216

and stochastic analysis.

2. <u>THE TWO-DIMENSIONAL EULER FLUID: STOCHASTIC SOLUTIONS OF THE EQUATIONS</u>
   <u>OF MOTION OF GIBBSIAN TYPE</u>

In this section we present some results essentially obtained in $[1] - [7]$.
The Euler equation for an incompressible fluid in $\mathbb{R}^2$ is given by

$$\frac{\partial}{\partial t} u = -(u.\nabla)u - \nabla p, \qquad \text{div } u = 0 \tag{2.1}$$

where $u(t,x) \equiv (u_1, u_2)(t,x)$ is the velocity vector, $\nabla$ is the gradient
operator, $\nabla \equiv (\partial_1, \partial_2)$, $\partial_i \equiv \frac{\partial}{\partial x_i}$, $u.\nabla \equiv u_1 \partial_1 + u_2 \partial_2$, and p is the pressure
term. This is the formal limit of the Navier-Stokes equation with viscosity
zero. For a domain $\Lambda \subset \mathbb{R}^2$ with piecewise $C^1$ boundary, a natural extension
of (2.1) is

$$\frac{\partial}{\partial t} u = -(u.\nabla)u - f, \text{rot } f = 0, \text{ div } u = 0. \tag{2.2}$$

When $\Lambda$ is simply connected, thus in particular when $\Lambda = \mathbb{R}^2$, we have
rot f = 0 iff there exists a function p such that $f = \nabla p$. In this case,
(2.2) reduces to (2.1).

   We shall study (2.2) with boundary condition n.u = 0 on $\partial\Lambda$, where n is
the unit normal to $\partial\Lambda$. This equation expresses the absence of flow through
the boundary $\partial\Lambda$. This boundary condition, together with div u = 0, implies,
in our two-dimensional case, that there exists a function $\phi$ ("stream
function") on $\Lambda$ such that

$$u = (-\partial_2 \phi, \partial_1 \phi). \tag{2.3}$$

We set $\nabla^\perp \equiv (-\partial_2, \partial_1)$, so that $\nabla^\perp \phi = u$. Inserting this into (2.2), we get

$$\partial_t \nabla^\perp \phi = - \sum_{i=1}^{2} \left[(\nabla^\perp)_i \phi\right] (\nabla)_i \nabla^\perp \phi - f. \tag{2.4}$$

$\nabla^\perp.u \equiv$ rot u is called the vorticity field.

   We have $\sum_i (\nabla^\perp)_i (\nabla^\perp)_i = \Delta$ and $\sum_i (\nabla^\perp)_i (\nabla)_i = 0$. Moreover, $\nabla^\perp.f =$ rot f = 0,
by (2.2). Applying $\nabla^\perp$ to (2.4) and using the latter formulae, we get easily

$$\partial_t \Delta\phi = (\nabla\phi). \nabla^\perp \Delta\phi = -(\nabla^\perp\phi). \nabla\Delta\phi. \tag{2.5}$$

The boundary condition

$$n.u = 0 \text{ on } \partial\Lambda \tag{2.6}$$

is equivalent to the fact that $\phi$ is constant on each component $\partial^k\Lambda$ of $\partial\Lambda$. Thus (2.2) with the boundary condition (2.6) is equivalent to (2.3), (2.5) with the boundary condition $\phi$ constant on each component $\partial^k\Lambda$ of $\partial\Lambda$. We shall call any such equation with the corresponding boundary conditions a "classical Euler equation".

Let $\partial\Lambda$ have components $\partial^k\Lambda$, $k = 0,\ldots,m$, with $\partial^o\Lambda$ the boundary of the unbounded component of $\mathbb{R}^2 - \Lambda$. Let $\alpha_k$, $k = 1,\ldots,m$, be a function such that $\Delta\alpha_k = 0$ in $\Lambda$ (harmonic function) and such that $\alpha_k = 1$ on $\partial^k\Lambda$, $\alpha_k = 0$ on $\partial\Lambda - \partial^k\Lambda$. It is easy to show, see [2] for details, that u is a smooth solution of the Euler equation in $\Lambda$ iff there exist m real constants $\beta_1,\ldots,\beta_m$ such that

$$u = \nabla^{\perp} (\psi + \sum_{k=1}^{m} \beta_k \alpha_k) \tag{2.7}$$

with $\psi$ a smooth solution of $-\partial_t \Delta\psi = B(\Delta\psi)$,

$$B(\Delta\psi) = (\nabla^{\perp}\psi) . \nabla\Delta\psi + (\nabla^{\perp} \sum_{k=1}^{m} \beta_k \alpha_k) . \nabla\Delta\psi \tag{2.7'}$$

where $\psi = 0$ on $\partial\Lambda$.

We define the energy functional H(u) by $H(u) = \dfrac{1}{2} \displaystyle\int_{\Lambda} u^2 dx$, for any $u \in L^2(\Lambda,dx)$. If u is a solution of the Euler equation then an easy computation, using the above remark, yields

$$H(u) = \hat{H}(\psi), \tag{2.8}$$

with $\hat{H}(\psi) \equiv -\dfrac{1}{2} \displaystyle\int \psi\Delta\psi \, dx$ and $\psi$ as in (2.7'), the relation between u and being that given in (2.7). It can also be seen that $\dfrac{d}{dt} H(u) = \dfrac{d}{dt} \hat{H}(\psi) = 0$, so H(u) is a constant of motion with respect to the Euler equation for u. Define, for any u such that rot $u \in L^2(\Lambda,dx)$,

$$S(u) \equiv \dfrac{1}{2} \displaystyle\int_{\Lambda} (\text{rot } u)^2 dx. \tag{2.9}$$

S is the so-called enstrophy functional. Using again (2.7), it is easy to see that $S(u) = \hat{S}(\psi)$ whenever u and $\psi$ are related as above. Moreover, it follows

218

that $\frac{d}{dt}$ S(u) = 0 for u a solution of the Euler equation, i.e. the enstrophy is a constant of motion for the classical Euler equations.

If $\partial\Lambda$ is connected, we have $\psi = \phi$ with $\phi$ the solution of the Euler equation (in this case $\beta_1 = \ldots = \beta_m = 0$ in (2.7)).

In the following we shall use the functionals $\hat{H}$, $\hat{S}$ to construct stochastic solutions of the Euler equation.

<u>Remark.</u>   In the case $\partial\Lambda$ connected, there are other constants of motion besides $\hat{H}(\phi)$, $\hat{S}(\phi)$, namely $\hat{S}_f(\phi) \equiv \int_\Lambda f(\Delta\phi(x))dx$, for any $f \in C(\mathbb{R})$ such that $\hat{S}_f(\phi) < \infty$.

In the following we shall, for simplicity, assume $\partial\Lambda$ connected (in this case $\phi = \psi$). We call the map $u(0) \to u(t)$ (resp. $\phi(0) \to \phi(t)$), with $u(t)$ (resp. $\phi(t)$) solutions of the classical Euler equation, the (classical) Euler flow (with initial condition $u(0)$ (resp. $\phi(0)$)).

It is well known (from classical results of Wolibner, Yudowich, Kato and others), that for u smooth, or at least of finite energy, there exist global-in-t solutions of the classical Euler equations (see e.g. [3], [8], [40] and references therein).

We shall associate to this classical flow a "stochastic flow". We assume $\Lambda$ bounded open. First of all we define a certain Gaussian measure associated with the enstrophy functional $\hat{S}(\psi)$. Let $\gamma > 0$ and let $\mu_\gamma$ be the Gaussian measure on the dual $\mathcal{D}'(\Lambda)$ of $\mathcal{D}(\Lambda) \equiv C_o^\infty(\Lambda)$ such that

$$\int_{\mathcal{D}'(\Lambda)} e^{i\gamma<g,\xi>} d\mu_\gamma(\xi) = e^{-\gamma/2 \, \|g\|_2^2}, \qquad (2.10)$$

for any $g \in \mathcal{D}(\Lambda)$. $<,>$ is the dualization of $\mathcal{D}(\Lambda)$ and $\mathcal{D}'(\Lambda)$ and $\|\cdot\|_2$ is the $L^2(\Lambda,dx)$-norm. $\mu_\gamma$ is thus the normal distribution associated with the Hilbert space $L^2(\Lambda,dx)$, with norm $\gamma^{1/2} \|\cdot\|_2$.

Let $\eta(x)$, $x \in \Lambda$ be the canonical generalized random field with underlying probability measure $\mu_\gamma$ (in the sense that
$\text{Prob}(Z_{f_1,\ldots,f_n}(a_1,\ldots,a_n)) = \mu_\gamma(Z_{f_1,\ldots,f_n}(a_1,\ldots,a_n))$, with
$Z_{f_1,\ldots,f_n}(a_1,\ldots,a_n)$ the cylinder set

$$\{\omega \mid <f_1,\eta>(\omega) \leq a_1,\ldots, <f_n,\eta>(\omega) \leq a_n\}, \quad a_i \in \mathbb{R}, f_i \in \mathcal{D}(\Lambda).$$

We shall look upon $\eta$ as a stochastic realization of the vorticity field rot u at time zero and hence, since for classical solutions $u,\phi$ of the Euler equation we have rot $u = \Delta\phi$, also of $\Delta\phi$ at time zero.

$\mu_\gamma$ gives to $\eta$ a "white noise" distribution. $\mu_\gamma$ is determined by the constant $\gamma$ and the enstrophy functional $S(u)$, inasmuch as formally

$$d\mu_\gamma(u) = Z^{-1} e^{-\gamma S(u)} \prod_{x \in \Lambda} d \text{ rot } u(x),$$

with Z the normalization making $d\mu_\gamma$ a probability measure (incidentally: a "direct" justification of such formal expressions is actually possible, using non-standard analysis, see [38]).

Let $\{f_n\}$ be the complete orthogonal system in $L^2(\Lambda)$ consisting of the eigenfunctions $f_n$ of the Dirichlet Laplacian $\Delta$ on $\Lambda$, to the eigenvalues $-\lambda_n$. We can realize $(\eta,\mu_\gamma)$ as $(X,P_\gamma)$, with $P_\gamma = \prod_n d\mu_n^\gamma$, $\mu_n^\gamma$ being one-dimensional Gaussian measures of mean zero and variance $\frac{1}{\gamma}$ and $<f,X> = \sum_n <f,f_n> X_n$, $X_n$ i.i.d. Gaussian random variables with distribution $\mu_n^\gamma$.

Define

$$:\hat{H}_N(X): \equiv \frac{1}{2} \sum_{n=0}^{N} \lambda_n^{-1} (X_n^2 - \frac{1}{\gamma}). \tag{2.11}$$

Then $E(:\hat{H}_N(X):) = 0$. Moreover, $:\hat{H}_N(X):$ converges in $L^2(P_\gamma)$ and by subsequences $P_\gamma$ -a.s. as $N \to \infty$ to a random variable $\hat{H}(X) \in L^2(\mu_\gamma)$ (since

$$E(|\sum_{n=m}^{m+\ell} \lambda_n^{-1} (X_n^2 - \frac{1}{\gamma})|^2) = \sum_{n=m}^{m+\ell} \frac{2}{\gamma^2} \frac{1}{\lambda^2} \to 0 \quad \text{as} \quad m \to \infty, \text{ because } \lambda_n \approx 4\pi^2/|\Lambda|,$$

by Weyl's estimate, as $n \to \infty$).

$:H(X):$ can be thought of as a renormalized energy associated with the random field $(X,P_\gamma)$ (hence with the random field $(\eta,\mu_\gamma)$). We recall the definition of the energy functional, in the case of smooth $\phi$,

$\hat{H}(\phi) = -\frac{1}{2} \int_\Lambda \phi\Delta\phi \, dx = -\frac{1}{2} \int_\Lambda \Delta\Delta^{-1}\phi \, \Delta\phi \, dx.$ Formally, then, with $\Delta\phi$ replaced by

$X$, $\hat{H}(X) = \frac{1}{2} \sum_n \frac{1}{\lambda_n} X_n^2$, which formally is $:\hat{H}(X): + \frac{1}{2\gamma} \sum_n (1/\lambda_n)$. Since

$:\hat{H}(X):$ is $P_\gamma$-a.s. finite and $\Sigma(1/\lambda_n) = \infty$, we see that $\hat{H}(X)$ is $P_\gamma$-a.s. equal to $+\infty$.

Thus the (time zero) stochastic vorticity field X (i.e. $\eta$) has diverging energy (but finite renormalized energy $:\hat{H}(X):$).

Remark.  Elements in the support of $P$  have thus infinite energy.  A detailed study of the support properties of $P$  in the related case of periodic boundary conditions for a square $\Lambda$ has been given in $[5] - [7]$, $[9]$.

We have, for any $\beta \geq 0$,

$$\exp\left[-\beta \; :\hat{H}(X):\right] \in L^1(P_\gamma), \quad \exp\left[-\beta \; :\hat{H}(X):\right] > 0,$$

$P_\gamma$-a.s., thus

$$dP_{\beta,\gamma}(X) \equiv \exp\left[-\beta \; :\hat{H}(X):\right]dP_\gamma(X)\Big/\int \exp\left[-\beta \; :\hat{H}(X):\right]dP_\gamma(X)$$

is a probability measure , absolutely continuous with respect to $P_\gamma$.

$P_{\beta,\gamma}$ is a mathematical realization of the formal object

$$Z^{-1} \exp(-\beta H(u)-\gamma S(u)) \quad \prod_{x\in\Lambda} d \; \text{rot} \; u(x),$$

with Z a normalization, i.e. of a Gibbsian measure with density given by the constants of motion H and S.

We summarize these results in the following:


Theorem 1.  Let $\Lambda$ be a bounded open subset of $\mathbb{R}^2$ and let $\gamma > 0$.  Then there exists a Gaussian measure $P_\gamma$ formally given by $Z^{-1} e^{-\gamma S(u)} \prod_{x\in\Lambda} d \; \text{rot} \; u(x)$, where $S(u) = \frac{1}{2}\int_\Lambda \text{rot} \; u^2 dx$ is the enstrophy functional of the Euler fluid in $\Lambda$, and Z is a normalization.  $P_\gamma$ can be realized as $d\mu_n^\gamma$, with $\mu_n^\gamma$ one-dimensional Gaussian measures of mean zero and variance $1/\gamma$.

The energy functional $H(u)$, formally given by $\frac{1}{2}\int u^2 dx$, is $P_\gamma$-a.s. infinite.  One can define a renormalized energy $:\hat{H}:$ in $L^2(P_\gamma)$, hence $P_\gamma$-a.s. finite.  $:\hat{H}(X):$, for $X \in \text{supp } P_\gamma$, is defined as the $L^2(P_\gamma)$-limit of

$$:\hat{H}_N(X): \equiv \frac{1}{2} \sum_{n=0}^{N} \lambda_n^{-1} \left(X_n^2 - \frac{1}{\gamma}\right),$$ with $\lambda_n$ the n-th eigenvalue of the negative of the Dirichlet Laplacian in $\Lambda$ and $X_n$ the n-th coordinate of X.

One has, for any $\beta \geq 0$, $\quad \exp\left[-\beta \; :\hat{H}(\cdot):\right] \in L^1(P_\gamma)$, hence

$$dP_{\beta,\gamma} \equiv \exp\left[-\beta \; :\hat{H}:\right]dP_\gamma\Big/\int \exp\left[-\beta \; :\hat{H}:\right]dP_\gamma$$

is a probability measure absolutely continuous with respect to $P_\gamma$.
Formally,

$$dP_{\beta,\gamma} = Z^{-1} \exp\left[-\beta H(u) - \gamma S(u)\right] \prod_{x \in \Lambda} d \text{ rot } u(x).$$

<u>Remark.</u>  $P_\gamma$ can also be defined, in the case $\Lambda = \mathbb{R}^2$, as the weak limit of
the above Dirichlet Gaussian measure $P_\gamma^\Lambda$ as $\Lambda \uparrow \mathbb{R}^2$.  In fact the $P_\gamma^\Lambda$ are
centered and the covariances $(-\Delta_{\partial\Lambda})^{-1}$ converge as $\Lambda \uparrow \mathbb{R}^2$ ($\Delta_{\partial\Lambda}$ is the
Dirichlet Laplacian in $\Lambda$).

<u>Remark.</u>  Gaussian Gibbs measures of the above type have also been discussed
for magnetohydrodynamics, see e.g. $[28]$.  There is also a large body of
numerical work concerning the possible relevance of the Gibbs measures to
the problems of "fully developed turbulence", see e.g. $[14]$, $[17]$, $[43]$.


The above results are contained essentially in $[2]$.  The case of $\Lambda$ a
square with periodic boundary conditions has been treated in $[1]$, $[5]$, $[6]$.
For related results, see also $[7]$, $[9]$.

The natural question now arises: in what sense are $P_{\beta,\gamma}$ invariant
measures for a possible stochastic flow associated with the Euler equation,
Such a flow $\eta(0) \rightarrow \eta(t)$ would map $\mathcal{D}'(\Lambda)$ into itself in such a way that any
cylinder set $A = Z_{f_1,\ldots,f_n}(a_1,\ldots,a_n)$ in the support of $\mu_\gamma$, as defined
above, would be mapped into a set $A_t$ again in the support of $\mu_\gamma$, with $A_t$
defined as A with $\eta \equiv \eta(0)$ replaced by $\eta(t)$.

Formally, $P_{\beta,\gamma}$ are invariant, since "they are built with the invariants
H and S of the Euler flow".  However, we should remember that $P_{\beta,\gamma}$ has
support on solutions of infinite energy, so in the definition of the flow
itself one cannot with certainty use results involving finite classical
energy.

The following results are obtained in $[4]$.


We shall first see that we can define an object corresponding to $B(\Delta\phi)$ in
(2.7') (here $\phi = \psi$), for $\phi$ such that $\Delta\phi = \eta$, as an $L^2(\mu_\gamma)$-limit of a
corresponding regularized quantity.  Let $\eta_\varepsilon$ be a regularization of $\eta$ such
that $<f;\eta_\varepsilon> \rightarrow <f,\eta>$ in $L^2(d\mu_\gamma)$ as $\varepsilon \rightarrow 0$, for all $f \in C_0^\infty(\Lambda)$, and $\eta_\varepsilon$ is a $C^\infty$
function (e.g. $\eta_\varepsilon = g_\varepsilon * \eta$, with $g_\varepsilon \in C_0^\infty(\Lambda)$, $g_\varepsilon \rightarrow \delta_0$, $\varepsilon \downarrow 0$).  Let

$\phi_\varepsilon = (\Delta_{\partial\Lambda})^{-1}\eta_\varepsilon$ (with $\Delta_{\partial\Lambda}$ the Laplacian with Dirichlet boundary conditions on $\Lambda$). Define, corresponding to (2.7)',

$$B(\eta_\varepsilon) = (\nabla^\perp\phi_\varepsilon) \cdot \nabla\eta_\varepsilon \qquad\qquad (2.12)$$

and

$$<f,B(\eta_\varepsilon)> = \int f(x)\ B(\eta_\varepsilon(x))dx.$$

Using a partial integration, we get

$$<f,B(\eta_\varepsilon)> = \sum_{i,j} \int (\nabla^\perp)_i\ \nabla_j f(x)\ \phi_\varepsilon(x)\ \nabla_i\ \nabla_j\ \phi_\varepsilon(x)dx$$

$$\qquad\qquad (2.13)$$

$$= \sum_{i,j} \iiint (\nabla^\perp)_i\ \nabla_j f(x)G(x,z)G_{ij}(z,y)\eta_\varepsilon(y)\eta_\varepsilon(z)dx\ dy\ dz$$

where $G(x,z)$ is the kernel in $L^2(\Lambda)$ of $(-\Delta_{\partial\Lambda})^{-1}$ and $G_{ij}(x,y) \equiv \dfrac{\partial^2}{\partial x_i\ \partial x_j} G(x,y)$.
A direct estimate, see $[2]$ for details, shows that $<f,B(\eta_\varepsilon)> \in L^2(\mu_\gamma)$ and
$<f,B(\eta_\varepsilon)>$ converges in $L^2(\mu_\gamma)$ as $\varepsilon \to 0$. Call $<f,B(\eta)>$ the limit.

We shall now use the realization $P_\gamma$ of $\mu_\gamma$ and denote points of supp $P_\gamma$ by
$X$ (the process $X$ being in canonical form). Define $<f,B(X)>$ in the same way
as we defined $<f,B(\eta)>$. Then $B_n(X) \equiv <f_n,\ B(X)>$ is in $L^2(P_\gamma)$ and we have

$$B_n(X) = - \sum_{\substack{k\neq\ell,\ell\neq n \\ k\neq n}} \lambda_k^{-1}\ \lambda_\ell^{-1}\ X_k\ X_\ell\ (\sum_{i,j}\ \nabla_i\ \nabla_j\ f_n,\ (\nabla)_i\ f_k\ \nabla_j\ f_\ell),\ (2.14)$$

which shows that $B_n(X)$ is independent of $X_n$.

Let $FC^1$ be the functions in supp $P_\gamma$ which depend only on a finite number
of coordinates, i.e.

$$F \in FC^1 \text{ iff } \tilde{F}(X) = F(<f_{i_1},X>,\ldots<f_{i_n},X>) \text{ for some } \tilde{F}\in C_b^1(\mathbb{R}^n),$$

$$i_1,\ldots,i_n \in \mathbb{N}.$$

$\{f_n\}$ is the complete orthonormal set of eigenfunctions of $-\Delta$ in $L^2(\Lambda,dx)$.
$FC^1$ is dense in $L^2(P_\gamma)$ $(\simeq L^2(\mu_\gamma))$. For $F \in FC^1$ we define the linear
operator $\tilde{B}$ by $(\tilde{B}F)(X) \equiv \sum_n B_n(X)\ \dfrac{\partial F}{\partial X_n}$ , with $X \in$ supp $P_\gamma$, $B_n(X)$ defined by
(2.14). Since $F$ depends only on a finite number of the $X_n$ and $B_n(\cdot)\in L^2(P_\gamma)$

we have $BF(.) \in L^2(P_\gamma)$. The operator $\overset{\sim}{B}$ is thus densely defined on $L^2(P_\gamma)$, with domain $FC^1$. It is, by definition, a derivation on $FC^1$. We call $\overset{\sim}{B}$ the Euler-Liouville operator.

We saw above that $B_n(X)$ is independent of $X_n$, so that $\overset{\sim}{B}$ is a divergence free vector field on $\mathcal{D}'(\Lambda)$, in the sense that

$$\text{div } \overset{\sim}{B} = \sum_n \frac{\partial B_n}{\partial X_n} = 0. \tag{2.15}$$

We shall look upon $\overset{\sim}{B}$ as the local infinitesimal generator of a (yet to be defined) stochastic Euler flow. We call a measure $\mu$ infinitesimally invariant (under the yet to be defined stochastic Euler flow) if

$$\int \overset{\sim}{B}F \, d\mu = 0 \quad \text{for all} \quad F \in FC^1. \tag{2.16}$$

This is equivalent to $\overset{\sim}{B}{}^*1 = 0$ where $^*$ is the adjoint in $L^2(\mu)$ and $1$ is the function identically $1$ on supp $\mu$.

Remark. Let us comment briefly on equation (2.16). It is an expression of the *stationary Hopf functional equation* associated with the Euler equation. In fact, Hopf's equation for the probability distribution associated with a stochastic Euler flow is by definition a family of probability measures $\mu_t$, $t \geq 0$, such that

$$\frac{\partial}{\partial t} \int_{\mathcal{D}'(\Lambda)} e^{i<f,\eta>} \, d\mu_t(\eta) = i \int_{\mathcal{D}'(\Lambda)} <f,B(\eta)> e^{i<f,\eta>} \, d\mu_t(\eta), \tag{2.17}$$

with the initial condition $\mu_o$, for any $f \in \mathcal{D}(\Lambda)$. $<,>$ is the distributional pairing. See $[39]$, $[37]$, $[6]$, $[2]$. Stationariness implies $\mu_t = \mu_o$ for all $t$ and the equation reduces to

$$\int_{\mathcal{D}'(\Lambda)} <f,B(\eta)> e^{i<f,\eta>} \, d\mu_o(\eta) = 0. \tag{2.18}$$

Replacing $(\eta,\mathcal{D}'(\Lambda))$ by the realization $(X, \prod_n \mathbb{R}^{(n)})$ and taking $f$ to be $f_n$, we see that (2.18) is equivalent to (2.16), for $\mu = \mu_o$. However, for $\mu = P_{\beta,\overset{\sim}{\gamma}}$,

$$\overset{\sim}{B}{}^* \supset \sum_n \left( \frac{\partial}{\partial X_n} \right)^* B_n, \tag{2.19}$$

224

with $B_n$ the operator in multiplication by the real-valued function $B_n(X)$ in $L^2(P_{\beta,\gamma})$. $P_{\beta,\gamma}$ is, by its definition, quasi-invariant under translations by elements $X_n$ and an easy computation shows that

$$( \frac{\partial}{\partial X_n} )^* \supset - \frac{\partial}{\partial X_n} + (\gamma + \frac{B}{\lambda_n}) X_n. \qquad (2.19')$$

Moreover, one can easily show that $\sum_{n=1}^{N} \frac{1}{\lambda_n} X_n B_n(X) = 0$, $P_{\beta,\gamma}$ -a.s.

This then implies, from (2.19), (2.19'), $\tilde{B}^* \supset - \tilde{B}$, which shows that $\tilde{B}$ is closable and $i\tilde{B}$ is symmetric as an operator in $L^2(P_{\beta,\gamma})$ (complex). Moreover, $\tilde{B}^*1 = 0$, i.e. $P_{\beta,\gamma}$, is infinitesimally invariant. Let $P_m$ be the closure in $L^2(P_{\beta,\gamma})$ of the polynomials in $X_n$ of degree $\leq m$ and set $P = U P_m$. Then $P \subset L^P(P_{\beta,\gamma})$ for all $p > 0$ and $B_n(\cdot) \in P_2$. $\tilde{B}$ extends in a natural way to $P \cup F C^1$ and $:\hat{H}(X):$ is then in its domain, in fact $\tilde{B}:\hat{H}(X): = 0$.

Now define, for $F \in L^2(P_{\beta,\gamma})$, $J F(X) \equiv \overline{F}(-X)$. Then $J$ is an antilinear map of the complex $L^2(P_{\beta,\gamma})$ such that $J^2 = 1$, hence it is a complex conjugation in $L^2(P_{\beta,\gamma})$. It is easy to show that $Ji\tilde{B} = i\tilde{B}J$, so that $i\tilde{B}$ is real with respect to $J$, hence, by a theorem of von Neumann, there are self-adjoint extensions of $i\tilde{B}$ in $L^2(P_{\beta,\gamma})$. Let $A$ be any such a self-adjoint extension. Then $e^{itA}$ is a strongly continuous unitary group $U_t$ in $L^2(P_{\beta,\gamma})$.

We summarize these results in the following:

<u>Theorem 2.</u>   Let $\Lambda$ be a bounded open subset of $\mathbb{R}^2$ with $\partial\Lambda$ connected. Then the operator $B$ appearing on the right-hand side of the Euler equation

$$u = \nabla^{\perp}\phi, \quad -\partial_t \Delta\phi = B(\Delta\phi),$$

$$B(\Delta\phi) = (\nabla^{\perp}\phi) \cdot \nabla\Delta\phi, \quad \text{with } \phi = 0 \text{ cn } \partial\Lambda,$$

is well defined also for rot $u = X \in$ supp $P_\gamma$. Here $P_\gamma$ is the Gaussian measure of Theorem 1, in the sense that, for any $f \in C_o^\infty(\Lambda)$,

$$<f,B(X)> = \sum_{n=1}^{\infty} <f,f_n> <f_n,B(X)>,$$

where $<f_n,B(X)> \equiv B_n(X)$ is given by (2.14), $f_n$ is the orthonormal set of eigenfunctions of the negative of the Laplacian $-\Delta_{\partial\Lambda}$ with Dirichlet boundary conditions on $\partial\Lambda$, $< , >$ being the pairing in distributional sense. We have

$B_n \in L^2(P_\gamma)$ and $\partial B_n / \partial X_n = 0$, $P_\gamma$-a.s. Let $FC^1$ be the cylinder functions in $L^2(P_\gamma)$, once continuously differentiable with bounded derivatives. $FC^1$ is dense in $L^2(P_\gamma)$ and the Euler-Liouville operator $\tilde{B}$ defined on $FC^1$ by

$$(\tilde{B}F)(X) = \sum_n B_n(X) \frac{\partial}{\partial X_n} F$$

is a divergence free derivation on $FC^1$. The measures $P_{\beta,\gamma}$ of Theorem 1 are infinitesimally invariant in the sense that

$$\int \tilde{B}F dP_{\beta,\gamma} = 0 \text{ for all } F \in FC^1,$$

i.e. they satisfy the stationary Hopf functional equation. $\tilde{B}$ is closable and $i\tilde{B}$ is symmetric in $L^2(P_{\beta,\gamma})$. $\tilde{B}$ extends in a natural way to $\underset{m}{U} P_m \cup FC^1$, where $P_m$ is the closure in $L^2(P_{\beta,\gamma})$ of the polynomials in $X_n$ of degree $\leq m$. We have $\tilde{B}:\hat{H}: = 0$. Moreover, $i\tilde{B}$ has at least one self-adjoint extension. For any self-adjoint extension $A$ of $i\tilde{B}$ we have that $U_t = e^{itA}$ is a unitary group in $L^2(P_{\beta,\gamma})$

Remark.    Formally, $U_t$ gives a stochastic $L^2$-flow associated with the Euler equation and the measures $P_{\beta,\gamma}$.

We might ask whether there is a self-adjoint extension such that, for the corresponding $U_t$, we have $(U_t F)(X) = F(\gamma_t(X))$, with $\gamma_t$ a group of measurable transformation of $(\underset{n}{\Pi} \mathbb{R}^{(n)}, P_{\beta,\gamma})$ leaving invariant $P_{\beta,\gamma}$. This is to our knowledge open; we shall now discuss a weaker result in this direction, which consists in constructing a flow $\gamma_t$ on supp $P_{\beta,\gamma}$ and a strongly continuous contraction group $U_t$ in $L^2(P_{\beta,\gamma})$. The method of constructing $\gamma_t$ is adapted from one used for smoother solutions by Boldrighini in [9], that was sketched by Albeverio and Høegh-Krohn in [10], [11] and carried through in [4] and independently, in the case of a square with periodic boundary conditions, by Caprino and De Gregorio in [7].

Define for any $N \in \mathbb{N}$, $f \in FC^1$, $\forall X \in$ supp $P_{\beta,\gamma}$:

$$(\tilde{B}^N F)(X) = \sum_n B_n^N(X) \ F_{,n}(X) \tag{2.20}$$

with $F_{,n}(X) \equiv \dfrac{\partial F}{\partial X_n}(X)$ and $B_n^N(X)$ defined as $B_n(X)$ but with the sum running only over the set $I_N$ of all indices such that $\lambda_k \leq N$, $\lambda_\ell \leq N$. We have that $B_n^N$ converges as $N \to \infty$ in $L^2(P_{\beta,\gamma})$ to $B_n$. Also, $\tilde{B}^N F$ converges as $N \to \infty$ in $L^2(P_{\beta,\gamma})$ to $\tilde{B}F$, for all $F \in FC^1$. The latter result follows from the former and (2.20), the sum being finite for $F \in FC^1$. The former result follows from a basic estimate on $\| B_n^{N'} - B_n^N \|_2$ for all $N' \geq N$, obtained from the definition of $B_n^N$, where $\| \ \|_2$ is the $L^2(P_{\beta,\gamma})$-norm. It is similar to results published in [1], [5], [6] for the periodic case, and is written out in [2], to which we refer for details. Let us also mention that

$$\| B_n \|_2^2 = \frac{1}{\gamma} \int_\Lambda \int_\Lambda \Big| \sum_{i,j} \int_\Lambda (\nabla^\perp)_i \nabla_j f_n(x) G(x,z) G_{ij}(z,y) dz \Big|^2 dx \, dy$$

$$= \frac{1}{\gamma} \| A \|_{HS}^2$$

where $\| A \|_{HS}$ is the Hilbert-Schmidt norm and

$$A \equiv \sum_{i,j} (-\Delta)^{-1} (\nabla^\perp)_i \nabla_j f_n, \ \nabla_i \nabla_j (-\Delta)^{-1} .$$

That $\| A \|_{HS}^2 < \infty$ is an easy estimate, using essentially

$$\| (-\Delta)^{-1} \|_{HS}^2 = \int_\Lambda \int_\Lambda |G(x,y)|^2 \, dx \, dy < \infty .$$

Let us now look at the equation for the unknown quantities $X_n^N(t)$:

$$\frac{d}{dt} X_n^N(t) = B_n^N(X_m^N(t), \ m \in I_N), \ n \in I_N \quad \left.\vphantom{\frac{d}{dt}}\right\} \tag{2.21}$$

with initial condition $X_n^N(0) = X_n$.

$B_n^N(X_m^N(t), \ m \in I_N)$, is defined by the right-hand side of (2.14) with $X_k X_\ell$ replaced by $X_k^N(t) X_\ell^N(t)$. Set also

$$X_n^N(t) \equiv X_n^N(0) = X_n \quad \text{for} \quad n \notin I_N . \tag{2.21'}$$

Since $B_n^N(X_m^N(t), \ m \in I_N)$ is given ($N < \infty$!) by a finite linear combination of products $X_k^N(t) \ X_\ell^N(t)$ with finite coefficients, it is not difficult to show, using essentially existence theorems for systems of ordinary differential equations, that (2.21), (2.21)' has a solution for all $t$, which we denote by $X_n^N(t)$.

Remark. A related statement has been discussed in detail for the situation with periodic boundary conditions in [7].

Define, for any $t \in \mathbb{R}_+$, $N \in \mathbb{N}$, $\gamma_t^N$ by $\gamma_t^N(X_n^N(0)) \equiv X_n^N(t)$ and extend it in the natural way to all $t \in \mathbb{R}$. $\gamma_t^N$ is a group of measurable transformations of $(\Pi_n \mathbb{R}^{(n)}, P_{\beta,\gamma})$, which leaves invariant $P_{\beta,\gamma}$. Define an operator $U_t^N$ in $L^2(P_{\beta,\gamma})$ by $(U_t^N F)(X) \equiv F(\gamma_t^N(X))$, $X \in \text{supp } P_{\beta,\gamma}$, $F \in FC^1 \cup P$, and extended by continuity to all of $L^2(P_{\beta,\gamma})$, where $\gamma_t^N(X)$ is such that $\langle f_n, \gamma_t^N(X) \rangle \equiv \gamma_t^N(X_n)$. It is easy to see that $U_t^N$ is unitary, with inverse

$$(U_t^N)^{-1} = (U_t^N)^* \quad \text{such that} \quad (U_t^N)^* F(X) = F(\gamma_{-t}^N(X)).$$

Moreover, we have, for any $F \in FC^1 \cup P$,

$$\frac{d}{dt}(U_t^N F)(X) = (\widetilde{B}^N F)(\gamma_t^N(X)), \tag{2.22}$$

with $(U_o^N F(X) = F(\gamma_o^N(X)) = F(X)$ and

$$(\widetilde{B}^N F)(\gamma_t^N(X)) = \sum_{n \in I_N} \widetilde{B}_n^N(X_m^N(t), m \in I_N) \frac{\partial}{\partial X_n^N(t)} F(\gamma_t^N(X)).$$

We note that

$$\| B_n^N(X_m^N(t), m \in I_N) \|_2 = \| B_n^N(X) \|_2 \tag{2.23}$$

and

$$\| \widetilde{B}^N F(\gamma_t^N(\cdot)) \|_2 = \| (\widetilde{B}^N F)(\cdot) \|_2 < \infty \,,$$

by the invariance of $P_{\beta,\gamma}$ under $\gamma_t^N$, the fact that $FC^1 \cup P \subset D(\widetilde{B}^N)$, and where we have also used the definitions of $B_n^N(X_m^N(t), m \in I_N)$ and $B_n^N(X)$.

Incidentally let us note that since $B_n^N(X)$ converges in $L^2(P_{\beta,\gamma})$ as $N \to \infty$, we can choose a subsequence, call it again $N$, such that $B_n^N(X)$ converges $P_{\beta,\gamma}$-a.x. to $B_n(X)$ as $N \to \infty$ and the subsequence and the exceptional set can be chosen independent of $n$.

Let us consider the family $F$ in $FC^1$ of functions $X \to F(X)$ of the form
$$F(X) = e^{i\alpha X_n}, \quad n \in \mathbb{N}, \ \alpha \in \mathbb{R}.$$
Linear combinations of finite products of such functions are dense in $FC^1$, in the sup-norm, and in $L^2(P_{\beta,\gamma})$. We have, for any $G \in L^2(P_{\beta,\gamma})$,

$$| (G, F \circ \gamma_t^N) | \leq \| G \|_2 \| F \circ \gamma_t^N \|_2 = \| G \|_2$$

where we have used the invariance of the $P_{\beta,\gamma}$-measure under $\gamma_t^N$.

On the other hand, by (2.22) for any $t' > t$, $t', t \in \mathbb{R}$ we have, with $F$ as above:

$$U_{t'}^N F - U_t^N F = e^{i\alpha X_n^N(t')} - e^{i\alpha X_n^N(t)}$$

$$= \int_t^{t'} \frac{d}{ds} e^{i\alpha X_n^N(s)} ds = \int_t^{t'} (\tilde{B}^N F)(\gamma_s^N(X)) ds, \tag{2.24}$$

hence

$$\| U_{t'}^N F - U_t^N F \|_2 \leq \int_t^{t'} \| \tilde{B}^N F(\gamma_s^N(\cdot)) \|_2 \, ds.$$

But

$$\tilde{B}^N F(\gamma_s^N(X)) = i\alpha B_n^N(X_m^N(t), m \in I_N) \, e^{i\alpha X_n^N(s)}, \tag{2.25}$$

hence, from (2.23),

$$\| \tilde{B}^N F(\gamma_s^N(\cdot)) \|_2 = |\alpha| \, \| B_n^N(\cdot) \|_2. \tag{2.26}$$

But, as remarked above, $B_n^N(X)$ converges to $B_n(X)$ in $L^2(P_{\beta,\gamma})$ as $N \to \infty$, hence in particular the right-hand side of (2.26) converges as $N \to \infty$ to $\| B_n(\cdot) \|_2$. Thus from (2.26) in particular $\| \tilde{B}^N F(\gamma_s^N(\cdot)) \|_2$ is uniformly bounded for all $N \in \mathbb{N}$, independently of $s \in \mathbb{R}$, i.e.

$$\| \tilde{B}^N F(\gamma_s^N(\cdot)) \|_2 \leq C_n \, |\alpha|, \tag{2.27}$$

with $C_n$ independent of $s \in \mathbb{R}$, $N \in \mathbb{N}$. Inserting this into (2.24), (2.25) we get

$$\| e^{i\alpha X_n^N(t')} - e^{i\alpha X_n^N(t)} \|_2 \leq (t'-t) \, |\alpha| \, C_n. \tag{2.28}$$

Also

$$\| \frac{\partial}{\partial t} e^{i\alpha X_n^N(t)} \|_2 \leq |\alpha| \, C_n. \tag{2.28}'$$

This implies that the family $\{F \circ \gamma_t^N, N \in \mathbb{N}, F \in F\}$ is an equicontinuous uniformly bounded (by 1) family of maps from $\mathbb{R}$ into $L^2(P_{\beta,\gamma})$. Hence $\{(G, F \circ \gamma_t^N), \ N \in \mathbb{N}, \ F \in \mathcal{F}, \ G \in L^2(P_{\beta,\gamma})\}$ is an equicontinuous uniformly bounded (by $\|G\|_2$) family of maps from $\mathbb{R}$ into $C$.

By weak compactness we can choose a subsequence $N$ such that $(G, F \circ \gamma_t^N)$ converges for all $G \in L^2(P_{\beta,\gamma})$ as $N \to \infty$ to $(G, \tilde{F}(t))$ for all $t \in Q$. We can now apply the Ascoli-Arzela theorem to the above family, on $[a,b]$, for any $-\infty < a < b < \infty$, taken to be a sequence indexed by $N$, obtaining then that there is a subsequence, with index denoted again by $N$, such that $(G, F \circ \gamma_t^N)$ converges uniformly for all $t \in [a,b]$ to the continuous bounded function $(G, \tilde{F}(t))$, for all $G \in L^2(P_{\beta,\gamma})$. Moreover, from (2.28)' we get

$$\left| \frac{\partial}{\partial t} (G, F \circ \gamma_t^N) \right| \leq \|G\|_2 \ \left\| \frac{\partial}{\partial t} F \circ \gamma_t^N \right\|_2 \leq |\alpha| \ C_n \|G\|_2$$

and from the above convergence we have that $\frac{\partial}{\partial t} (G, \tilde{F}(t))$ exists, is the limit of $\frac{\partial}{\partial t} (G, F \circ \gamma_t^N)$ as (a subsequence of) $N \to \infty$, and

$$\left| \frac{\partial}{\partial t} (G, \tilde{F}(t)) \right| \leq |\alpha| \ C_n \ \|G\|_2 \ .$$

Furthermore, from equation (2.24) we have for $t = 0$, $t' = s$,

$$(G, U_s^N F) = (G, F) + \left( G, \int_0^s (\tilde{B}^N F)(\gamma_{s'}^N, (\cdot)) ds' \right).$$

Going to the limit $N \to \infty$, we get

$$\tilde{F}(s) = F + \underset{N \to \infty}{w-\lim} \int_0^s \tilde{B}^N F \ (\gamma_{s'}^N, (\cdot)) ds'. \tag{2.29}$$

By construction, $\frac{\partial}{\partial \alpha} (G, e^{i\alpha X_n^N(t)}) \big|_{\alpha=0} = i(G, X_n^N(t))$. As $N \to \infty$ by similar arguments this converges to $(G, X_n(t))$ for some $X_n(t) \in L^2(P_{\beta,\gamma})$.

The convergence of $(G, e^{i\alpha X_n^N(t)})$ as $N \to \infty$ is uniform in $\alpha$, by a similar argument (since

$$\left| (G, e^{i\alpha X_n^N(t)} - e^{i\beta X_n^N(t)}) \right| \leq \|G\|_2 \|X_n^N(t)\|_2 |\beta - \alpha| \leq \|G\|_2 \|X_n^N(0)\|_2 |\beta - \alpha|$$

and $(G, e^{i\alpha X_n^N(t)})$ is a uniformly bounded family from the set $\mathbb{R}$ of the $\alpha$'s into C).  Thus

$$\lim_{N\to\infty} \frac{\partial}{\partial\alpha} (G, e^{i\alpha X_n^N(t)})\Big|_{\alpha=0} = i(G, X_n(t))$$

$$= \frac{\partial}{\partial\alpha} (G, \tilde{F}(t))\Big|_{\alpha=0} .$$

We summarize these results in the following:

<u>Theorem 3.</u>    Let $\Lambda$ be a bounded open subset of $\mathbb{R}^2$ with $\partial\Lambda$ connected.  Let $B_n$ be as in Theorem 2. Define $B_n^N, \tilde{B}_n^N$ as the Galerkin approximation of $B_n, \tilde{B}$ by "cutting off modes corresponding to energies $> N$", defined by (2.20).  $B_n^N$ converges in $L^2(P_{\beta,\gamma})$ as $N \to \infty$ to $B_n$.  Define $X_n^N(t)$ as the solution of the corresponding Galerkin approximation (2.21) of the Euler equation and set $\gamma_t^N(X_n^N(0)) \equiv X_n^N(t)$,

$$(U_t^N F)(X) \equiv F(\gamma_t^N(X)), \quad X \in \text{supp } P_{\beta,\gamma}, \quad F \in FC^1 \cup P.$$

Then $U_t^N$ is unitary, with generator extending $i\tilde{B}^N$, and $P_{\beta,\gamma}$ is invariant under $\gamma_t^N$.  The family $\{F \circ \gamma_t^N, N \in \mathbb{N}, F \in F\}$ with $F$ the set of functions of the form $F(X) = e^{i\alpha X_n}$, for some $n \in \mathbb{N}$, $\alpha \in \mathbb{R}$, is equicontinuous, uniformly bounded by 1, from $\mathbb{R}$ into $L^2(P_{\beta,\gamma})$.

There is a subsequence of $N$ such that $F \circ \gamma_t^N$ converges weakly in $L^2(P_{\beta,\gamma})$, in a uniform way for all $t$ in compacts of $\mathbb{R}$, to a limit $\tilde{F}(t)$ in $L^2(P_{\beta,\gamma})$. One has for all $G \in L^2(P_{\beta,\gamma})$ that

$$\frac{\partial}{\partial t} (G, \tilde{F}(t)) = \lim_{N\to\infty} \frac{\partial}{\partial t} (G, F \circ \gamma_t^N)$$

and

$$\Big|\frac{\partial}{\partial t} (G, \tilde{F}(t))\Big| \leq |\alpha| \, C_n \, \|G\|_2 .$$

$\tilde{F}(t)$ solves the limit Euler equation in the sense that

$$F(t) = F(0) + \text{w-}\lim_{N\to\infty} \int_0^t \tilde{B}^N F (\gamma_s^N(\cdot))ds,$$

with $F(0) = 1$.  $X_n^N(t)$ converges weakly by subsequences to $L^2(P_{\beta,\gamma})$,

231

uniformly in $\alpha$ to a limit $X_n(t) \in L^2(P_{\beta,\gamma})$ and one has $\frac{\partial}{\partial \alpha} \tilde{F}(t)\big|_{\alpha=0} = iX_n(t)$, weakly in $L^2(P_{\beta,\gamma})$. $\square$

Define $U_t$ as a mapping in $L^2(P_{\beta,\gamma})$ by $U_t\tilde{F}(s) \equiv \tilde{F}(s+t)$. Then $(U_t, t \in \mathbb{R})$ is a one-parameter group of abelian transformations of $L^2(P_{\beta,\gamma})$ which are contractions (hence it extends in particular to the whole $L^2(P_{\beta,\gamma})$). $U_t$ is weakly continuous and differentiable, as a map from $\mathbb{R}$ into the bounded linear operators on $L^2(P_{\beta,\gamma})$, hence in particular strongly continuous. From (2.29) we have easily, using the strong convergence of $\tilde{B}^N F$ to $\tilde{B}F$, that the infinitesimal generator of $U_t$, as a semigroup, is a closed extension $\hat{B}$ of $\tilde{B}$, thus $U_t = e^{t\hat{B}}$, $t \geq 0$. We can then apply results of $\begin{bmatrix} 55 \end{bmatrix}$ to decompose $U_t$ to give a unitary and a completely nonunitary part. Finally, we can define a one-parameter weakly continuous differentiable group of maps $\gamma_t$ from $\mathbb{R}$ into $L^2(P_{\beta,\gamma})$ by $\gamma_t X_n(s) = X_n(s+t)$. We summarize these results in the following:

<u>Theorem 4.</u>   There exists a one-parameter strongly continuous group $U_t$ of contractions of $L^2(P_{\beta,\gamma})$, defined by $U_t\tilde{F}(s) \equiv \tilde{F}(s+t)$, with $\tilde{F}(t)$ as in Theorem 3, hence such that $U_t$ is the limit of the Euler evolution groups $U_t^N$. Its infinitesimal generator is a closed extension $\hat{B}$ of the Euler Liouville operator $\tilde{B}$ of Theorem 2.

$e^{t\hat{B}}$, $(e^{t\hat{B}})^*$, $t \in \mathbb{R}_+$, are contraction semigroups in $L^2(P_{\beta,\gamma})$. $\hat{B}$ is maximally densely defined dissipative (such that $\mathrm{Re}(\hat{B}G,G) \leq 0 \; \forall G \in L^2(P_{\beta,\gamma})$) and there exists a unique orthogonal decomposition $L^2(P_{\beta,\gamma}) = H_1 \oplus H_2$, with $U_t \upharpoonright H_1$ unitary and $U_t \upharpoonright H_2$ completely nonunitary (i.e., $\{0\}$ is the only subspace of $H_2$ on which $U_t$ is unitary). The maps $\gamma_t(X_n(s) \equiv X_n(s+t)$ form a weakly continuous and differentiable group from $\mathbb{R}$ into $L^2(P_{\beta,\gamma})$.

<u>Remark.</u>   By the construction, $(U_t, \gamma_t)$ is the limit of the cut-off Euler flow $(U_t^N, \gamma_t^N)$. However, it is open whether $(U_t F)(X) = F(\gamma_t(X))$, and in particular $U_t$ is unitary and $P_{\beta,\gamma}$ invariant under $\gamma_t$, as was the case for $(U_t^N, \gamma_t^N)$. In $\begin{bmatrix} 4 \end{bmatrix}$ this question is discussed further. In particular, the condition that $i\tilde{B}$ is essentially self-adjoint on $\mathcal{D} = FC^\infty$ (or any domain on which $\tilde{B}^N$ converges to $\tilde{B}$ as $N \to \infty$) yields that, with A the closure of $i\tilde{B}$, $U_t = e^{iAt}$ is the unique unitary group in $L^2(P_{\beta,\gamma})$ defined as the limit of $U_t^N$,

232

and one has $(e^{iAt}F)(X) = F(\gamma_t \circ X)$, with $\gamma_t$ leaving $P_{\beta,\gamma}$ and $:\hat{H}:$ invariant. In this case one has also that $\tilde{F}(t)$ in Theorem 3 is the strong limit of $F \circ \gamma_t^N$. However, the above essential self-adjointness of $i\tilde{B}$ is an open problem.

## 3. POISSON STATISTICAL SOLUTIONS OF THE EULER EQUATION, VORTEX MODEL, TWO-DIMENSIONAL COULOMB SYSTEMS AND PLASMAS

Let us now consider the classical Euler fluid with initial vorticity rot u a sum of Dirac measures, i.e., with u a function of $x \in \Lambda$:

$$\text{rot } u \, (dx) = \sum_{i=1}^{n} a_i \, \delta_{x_i} (x) \tag{3.1}$$

where $\delta_{x_i}$ is the Dirac measure concentrated at the points $x_i \in \Lambda$, $\Lambda$ is an open region of $\mathbb{R}^2$ with smooth boundary, and $a_i$ are real numbers denoting the intensities of the vortices localized in $x_i$.

From (3.1) we get

$$u(x) = \sum_{i=1}^{n} a_i \, \nabla^{\perp} (\Delta)^{-1} (x,x_i), \tag{3.2}$$

where $\Delta$ is the Dirichlet Laplacian in $\Lambda$ and $\Delta^{-1}(x,y)$ the corresponding fundamental solution. Due to the logarithmic singularity of this kernel, u diverges as $|x-x_i|^{-1}$ at the point $x_i$. This gives problems in studying (weak) solutions of the Euler equation of motion with initial condition (3.1). These problems are very well discussed in $\boxed{11}$, to which we refer. Here we shall only briefly mention the relation with the so-called vortex model, the two-dimensional Coulomb gas of statistical mechanics, and the two-dimensional guiding center plasma.

Let us start with the vortex model. For simplicity, we consider the case $\Lambda = \mathbb{R}^2$. By definition the vortex model (Lin-Onsager model) describes n degrees of freedom $x_i(t)$, $t \geq 0$, $i = 1,\ldots,n$ in $\mathbb{R}^2$ as solutions of the system of ordinary differential equations

$$\dot{x}_i = (\nabla^{\perp})_i \sum_{\substack{j=1, \\ j \neq i}}^{n} a_j \, G(x_i(t), x_j(t)) \tag{3.3}$$

where $G = (-\Delta)^{-1}$ is the Newton kernel, with initial condition $x_i(0) = x_i$. If

$x_i(t)$ is a solution of (3.3), then with $\text{rot } u(t,dx) = \sum_{i=1}^{n} a_i \, \delta_{x_i(t)}(dx)$ and writing $\text{rot } u(t,f) \equiv \int \text{rot } u(t,dx) f(x) dx$, $f \in C_o^\infty(\mathbb{R}^2)$, in the sense of distributions,

$$\frac{\partial}{\partial t} \text{rot } u(t,f) = \text{rot } u(t, u_o \cdot \nabla f),$$

with $u_o(t,x) = \int \nabla_x^\perp G(x,x') \, \chi_{\{x \neq x'\}} \, \text{rot } u(t,dx')$, an equation equivalent to the Euler equation for smooth $u(t,x)$.

In this sense, then, but also in much stronger form, cf. in particular recent results by Marchioro and Pulvirenti [12], the vortex model is related to the Euler equation. (Below we shall see yet another form of this relation, when looking at Poisson invariant measures associated with the classical Euler equations for an inviscid fluid). The vortex model is a classical Hamiltonian system with Hamiltonian function

$$H = \sum_{i \neq j}^{n} a_i a_j \, G(x_i - x_j).$$

The classical dynamics of this Hamiltonian system has been studied recently in particular by Dürr, Marchioro and Pulvirenti, see [18], [12] and below.

The classical equilibrium statistical mechanics of a system of a large number of vortices is by definition the study of canonical Gibbs measures of the form $Z^{-1} e^{-\beta H} \Pi_i dx_i$, hence it coincides with the study of a classical Coulomb system of charged particles in two dimensions, with charges $a_i$ interacting with the two-body potential $G(x_i - x_j)$. Due to the logarithmic singularity of the interaction, this system has peculiar features which will be discussed further below.

Let us now discuss briefly how the vortex model, with arbitrary number of vortices, is related to statistical solutions of the Euler equations of Poisson type. A measure $\mu$ on the space $\mathcal{D}'(\Lambda)$ of generalized functions on a domain $\Lambda$ of $\mathbb{R}^2$ is a generalized Poisson measure if its Fourier transform (characteristic functional) has the form

$$L(f) = \exp \left[ \int_\Lambda \int_\mathbb{R} (e^{i\alpha f(x)} - 1) d\nu(\alpha) dx \right]$$

for $f$, a test function, and $\nu$, some bounded measure on $\mathbb{R}$. Expanding

234

the first exponential, we get

$$L(f) = e^{-|\Lambda|\,\|\nu\|} \sum_{n=0}^{\infty} \frac{1}{n!} \int_{\Lambda}\cdots\int_{\Lambda}\int_{\mathbb{R}}\cdots\int_{\mathbb{R}} e^{i\sum_{j=1}^{n}\alpha_j f(x_j)} \prod_{j=1}^{n} d\nu(\alpha_j)\,dx_j,$$

with $\|\nu\| \equiv$ total variation of $\nu$ and $|\Lambda| \equiv$ volume of $\Lambda$.  Using

$\sum_{j=1}^{n} \alpha_j f(x_j) = \langle f, \sum_{j=1}^{n} \alpha_j \delta_{x_j}\rangle$, in the sense of distributions, we see that $\mu$

is a "grand canonical" measure with support on the family $\Omega$ of all linear

combinations $\ell_{\underline{\alpha},\underline{x}}^{n} \equiv \sum_{j=1}^{n} \alpha_j \delta_{x_j}$, with $\underline{\alpha} = (\alpha_i, i=1,\ldots,n)$, $\underline{x}=(x_1,\ldots,x_n)$, $n\in\mathbb{N}$,

with

$$\mu\{\sum_{j=1}^{k} \alpha_j \delta_{x_j} \mid k=n,\ \alpha_j \in A_j,\ x_j \in B_j\} = \frac{e^{-|\Lambda|\,\|\nu\|}}{n!} \prod_{j=1}^{n} (\nu(A_j)|B_j|),$$

where $A_j$ are Borel subsets of $\mathbb{R}$ and $B_j$ are Borel subsets of $\Lambda$.  Let
$\xi(x) = \mathrm{rot}\, u(x)$ be the stochastic field so that

$$\mathrm{Prob}(\langle\xi,f_n\rangle \leq a_1,\ldots,\langle\xi,f_n\rangle \leq a_n)=\mu(\langle\xi,f_n\rangle \leq a_1,\ldots,\langle\xi,f_n\rangle \leq a_n),$$

$$\forall a_i \in \mathbb{R},\ f_i \in C_0^{\infty}(\Lambda).$$

$\xi$ is a stochastic measure [6].

Let us now assume that $\nu$ has discrete support.  It has been proven by
Dürr and Pulvirenti [18] (see also [12], [44], [56]) that if $\Lambda$ is bounded
there exists a global-in-time solution $x_i(t)$ of the equation of motion of
$n$-vortices (3.1), defined for Lebesgue a.e. initial condition $x_i(0)$.  We
shall call this the vortex flow.

Since the Lebesgue measure is a Liouville measure for the Hamiltonian
vortex system, we have that $\mu$ is invariant on the $k=n$ sector of its support
and hence $\mu$ is invariant under the vortex flow.  In what sense is $\mu$ also
invariant under the Euler flow (associated with classical solutions of the
Euler equation)?  This has been shown, in the case of $\Lambda$ a square with
periodic boundary conditions and $\nu$ with support on a finite subset of $\mathbb{R}_+$,
by Boldrighini and Frigio, in the sense that $\mu$ satisfies the stationary
Hopf equation ("infinitesimal invariance") (in a sense corresponding to that
discussed in Section 2).  In fact, Boldrighini and Frigio extend this result
to other infinitely divisible vorticity distributions.  (Other types of

invariant measures for hydrodynamics are discussed, e.g., in [46], [47].)
Moreover, they show that for the Poisson distribution $\mu$ with above supports
for $\nu$ there exist analytic in t solutions of the Euler equations in the form
$\frac{\partial}{\partial t} <\text{rot } u,f> = <B(\text{rot } u),f>$, with initial condition in a set of $\mu$-measure 1.

As we mentioned above, the vortex model is related to the classical
Coulomb problem because of the form of $H$ above. Let us now say a word on
the relation between the Coulomb model and the Sine-Gordon model of quantum
fields. A quantum field theory with Sine-Gordon interaction has been shown
in [52], [53] (see also for further references, e.g., [57]) to be related to
the Coulomb model by the so-called Sine-Gordon transformation. Let $\mu_o$ be the
free Euclidean Markov field on $\mathbb{R}^d$, i.e. the Gaussian measure on $S'(\mathbb{R}^d)$
with mean zero and covariance $(-\Delta + m^2)^{-1}$ for some $m > 0$ ($m \geq 0$ if $d \geq 3$).
The generating functional of Schwinger functions of this model is by
definition

$$E(e^{i<\xi,f>}) = Z_\Lambda^{-1} E_{\mu_o}(e^{i<\xi,f>} e^{-\lambda_\varepsilon \int_{\mathbb{R}} \int_\Lambda e^{i\alpha\xi_\varepsilon(x)} dx \, d\nu(\alpha)}),$$

with

$$Z_\Lambda \equiv E_{\mu_o}(e^{-\lambda_\varepsilon \int_{\mathbb{R}} \int_\Lambda e^{i\alpha\xi_\varepsilon(x)} dx \, d\nu(\alpha)}), \qquad f \in \mathcal{D}(\mathbb{R}^d),$$

where $\xi_\varepsilon$ is the convolution of $\xi$ with a suitable smooth function $\chi_\varepsilon$ such
that $\chi_\varepsilon \to \delta$ as $\delta \downarrow 0$, $f \in C_o^\infty(\mathbb{R})$, and $\nu$ is a positive measure on $\mathbb{R}$.
Expanding the second exponential we get, with $G_\varepsilon \equiv \chi_\varepsilon * G * \chi_\varepsilon$, $\hat{G}_\varepsilon \equiv \chi_\varepsilon * G$,

$$E(e^{i<\xi,f>}) = e^{-\frac{1}{2}<f,Gf>} Z_\Lambda^{-1} \sum_{n=0}^\infty \frac{(-\lambda_\varepsilon)^n}{n!} \int_{\mathbb{R}^n} \int_{\Lambda^n} e^{-\frac{1}{2}\sum_{i=1}^n \alpha_i^2 G_\varepsilon(0)}$$

$$e^{-\frac{1}{2}\sum_{i\neq j}^n \alpha_i\alpha_j G_\varepsilon(x_i-x_j)} e^{-\sum_{i=1}^n \alpha_i<f,\hat{G}_\varepsilon(\cdot-x_i)>} \prod_{i=1}^n dx_i \, d\nu(\alpha_i).$$

We then see that this is essentially the grand canonical measure for a
classical statistical mechanical system with interacting charged particles
having charges distributed according to $\nu$ and regularized Yukawa ($m > 0$),
resp. Coulomb ($m = 0$), interaction $G_\varepsilon(x_i - x_j)$. The same relation holds as
$\varepsilon \downarrow 0$, $m \downarrow 0$, $d = 2$ for suitable choice of $\lambda_\varepsilon$ with supp $\nu \subset (-\sqrt{4\pi}, \sqrt{4\pi})$ [54],
which then gives the Coulomb gas ([58],[61]).

236

Let us now make a short digression to remark how the connection between the models of a Euler fluid and of a guiding center plasma is established. The guiding center plasma, see e.g. [49], [51], [25], describes a fluid of charged particles of charge density $n(t,x)$, $(x,y) \in \mathbb{R}^2$, with time evolution given (in suitable units) by $\frac{\partial}{\partial t} n + v \cdot \nabla n = 0$. $v$ is the velocity vector, given by the vector in the x,y-plane, $v = \nabla \phi \times e_z$; $e_z$ is the unit vector in the direction $z(e_z$ having then the interpretation of a constant magnetic field in the direction z); $\phi$ is a scalar function such that $\Delta \phi = n$. We have then $v = \nabla^{\perp} \phi$ and the above equation coincides with the Euler equation (2.5).

Thus we see plasmas and the two-dimensional classical Coulomb gas system arising in several connections involving a Euler fluid. Let us therefore close with some remarks on such systems of charged particles. Their mathematical study has been pursued quite intensively in recent years, from different points of view. Although some results have been achieved, a complete proof of the existence of a phase transition is, to our knowledge, still lacking, due to the long range and/or to the singularity of the interaction. Furthermore, only partial results have been obtained concerning the formulation of the equilibrium properties in terms of Gibbs random fields (the Dobruchin, Lanford-Ruelle approach). (See, however, e.g. [45]; for the relevance of phase transitions in meteorology, see e.g. [13], [36] and references therein; for discrete Coulomb systems, see e.g. [42].)

Recently, a great deal of attention has been devoted also to a related model, the so-called two-dimensional one-component plasma or jellium (see e.g. [15], [16], [19], [20]-[35], [50], and references therein). Its classical Hamiltonian function is given in a region $\Lambda$ by

$$H = \alpha^2 \sum_{i<j}^{N} G(x_i,x_j) - \rho \sum_{i=1}^{N} \int_{\Lambda} dy\, G(x_i,y)$$

$$+ \frac{1}{2} \rho^2 \int_{\Lambda}\int_{\Lambda} dy\, dz\, G(y,z),$$

where $G = (-\Delta)^{-1}$ is the Newton kernel in dimension 2, $\alpha > 0$ is a positive constant ("charge at $x_i$"), $x_i \in \Lambda$, $\rho \equiv \alpha^2 N/|\Lambda|$. This system has been shown to be thermodynamically stable in the sense of the free energy by Merlini and Sari [22] (the corresponding model in three dimensions was discussed by Lieb and Narnhofer [16]).

Some recent progress in the study of the statistical mechanical problems of this model (study of Gibbs measures associated with it and associated Gibbs fields) includes exact solutions for a special value of $\beta\alpha^2$ ($\beta$ the inverse temperature), see e.g. [19], [21], [34] as well as study of correlation functions, see e.g. [45], [60].

The problem of the existence of a phase transition for sufficiently low temperature is still open. An approach introducing Dobrushin boundary conditions (somewhat as in Ising systems) to characterize different phases has been initiated in [32], [59]. Let us just quote one of the results obtained. Let us put fixed point charges $x_i$ in a crystalline configuration $Y = \rho^{-1/2} \, \mathbb{Z}^2 \cap (\Lambda_o - \Lambda)$, in a square $\Lambda_o$ of sides $(N_o/\rho)^{1/2}$ surrounding the square $\Lambda$ of side $(N/\rho)^{1/2}$, $N_o > N$. Under a suitable boundedness condition on the one-particle correlations it is shown that the thermodynamical limit of the corresponding free energy for these Dobrushin boundary conditions is the same as that without Dobrushin boundary conditions. For this and other related results see [32], [59].

A study of the relation between decay of a truncated pair correlation function at infinity and absence of long range order (indication of absence of crystalline phase) is given in [33].

There is a large body of literature on computations for thermodynamic functions, see e.g. [15], [19], [21], [27], [29], [31], [34]. Recently a new powerful method has been developed using exact calculations for systems of few particles and applied to the above model and related models, see [27], [31], [34], [35]. This provides new challenges for an analytic approach to phase transitions in Coulomb systems. A study of such phase transitions, as briefly mentioned above in connection with the relation between Euler flow, vortex-model and Coulomb systems, would also lead to interesting developments in the study of stochastic flows associated with Euler fluids.

In conclusion, we hope to have given at least some illustrations of the usefulness of measures in infinite dimensional space and other probabilistic objects as unifying tools in apparently distinct domains such as hydro-dynamics, classical statistical mechanics of Coulomb systems and the study of plasmas.

Clearly, much of the material we have reported is only at a preliminary
stage and many problems remain open, a challenge for the future.

ACKNOWLEDGEMENTS

It is a pleasure to acknowledge most valuable discussions with Detlev Dürr,
Margarida De Faria and Steinar Johannesen. The financial support of the
Deutsche Forschungsgemeinschaft (Research Program "Stochastic Fields"), the
Oslo Mathematics Seminar NAVF, the USP-Mathematisierung, and the ZiF of
University of Bielefeld (Projet No 2) are also gratefully acknowledged.


REFERENCES

[1]  S. Albeverio, M. De Faria, R. Høegh-Krohn, Stationary measures for the
         periodic Euler flow in two dimensions, *J. Stat. Phys. 20*,
         585-595 (1979).

[2]  S. Albeverio, R. Høegh-Krohn, Stochastic Euler flows with stationary
         distribution, Bochum Preprint.

[3]  M. De Faria,  Measures de Gibbs et équation d'Euler de l'hydrodynamique,
         Thèse de III$^e$ Cycle, Université de Provence, Marseille (1979).

[4]  S. Albeverio, R. Høegh-Krohn, M. De Faria, A remark on the existence of
         a stochastic flow associated with a Euler fluid in two dimensions
         (in preparation).

[5]  C. Boldrighini, S. Frigio, Equilibrium states for the two dimensional
         incompressible Euler fluid, *Atti Sem. Mat. Fis. Univ. Modena 27*,
         106-125 (1978).

[6]  C. Boldrighini, S. Frigio, Equilibrium states for a plane
         incompressible perfect fluid, *Commun. Math. Phys. 72*, 55-76
         (1980).

[7]  S. Caprino, S. De Gregorio,  On the statistical solutions of the two-
         dimensional periodic Euler equation, Roma Preprint (1984), *Math.
         Methods in the Appl. Sciences* (to appear).

[8]  R. Temam, Ed., *Turbulence and Navier-Stokes Equation*, Orsay 1975,
         Lect. Notes Maths. 565, Springer, Berlin (1976).

[9] C. Boldrighini, *Introduzione alla Fluidodinamica*, Quaderno Ric. Mat. Cons. Naz. Scienze, Rome (1979).

[10] S. Albeverio, R. Høegh-Krohn, Stochastic methods in quantum field theory and hydrodynamics, *Phys. Rept. 77*, 193-214 (1981).

[11] S. Albeverio, R. Høegh-Krohn, Some Markov processes and Markov fields in quantum theory, group theory, hydrodynamics and $C^*$-algebras, in *Stochastic Integrals*, Proc. LMS Durham Symposium 1980, Ed. D. Williams, Lect. Notes Maths. 851, Springer, Berlin (1981) 497-540.

[12] C. Marchioro, M. Pulvirenti, *Vortex Methods in two-dimensional Fluid Mechanics*, Edizioni KLIM, Roma; and Lect. Notes Phys. 203, Springer Verlag (1984).

[13] S. Albeverio, R. Høegh-Krohn, Let us talk about the weather - A model for large scale atmospheric features, *Proc. Confer. Les Embiez* (1980).

[14] R.H. Kraichnan, D. Montgomery, Two-dimensional turbulence, *Rep. Progr. Phys. 43*, 547-619 (1980).

[15] Ph. Choquard, Selected topics in the equilibrium statistical mechanics of Coulomb systems, in *Strongly Coupled Plasma*, Ed. G. Kalman, P. Carini, Plenum Press, N.Y. (1978) 347-406.

[16] E.H. Lieb, H. Narnhofer, The thermodynamic limit for jellium, *J. Stat. Phys. 12*, 291-310 (1975).

[17] H.A. Rose, P.L. Sulem, Fully developed turbulence and statistical mechanics, *J. de Phys. 39*, 441-483 (1978).

[18] D. Dürr, M. Pulvirenti, On the vortex flow in bounded domains, *Commun. Math. Phys. 83*, 265-273 (1983).

[19] A. Alastuey, B. Jancovici, On the classical two-dimensional one-component Coulomb plasma, *J. Phys. 42*, 1-12 (1981).

[20] D.C. Brydges, P. Federbush, Debye screening, *Commun. Math. Phys. 73*, 197-246 (1980).

[21] B. Jancovici, Exact results for the two-dimensional one-component plasma, *Phys. Rev. Lett. 46*, 386-388 (1981).

[22] R.R. Sari, D. Merlini, On the $\nu$-dimensional one-component classical
    plasma: the thermodynamic limit problem revisited, *J. Stat. Phys.*
    *14*, 91-100 (1976).

[23] J. Fröhlich, D. Ruelle, Statistical mechanics of vortices in an
    inviscid two-dimensional fluid, *Comm. Math. Phys. 87*, 1-36,
    (1982).

[24] R. Calinon, D. Merlini, Inhomogeneous stationary states of two
    dimensional magnetofluids, *J. Phys. of Plasma 30*, 90-    (1983).

[25] R. Calinon, D. Merlini, Equilibrium statistical mechanics treatment of
    a "modified" 2-dimensional guiding center plasma, *Phys. Fluids*
    *26*, 3508-3514 (1983).

[26] D. Merlini, Some equilibrium properties of the 2D strongly coupled one
    component plasma, in [15],   587-595.

[27] S. Johannesen, D. Merlini,  On the thermodynamics of the 2-dimensional
    jellium, *J. Phys. A 16*, 1449-1463 (1983).

[28] R. Calinon, D. Merlini, Stationary states of the 2-dimensional magneto
    hydrodynamic turbulence: non-dissipative limit, *J. Plasma Phys.*
    *22*, 385-396 (1979).

[29] R.R. Sari, D. Merlini, R. Calinon,  On the ground state of the
    one-dimensional classical plasma,  *J. Phys. A 9*, 1539-1551 (1976).

[30] J.Z. Imbrie, Debye screening for jellium and other Coulomb systems,
    *Commun. Math. Phys. 87*, 515-565 (1983).

[31] S. Johannesen, D. Merlini, The thermodynamic limit with a few
    particles: the two-dimensional one-component plasma, *J. Stat.*
    *Phys.*

[32] S. Albeverio, D. Dürr, D. Merlini, Remarks on the independence of the
    free energy from crystalline boundary conditions in the two-
    dimensional one component plasma, *J. Stat. Phys. 31*, 389-407
    (1983).

[33] F. Martinelli, D. Merlini, A refined Mermin argument for the two-
    dimensional jellium, *J. Stat. Phys. 34*, 313-318 (1984).

[34] T. Arede, S. Johannesen, D. Merlini, On the thermodynamic limit of
the Mehta-Dyson one dimensional model with long range
interaction, *J. Phys. A, 17,* 2505-2515 (1984).

[35] S. Johannesen, D. Merlini, On "critical points" in the two-dimensional
jellium, *Phys. Lett. 93A,* 21-23 (1982).

[36] H. Aref, Motion of three vortices, *Phys. Fl. 22,* 393-400 (1979).

[37] A.S. Monin, A.M. Yaglom, *Statistical Fluid Mechanics: Mechanics of
Turbulence,* I, II, The MIT Press (1975).

[38] S. Albeverio, J.E. Fenstad, R. Høegh-Krohn, T. Lindstrøm, *Nonstandard
Methods in Stochastic Analysis and Mathematical Physics,*
Academic Press (to appear).

[39] E. Hopf, Statistical hydromechanics and functional calculus, *J. Rat.
Mech. Anal.* 1, 87-123 (1952).

[40] P. Bernard, T. Ratiu, Eds., *Turbulence Seminar, Berkeley 1976/77,*
Lecture Notes Maths. <u>615</u>, Springer, Berlin (1977).

[41] G. Gallavotti, Problèmes ergodiques de la mécanique classique, Cours
de $3^e$ Cycle, Physique, EPF-Lausanne (1976).

[42] K.R. Ito, Kosterlitz-Thouless transition and Mayer expansion, ZiF-
Preprint (1984).

[43] R.H. Kraichnan, Realizability inequalities and closed moment
equations, in *Nonlinear Dynamics,* Ed. R.H.G. Helleman, N.Y.
Acad. Sci. (1980) 37-46.

[44] E.A. Novikov, Stochastization and collapse of vortex systems, in
[43], 47-54.

[45] J.R. Fontaine, Ph. A. Martin, Equilibrium equations and symmetries
of classical Coulomb systems, *J. Stat. Phys., 36* (1984), 163-179.

[46] A.M. Vershik, D.A. Ladyzenskaja, On the evolution of measures defined
by the Navier-Stokes equations, and on the solvability of the
Cauchy problem for Hopfs statistical equation, *Dokl. Ak. Nauk
SSR 226,* 18-21 (1976).

[47] A.M. Vishik, A.I. Komech, A.V. Fursikov, Some mathematical problems of statistical hydromechanics, *Russ. Math. Surv. 34*, 149-234 (1979).

[48] E. Lieb, H. Narnhofer, The thermodynamic limit for jellium, *J. Stat. Phys. 12*, 291-310 (1975).

[49] C.E. Seyler, jr., Y. Sahn, D. Montgomery, G. Knorr, Two-dimensional turbulence in inviscid fluids or guiding center plasmas, *Phys. Fl. 18*, 803-313 (1975).

[50] P.J. Forrester, Interpretation of an exactly solvable two-component plasma, *J. Stat. Phys. 35*, 77-87 (1984).

[51] M. Mond, G. Knorr, Elementary derivation of the kinetic equation for the two-dimensional guiding centre plasma. *J. Plasma Phys. 19*, 121-133 (1978).

[52] S. Albeverio, R. Høegh-Krohn, Uniqueness of the physical vacuum and the Wightman functions in the infinite volume limit for some non polynomial interactions, *Commun. Math. Phys. 30*, 171-200 (1973).

[53] S. Albeverio, R. Høegh-Krohn, Homogeneous random fields and statistical mechanics, *J. Funct. Anal. 19*, 242-272 (1975).

[54] S. Albeverio, R. Høegh-Krohn, The Wightman axioms and the mass gap for strong interactions of exponential type in two-dimensional space-time, *J. Funct. Anal. 16*, 39-82 (1974).

[55] E.B. Davies, *One-parameter Semigroups*, Acad. Press (1980).

[56] C. Marchioro, E. Pagani, Evolution of two concentrated vortices in a two-dimensional bounded domain, Trento Preprint (1984).

[57] S. Albeverio, R. Høegh-Krohn, Diffusion fields, quantum fields, and fields with values in Lie groups, in *Stochastic Analysis and Applications*, Ed. M. Pinsky, M. Dekker, New York (1984) 1-98.

[58] J. Fröhlich, Classical and quantum statistical mechanics in one and two dimensions: two component Yukawa and Coulomb systems, *Commun. Math. Phys. 47*, 233-268 (1976).

[59] D. Merlini (in preparation).

[60]  Ch. Gruber, Ch. Lugrin, Ph. A. Martin, Equilibrium properties of
       classical systems with long range forces.  BBGKY-equation,
       neutrality, screening and sum rules, *J. Stat. Phys. 22*,
       193-      (1980).

[61]  G. Gallavotti, F. Nicolò,  The screening phase transition in the two
       dimensional Coulomb gas, BiBoS-Preprint, 1985 (to appear in
       *Trends and Perspectives in the Eighties*, Edts. S. Albeverio,
       Ph. Blanchard, World Publ., Singapore, 1985).


S. Albeverio,
Mathematisches Institut,
Ruhr-Universität Bochum.

Bielefeld-Bochum Research Center Stochastics,
Volkswagenstiftung.


R. Høegh-Krohn,
Matematisk Institutt,
Universitetet i Oslo.

Université de Provence et Centre de
  Physique Theorique, CNRS,
Marseille-Luminy.


D. Merlini,
Mathematisches Institut,
Ruhr-Universität  Bochum.
Liceo Cantonale, Bellinzona.